Fundamentals of Electronic Warfare

For a listing of recent titles in the
Artech House Radar Library, turn to the back of this book.

Fundamentals of Electronic Warfare

Sergei A. Vakin
Lev N. Shustov
Robert H. Dunwell

Artech House
Boston • London
www.artechhouse.com

Library of Congress Cataloging-in-Publication Data
Vakin, Sergei A.
Fundamentals of electronic warfare / Sergei A. Vakin, Lev N. Shustov, Robert H. Dunwell.
p. cm. — (Artech House radar library)
Includes bibliographical references and index.
ISBN 1-58053-052-4 (alk. paper)
1. Electronics in military engineering. 2. Command and control systems. 3. Electronic intelligence. 4. Electronic surveillance. I. Shustov, Lev N. II. Dunwell, Robert H. III. Title. IV. Series.
UG485.V25 2001
623'.043—dc21 2001022186

British Library Cataloguing in Publication Data
Vakin, Sergei A.
Fundamentals of electronic warfare. – (The Artech House radar library)
1. Electronics in military engineering—Mathematics
2. Command and control systems—Mathematics
I. Title. II. Shustov, Lev N. III. Dunwell, Robert H.
623'.043'0151

ISBN 1-58053-052-4

Cover design by Yekaterina Ratner

685 Canton Street
Norwood, MA 02062

International Standard Book Number: 1-58053-052-4
Library of Congress Catalog Card Number: 2001022186

10 9 8 7 6 5 4 3 2 1

Table of Contents

Introduction

At present and for the foreseeable future, electronic systems serve and will continue to serve as the foundation of systems for the control of forces and weapons (military equipment) in all branches of the armed forces. In antiaircraft defense (AAD) forces, radar permits the gathering of information about the situation in the air. Radar also ensures the guidance of weapons to airborne targets.

Airforce high-precision weapon (HPW) installations, including synthetic antenna radar (SAR) and high-precision satellite radio navigation systems, detect land-based targets of small external dimensions at a distance of several hundred kilometers and determine their coordinates with a mean square error (MSE) of the order of ones of meters, permitting the delivery of massive strikes on the enemy in the very depths of its forces' operational formations [1].

Analogous examples of the high effectiveness of weapons attained as a result of using appropriate electronic systems can be offered for other branches of the armed forces as well.

Ensuring high operational and tactical indicators for weapons, electronic systems are at the same time one of the most vulnerable elements in control systems, since they can be detected based on their radiation and they are subject to countermeasures using technical methods (electronic warfare systems).

The dialectics of combat, measures, and countermeasures in this area of military engineering have given rise to electronic warfare (EW). Although EW was considered initially to be a type of operations and combat support, subsequently it has grown into an element for sustaining operations and combat.

At the present time, EW is defined as a set of measures and actions performed by the conflicting sides to detect and electronically attack enemy electronic systems for the control of forces and weapons, including high-precision weapons, as well as to electronically defend one's own electronic systems and other targets from technical intelligence (electronic intelligence, ELINT), jamming and nondeliberate interference. The latter presupposes ensuring electromagnetic compatibility (EMC).

Electronic attack can be achieved by

- Jamming (active, passive, false targets);
- Reduction of radar and thermal detectability;
- Changing the electrical properties of the environment (conditions for the propagation of electromagnetic waves).

ELINT, interacting with other types of technical intelligence [2], must solve a two-fold problem: the detection and analysis of radiation from electronic systems in the interests of jamming them; and the detection and analysis of jamming radiation in the interests of improving the electronic defense of one's own electronic systems in the dynamics of EW.

ELINT is also performed with the goal of gathering source information for the subsequent development and synthesis of optimum structures and algorithms for electronic systems intended to operate in a given jamming environment. Information arriving from ELINT and other technical intelligence systems in many ways provides the basis for the computer data and knowledge bases that form a part of the control systems for both EW and the electronic defense from jamming and ELINT systems.

The appearance of EW was motivated dialectically by the dynamics of combat between offensive and defensive systems. In the given case, we mean the battle between English bombers and Nazi German AAD in World War II (WWII). Before that time, there were individual cases of jamming using available devices, but they were neither massive nor organized in nature, and they did not lead to qualitative changes in the tactics of combat operations. Massive use of passive jammers by English bombers started on June 24, 1943 during the nighttime bombing raid on Hamburg and continued systematically in subsequent raids. The jamming targets were Würzburg gun-laying radar stations, which had an emitted signal wavelength of about 50 cm. In this case, the passive jamming consisted of manually launched tinfoil strips (dipoles) of the order of 25 cm in length (i.e., equal to approximately half the radar wavelength). As a result of this jamming, bomber losses decreased by about half for several months. This first use of jamming was preceded by lengthy reconnaissance of the jamming targets. The decision to employ jamming was made after losses reached the maximum permissible level.

To a large extent, this jamming was so highly effective due to the wavelength selected for the Würzburg radar; the absence in it of circuits to protect against passive jamming, which appeared only at the end of the war; and the absence of the possibility to change wavelength. In October 1943, U.S. bombers began employing active jamming against Würzburg radar using carpet-type jammers, which turned out to be highly effective because it was not possible to alter the carrier frequency.

The period corresponding to WWII and the first postwar years can be considered the initial phase in the development of systems and techniques for electronic warfare.

Their subsequent development also took place according to the laws of dialectics. Moreover, the combat target on one hand was radar, and on the other, EW systems. It is possible to distinguish approximately four basic phases of significant change in the structure and algorithms of both radar and EW systems. The period of the 1950s corresponds to the development of noncoherent radar with rapid alteration of the carrier frequency and antijamming devices built using pulse-to-pulse cancellation. In jammers, electronic frequency alteration was introduced. Special devices for the launching of chaff were developed.

In the 1960s and in the first half of the 1970s, pulse-coherent radar appeared, capable of filtering out passive jamming and permitting the detection of low-flying aircraft against the background of reflections from the surface of the Earth. Defense from active jamming was ensured in them by increasing the specific energy potential by approximately an order of magnitude and rapidly altering the carrier frequency. EW systems changed accordingly.

During the second half of the 1970s and at the beginning of the 1980s, there appeared radar that worked in two modes: pulse-coherent with narrowband signals, and with broadband signals having a big base. Increasing the base enabled a corresponding increase in the radar energy potential. The use of two modes made it difficult to effectively organize both active and passive jamming. In narrowband mode, the effect of broadband jamming is attenuated, whereas broadband mode enables a reduction in the effectiveness of narrowband jamming. EW systems and techniques were modified accordingly.

Besides the appearance of high-precision weapons systems, the fourth phase of EW is characterized by the use of radar with phased antenna arrays that can adjust to the jamming environment; a transition to multiposition circuits for signal processing; and the decrease in radar radiation detectability by making the transition to noise-like signals (NLS).

The capabilities of EW systems and techniques are increasing. Massive use of these systems has become widespread. After the Gulf War, EW became considered not only a type of operations and combat support, but a means of sustaining operations and combat.

In the first part of this monograph, we present the fundamentals of electronic warfare based on systems methodology. The various targets of EW operations are considered in sequence. Mathematical models of electronic systems viewed as EW targets are presented. This includes mathematical models of an optimum detector and radar tracker; models for optimum reception in radio communications systems and radio navigation; and models of recognition schemes.

Using the methods of the mass service theory and the theory of automated control systems, we present mathematical models of automated systems for the control of AAD forces and weapons as EW targets.

Using the methods of information theory and statistical radio technology, models of jamming signals have been developed. Elements of the theory of jamming and jamming systems and techniques are explained using games theory and the dynamics of averages.

Effectiveness indicators for jamming signals and EW systems and techniques are defined with reference to information, energy, operations and tactics, and military and economic parameters. Methods of calculating jamming corridors for active radar jamming with various antijamming levels (the level of information stability) are developed and presented, based on the energy criterion. Using this same energy criterion, methods of calculating radar jamming corridors for passive and active–passive jamming are developed and presented, and jamming methods based on the use of false targets, radar decoys, and thermal traps are analyzed.

Jamming possibilities based on reducing the radar detectability of aircraft and changing the electrical properties of the environment are evaluated.

The authors plan to publish a second volume of this monograph with the proposed title *EW Techniques and Electronic Intelligence*. In the second volume, we intend to analyze

- Jamming of radio communications and radio navigation systems;
- Techniques and effectiveness of jamming radar in ACS for control of AAD forces;

- Techniques and effectiveness of jamming radar in weapon control systems;
- Techniques and effectiveness of jamming to screen a single aircraft;
- ELINT methods in radio electronic warfare.

References

[1] Simonyan, R. G., *O nastupatel'nykh operatsiyakh v armii SSHA* (*About Offensive Operations in the US Army*). Moscow: *Voennaya mysl'* (Military Thought), Vol. 1, 1989, pp. 69–77.

[2] Brusnitsin, N. A., *Global'naya tekhnicheskaya razvedka SSHA* (*US Global Technical Intelligence*). Moscow: *Voennaya mysl'* (Military Thought), Vol. 10, 1990, pp. 57–65.

Acknowledgements

The authors sincerely thank S. S. Gaysaryan, G. Yu. Klyueva, M. V. Kosenko, A. E. Ovchinnikov, N. V. Shustova, T. S. Vakina, and the entire team at our department for their hard work in preparing this manuscript for publication. The authors express special gratitude to G. V. Zaporezhec, who took upon himself the task of reviewing this monograph.

1

Targets of Electronic Warfare Operations

1.1 A General Description of Targets of Electronic Warfare Operations

In the dynamics of electronic warfare (EW), objects and methods of operation are by and large determined by the specific conditions of the electronic environment and the nature of the problems being solved. Let us examine the basic targets of EW operations in more detail, based on an initial definition.

As was mentioned earlier, EW represents a set of measures and actions for the detection and jamming of electronic systems for the control of forces and weapons (including high-precision weapons), as well as for the electronic protection of one's own electronic systems against technical reconnaissance (electronic intelligence, or ELINT), jamming, and interference.

Typically, EW methods of operation can be either offensive or defensive in nature. When we speak of offensive operations, first and foremost we have in mind operations conducted using jamming or employing weapon systems that automatically home in on radiation sources. Both of these offensive operation methods are used not only to attack (or destroy) electronic systems for the control of forces and weapons, but also to defend one's own electronic systems against ELINT and jamming. In the first instance, the jamming targets are the receivers of radar, communications, and radio navigation systems. In the second case, the jamming targets are ELINT receivers that form a part of active jamming and ELINT stations, as well as of reconnaissance and strike systems.

Typical defensive EW operations involve protecting the respective systems from jamming and ELINT by increasing their equivalent specific energy potential, modifying their time, frequency, space and other characteristics, or by making them more secure (masking, deception).

As a rule, the most effective solution to the problem of protecting against jamming is achieved by combining both purely defensive and offensive operations. More precisely, this is achieved by destroying the jammer or by electronically jamming ELINT systems and active jammers. Thus, it may be said that jamming and ELINT, taken together, by and large comprise the basis of EW. Other considerations also serve to corroborate this statement. In the operational and tactical plane, according to contemporary views, jamming is considered to be an element for sustaining operations and combat activities, while EW defensive measures (antijamming and ELINT) represent a form of operations (combat) support. On a tactical level, the definition of jamming as a battle element can be substantiated using the example of airforce subdivisions conducting military operations against antiaircraft defense (AAD) missile systems on the front line. When aircraft battle formations are screened using jamming, they can strike antiaircraft positions without suffering significant losses. In this case, airborne jamming systems represent an essential battle element. They directly participate in the delivery of material damage, having beforehand caused information damage to the electronic antiaircraft control systems. By information damage, we mean the size of the coverage area eliminated from the region where the attacked electronic system was acquiring information while working in normal operating mode during the dynamics of EW. Therefore, jamming is a weapon and not merely a means of support.

On the operations level, the thesis we have mentioned about jamming is borne out by the example of military actions taken during a defensive operation to fend off strikes by an enemy who is conducting offensive air and ground operations. In this case, jamming systems may be given the task of thwarting a massive strike using high-precision weapons [1, 2]. By delivering information damage to synthetic aperture radar, ELINT stations within intelligence and strike systems, and high-precision radio navigation systems, jamming significantly decreases the probability that groups of aircraft and missiles, AAD systems, and other targets belonging to the defending side will be hit. Thus, their average combat life becomes significantly longer. In this instance, jamming prevents potential material damage by delivering information damage to the enemy's electronic systems for control of forces and weapons, thereby creating conditions for a counterstrike.

As in any other type of armed conflict, the counteraction between offense and defense serves as the motivating force for the development of EW systems and methods. In the case at hand, we are dealing with the action–counteraction of systems and methods for jamming and methods for electronic defense.

Everything mentioned above confirms and extends the thesis that systems and methods for jamming and ELINT comprise the basis of EW.

An examination of systems and methods for jamming electronic targets of various kinds, together with the specific methods and devices for their electronic protection, make the stages and essence of the content of EW dynamics quite clear. This allows us to accept the targets of EW operations as being the targets of jamming actions and to set forth the fundamentals of EW based on, in the main, a study of the jamming dynamics for various conditions of the electronic environment. Of course, an examination that is limited to the dynamics of jamming and reconnaissance can by no means exhaust the vast range of specific techniques and tactical methods for electronic protection and technical counterintelligence. They should be analyzed by suitable specialists in each specific subject area.

In the material that follows, we consider as the basic targets of jamming and ELINT: various kinds of radar; radio communications and data exchange systems as used in control systems for AAD fighter planes and tactical aircraft; command radio lines for the control of AAD missiles; high-precision radio navigation systems; and ELINT systems. We also analyze automated systems for the control of AAD forces and weapons as jamming targets.

Radar serves as the basis for modern all-weather AAD information support systems. At the same time, as a target of electronic warfare, it is one of the most vulnerable elements. This is due to the following circumstances.

Radar processes signals reflected from airplanes and other targets. To detect the latter against its own noise background, the energy of the signal emitted must be quite large. In principle, this permits us to detect radar radiation at distances far exceeding its effective range of operation, especially in those cases where we are dealing with radar for locating airplanes (targets) with low radar cross-sections (LO vehicles).

Radar can be detected not only based on radiation from the reconnaissance sensor. It is also possible to determine its coordinates, carrier frequency, spectral and other characteristics using appropriate receivers and range finders. This permits the subsequent determination of the type of radar and the deployment of appropriate firepower or jammers against it.

The energy of the reflected signal at the output to the radar receiver is inversely proportional to the fourth power of the distance to the object being detected. At the same time, for example, the spectral density of the masking jamming signal at the input to the same receiver is inversely proportional to the square of the distance from the radar to the jammer. Compared to other systems (radio communications, radio navigation, etc.), this situation presents great potential opportunities to electronically attack radars using active jamming.

Since the resolution of radar and its abilities to recognize the targets being detected are limited, in a significant number of cases there are also possibilities of jamming using passive jamming, false targets, and radar decoys.

Measures are being taken in modern and, to an even greater extent, future radar systems to increase their level of electronic protection and, correspondingly, to decrease the possibility of their radiation being detected and, as a result, of them being electronically attacked. Regardless, in view of their important role, they continue to remain the basic targets of jamming in the AAD system.

In a number of cases, despite definite difficulties with jamming, radio communications systems are the only channel in the battle control system for AAD fighters and tactical aircraft. Therefore, by attacking it, it is possible to significantly reduce the possibilities of centralized control, or to disrupt it totally. An analogous situation also occurs in command radio lines for the control of antiaircraft missiles, high-precision radio navigation systems, and electronic reconnaissance systems within intelligence and firepower installations. Therefore, these systems must also be included as targets of EW.

The examination, as targets of EW, of automated systems for the control of forces and weapons is required to determine the minimum degree of jamming necessary for the electronic systems that comprise them.

1.2 Mathematical Models of Electronic Systems as Targets of Electronic Warfare

1.2.1 Fundamental Principles

The detailed development of mathematical models of electronic systems, viewed as targets of EW, is a result of the requirement, in the material that follows, to determine threshold values for the effects of jamming, which are directly connected with the minimum required degree of jamming. In other words, these models permit us to determine, with a sufficient level of substantiation, criteria for jamming. These same circumstances permit us, in the materials below, to limit ourselves basically to additive jamming signals.

When viewing radar and other electronic devices as targets of EW (jamming), we should keep in mind that jamming signals directly affect radio receivers. A priori, as a rule, the jammer or other electronic attack system does not have full information about the target of the attack. In order to eliminate the element of chance for the victim receiver, it should be attempted to orient oneself by assuming that the victim system uses optimum or quasi-optimum processing methods for the signals received and within a certain standard (conditional) electronic environment. Depending on how the radar or other electronic device operates, differing variants of processing optimization may occur, thus resulting in various mathematical models. Let us look at the most typical ones.

1.2.2 Mathematical Models of an Optimum Radar Receiver Working in Search Mode

The basic purpose of a radar operating in volume-scan (search) mode is the detection and recognition of targets and their subsequent acquisition for tracking. Mathematical models for a radar receiver operating in the search mode may be viewed as a diagram that consists of an optimum detector and an optimum system for making decisions in conditions of uncertainty [3, 4]. In essence, receivers must, in the conditions being analyzed, permit us to make a decision about the validity or invalidity of one of two statistical hypotheses about the value of a certain parameter λ. In other words, it is necessary to solve a problem involving the statistical theory of detection and discrimination of signals, which is a part of the theory of parametric hypotheses in mathematical statistics [5].

According to methods for verification of parametric hypotheses developed to the present point in time, an optimum decision is made based on the examination of a likelihood ratio, which, in many cases, comprises sufficient statistics to make a decision about the parameter λ for the given sample function. In other words, the likelihood ratio contains all information about the parameter λ available in the sample function. Let us clarify what has been mentioned above using the specific example of reaching a decision about the parameter λ based on the results of observing the voltage (current) $y(t)$ at the output to a radar receiving antenna over a finite time interval $(0, T)$, which may be written in the following way:

$$y(t) = \lambda s(t) + n(t) \qquad 0 \leqslant t \leqslant T \tag{1.1}$$

where λ is the parameter we are interested in, $s(t)$ is the anticipated signal, and $n(t)$ is the noise, which is the sample function of a random process.

Let us assume that, with relationship to parameter λ, only two alternative hypotheses can occur; more precisely, $\lambda = 0$ when the sample function $y(t)$ represents only noise (hypothesis H_0), and $\lambda = 1$ if the sample function contains the sum of signal and noise (H_1). Let us further assume that hypotheses H_0 and H_1 have corresponding multidimensional probability densities $p_0(y)$ and $p_1(y)$. In the given case, the statistical problem of making a decision consists of determining which of the hypotheses is taking place based on the results of observing the sample function through the point in time $t = T$: $H_0(\lambda = 0)$ or $H_1(\lambda = 1)$. As was noted before, the decision is made based on an examination of the likelihood ratio

$$l(y) = \frac{p_1(y)}{p_0(y)} \quad t = T \tag{1.2}$$

by comparing it to some threshold value l_0.

If $l(y) < l_0, t = T$, then the decision is made that signal $s(t)$ is absent from the sample function being analyzed ($\lambda = 0$). In the case where $l(y) > l_0$, the valid hypothesis is assumed to be H_1 (i.e., signal $s(t)$ is present in the sample function).

There are two types of errors possible in the given situation. The first type of error is when a valid hypothesis H_1 is assumed, whereas in actuality the valid hypothesis is H_0 (i.e., a false alarm has occurred). The second type of error is when H_0 is taken to be valid, whereas the valid hypothesis is H_1 (i.e., the radar has missed its target).

Depending on information available and the criterion selected, an optimum receiver can have various structures and characteristics. In the radar operating modes considered, the Neyman–Pearson criterion is normally used, according to which, for a fixed probability of a false alarm P_{FA}, the probability of a missed target is minimized and the probability of a correct detection P_{DET} is maximized. If $n(t)$ is a sample function of white Gaussian noise and we are dealing with the detection of a radio signal, then it is possible to consider the likelihood ratio $l(y)$ as being equivalent to the parameter q, which is defined by the ratio [3, 4]

$$q = \sqrt{\frac{2E}{N_0}} \tag{1.3}$$

Here, E is the signal energy and N_0 is the one-sided spectral noise density.

The parameter q is compared to the threshold h, which is determined by the value accepted as the probability of a false alarm P_{FA}.

In the case where the parameters of the signal are fully known and the noise is white, the probabilities of a correct detection (P_{DET}) and a false alarm (P_{FA}) are defined by the following ratios:

$$P_{DET} = 0.5 - \Phi_0\left(\frac{h}{q} - q\right) \tag{1.4}$$

$$P_{FA} = 0.5 - \Phi_0\left(\frac{h}{q}\right) \tag{1.5}$$

$$\Phi_0(x) = \frac{1}{\sqrt{2\pi}} \int_0^x e^{-\frac{t^2}{2}}\, dt \tag{1.6}$$

Here, (1.6) is the probability integral [6].

In the conditions under consideration and for a given probability of a false alarm, it follows from (1.4) and (1.5) that the probability of correct detection is determined only by the ratio of the signal energy to the spectral noise density at the moment of time $t = T$ and does not depend on the signal waveform. The latter may be intermittent or modulated during the observation interval. At the same time, it is important to eliminate the loss of signal energy during the process of its transformation through the observation interval $(0, T)$. The optimum device for signal and noise processing is a matched filter (for example, a comb filter) or a correlation receiver.

The analysis of formulas (1.3), (1.4), and (1.5) shows that, given a fixed signal energy E, you can make the probability of correct radio signal recognition P_{DET} as small as you like by increasing the spectral noise density N_0, even in the case where its parameters are fully known and the receiver is optimum.

In actually implemented radar systems, including AAD radar systems, the parameters of the signal being detected are known to a certain degree of accuracy and can be represented simply by their probability densities or distribution functions. Detection of a fluctuating signal $s(t)$, optimum according to the Neyman–Pearson criterion, with a random phase ϕ evenly distributed in the interval $(-\pi, \pi)$ and a random envelope U

distributed according to Rayleigh's Law [3], is the most typical for radars:

$$p(U) = \frac{U}{D_s} \exp\left(-\frac{U^2}{2D_s}\right) \tag{1.7}$$

Here, D_s is the variance of the signal s, which in the given case is a quasi-sinusoidal wave.

For the conditions under consideration, the probability of a correct detection and a false alarm turn out to be linked by the following relationship:

$$P_{DET} = P_{FA}^{\frac{1}{1+0.5q^2}} \tag{1.8}$$

where q is defined in (1.3).

In this case, the structure of an optimum correlation receiver presupposes a quadrature channel with an appropriate threshold device.

The mathematical models given are a long way from exhausting the problems of the radio electronic protection of radars. In the main, they permit the determination of a threshold value q. Specific techniques and tactical methods of countering jamming generated during the dynamics of EW are considered below. As a rule, they are linked to specific methods of electronic attack.

1.2.3 Mathematical Models of an Optimum Receiver for an Electronic Tracker

In the given case, we are dealing with the examination, as targets of EW, of receivers that perform optimum evaluation of the parameter λ of a signal $s(t, \lambda)$, which is being received against a background of white noise. This can be a point evaluation, performed at a fixed moment of time, according to the results of measuring λ for a specific sample function. It can be performed over a certain period of time in the case where the parameter $\lambda(t)$ is a random process and represents an information transmission. The evaluation of $\lambda(t)$ under these conditions is performed using optimum filtering. The latter, as a particular case, also includes optimum precision evaluation. In the general case, the parameter $\lambda(t)$ being evaluated can represent a random vector.

The operational dynamics of a radar performing filtering of a random vector λ that is carrying information about target coordinates and their derivatives can be described using a canonical system of stochastic differential equations of state (of transmission) [7, 8]:

$$\frac{d\lambda}{dt} = g(t, \lambda) + n_\lambda(t) \quad \lambda(0) = \lambda_0 \tag{1.9}$$

Here, λ is a vector-column of dimension n; $g(t, \lambda)$ is a vector function-column of dimension n; $n_\lambda(t)$ is a vector-column of centered white noise forming the transmission λ; and λ_0 is the initial value.

The noise $n_\lambda(t)$ is called shaping white noise. The filtering device, described by the stochastic differential equation (1.9), is called the shaping filter of the transmission.

The process $y(t)$, observed at the output to the shaping filter, is defined by the following equation in vector format:

$$y(t) = s(t, \lambda(t)) + n_0(t) \tag{1.10}$$

where $s(t, \lambda(t))$ is a vector function-column of dimension m comprising the signal being observed, and $n_0(t)$ is a vector-column of centered Gaussian white observation noise with a matrix correlation function

$$M\left[n_0(t_1)n_0^T(t_2)\right] = N_0\delta(t_2 - t_1) \tag{1.11}$$

Here, N_0 is a symmetrical matrix of dimension n, the elements of which are one-sided spectral densities for noise observed divided by two; $\delta(x)$ is the delta function; and T is the transposition sign.

In the particular case of linear filtering, the equations of state (of transmission) and observation take on the following format:

$$\frac{d\lambda}{dt} = A(t)\lambda + n_\lambda(t) \tag{1.12}$$

$$y(t) = H(t)\lambda + n_0(t) \tag{1.13}$$

$A(t)$ is the coefficient matrix $(n \times n)$; $n_\lambda(t)$ is the shaping noise vector; $n_0(t)$ is the white observation noise; and $H(t)$ is the measurement (observation) matrix $m \times m$.

The random processes given in the stochastic differential equations (1.9) and (1.12) are Markovian. Their a posteriori probability density $p_{ps}(t, \lambda)$ is determined by the Fokker–Planck–Kolmogorov (FPK) equation, the drift and diffusion coefficients of which may be found with the aid of stochastic differential equations (1.9) or (1.12).

Using the methods of nonlinear and linear filtering, the FPK equation and stochastic differential equations (1.9) and (1.12) permit us, in principle, to synthesize a filter for the optimum evaluation of the vector $\lambda(t)$ in a jamming environment defined by white noise with a spectral density matrix N_0. The effectiveness of the filter can be quite high if the spectral density of observation noise is limited.

In practice, when useful signal energy is fixed and spectral density levels of observation noise are high, optimum filtering methods, just like optimum detection and recognition methods, have little effect. We will analyze this situation in greater detail using an example that analyzes one-dimensional equations for optimum nonlinear and linear filtering in continuous time.

In the case of nonlinear filtering, the integro-differential Stratonovich equation determining the a posteriori probability density $p(t, \lambda) = p_{ps}(t, \lambda)$ is written in the following way [3, 7, 8]:

$$\frac{\partial p(t,\lambda)}{\partial t} = \frac{1}{2}\frac{\partial^2}{\partial \lambda^2}(b(\lambda)p(t,\lambda)) - \frac{\partial}{\partial \lambda}(a(\lambda)p(t,\lambda)) + \left(F(t,\lambda) - \int_{\lambda} F(t,\lambda)p(t,\lambda)d\lambda\right)p(t,\lambda) \tag{1.14}$$

Here, a and b are the drift and diffusion coefficients of the FPK equation, as defined by stochastic differential equation (1.9):

$$F(t,\lambda) = \frac{s(t,\lambda)}{N_0}\left(y(t) - \frac{1}{2}s(t,\lambda)\right) \tag{1.15}$$

$$\lambda(0) = \lambda_0; \; p(0,\lambda) = p_{pr}$$

$$p(0,\lambda_0) = p_{pr0}(\lambda)$$

p_{pr} is the a priori probability density.

As the spectral noise density N_0 increases and the signal-to-noise ratio decreases, the last term introduces an increase in the degree of randomness into (1.14), which, in the end, causes the formation of a Wiener diffusion process for an a posteriori probability at the output of the optimum filter.

This is more evident in the case of linear filtering, as well as in the instance of an approximate solution of the filtering problem using the current linearization method. According to (1.12) and (1.13), the equations of state and observation for linear filtering can be written down in the following way:

$$\frac{d\lambda}{dt} = -\alpha\lambda + n_\lambda \tag{1.16}$$

$$y(t) = H\lambda + n_0 \tag{1.17}$$

where α is the dimensionality coefficient t^{-1}.

The equation for determining the optimum evaluation of $\hat{\lambda}$ at the output to a linear Kalman filter is written down in the following way [3, 9]:

$$\frac{d\hat{\lambda}}{dt} = -\alpha(t)\hat{\lambda} + H(t)\frac{2R(t)}{N_0}\left(y(t) - H(t)\hat{\lambda}\right) \tag{1.18}$$

where $R(t) = \sigma^2(t)$ represents the observation noise variance n_0.

If the spectral noise density is high enough and the signal-to-noise ratio is low, then the differential equation for optimum evaluation becomes analogous to the initial stochastic equation (1.16).

In the case when the optimum filtering problem is solved by the current linearization method, presupposing linearity of the equation of state and representation of the nonlinear equation of observation (1.10) in a linearized format:

$$y(t) = s(t, \hat{\lambda}) + \frac{\partial}{\partial\lambda}s(t, \hat{\lambda})(\lambda - \hat{\lambda}) + n_0$$

the equation for the quasi-optimum evaluation of parameter λ (at the output to the filter) assumes the following format [9]:

$$\frac{\partial\hat{\lambda}}{\partial t} = -\alpha\hat{\lambda} + \frac{2}{N_0}R(t)\frac{\partial s(t, \hat{\lambda})}{\partial\lambda}\left(y(t) - s(t, \hat{\lambda})\right) \tag{1.19}$$

Just as in the previous case, as the spectral noise density N_0 increases and the signal-to-jamming ratio decreases, the statistical characteristics of the right-hand side of (1.19) to an ever increasing degree approach a random Gaussian process and the equation for quasi-optimum evaluation at the output to a linearized filter approaches the stochastic equation of state (1.16). A similar analysis can be performed, for example, for each independent radar channel (range measurement, angle measurement, velocity tracking, etc.).

Thus, for each electronic system, regardless of how signals are formed and processed, there exists a domain of critical spectral noise density values N_0 for which operation of the electronic system as a tracker and filter and, particularly, as a means of solving the problem of coordinate evaluation for true and false targets, becomes impossible. The critical values for spectral density can manifest themselves in various ways. In particular, it is possible to speak of a critical spectral density, where tracking in an electronic system breaks down. It is possible to speak about a critical spectral density for white noise or about critical values for the variance of reference values in a narrowband noise sample function, where evaluation variances for corresponding signal parameters and target coordinates are greater than a certain threshold value.

Critical values for spectral noise density, where tracking breaks down, can be determined and then subsequently determined for specific structural variants of trackers. Threshold values for the evaluation variances of signal parameters and coordinate measurements can be found with the help of the Cramer–Rao inequality, which defines the lower limit for evaluation variances of parameters being measured.

In the case of a continuous probability density and a one-dimension evaluation parameter λ, the Cramer–Rao inequality can be written down in the following way [4, 10, 11]:

$$M\left[(\lambda^* - \lambda)^2\right] \geqslant \frac{\left(1 + \dfrac{\partial b}{\partial \lambda}\right)^2}{nM\left[\left(\dfrac{\partial \ln p(x, \lambda)}{\partial \lambda}\right)^2\right]} \tag{1.20}$$

where $x = (x_1, \ldots, x_n)$ is the sample function, represented by n independent random values having one and the same probability density; $p(x, \lambda)$ is the probability density x, dependent on the parameter λ being evaluated, which, in the given case, is a probability function; and b is the evaluation displacement, $\lambda^* = \lambda + b(\lambda)$. The symbol $M[-]$ designates the operation for determining the mathematical expectation.

The mathematical expectation for the square of the logarithmic derivative of $p(x, \lambda)$ with respect to λ is, by definition, the Fisher information quantity I_λ:

$$I_\lambda = \int_{-\infty}^{+\infty} \left(\frac{\partial \ln p(x, \lambda)}{\partial \lambda} \right)^2 p(x, \lambda) dx \tag{1.21}$$

The inequality (1.20) becomes an equality if two conditions are satisfied:

1. $\lambda^*(x)$ represents sufficient statistics for the evaluation of parameter λ;
2. The following equality holds true:

$$\frac{\partial}{\partial \lambda} \ln p(x, \lambda) = k(\lambda)(\lambda^* - \lambda) \tag{1.22}$$

In this case, if the evaluation of $\lambda^*(x)$ is not displaced, then it is termed effective. Correspondingly, the variance for an effective evaluation is minimal. The jamming side, when generating a jamming signal, can predefine the parameters for the probability density $p(x, \lambda)$ of the sample function $(x_1, \ldots, x_n)$ so that the minimum variance value for evaluation of the parameter λ is not less than a certain given value. According to the Cramer–Rao inequality, it can be stated, in this case, that no amount of subsequent optimum processing can make the evaluation variance of λ less. Let us clarify this with a specific example.

Let us assume that the probability density $p(x, \lambda)$ is Gaussian:

$$p(x, \lambda) = \frac{1}{\sqrt{2\pi\sigma^2}} \exp\left(-\frac{(x - \lambda)^2}{2\sigma^2} \right) \tag{1.23}$$

where λ represents the average value of x determined from the random sample $(x_1, \ldots, x_n)$. According to (1.20) and (1.21) we find the Fisher information quantity and then the lower variance limit for the evaluation $D[\lambda^*]$:

$$I_\lambda = \int_{-\infty}^{+\infty} \left(\frac{x - \lambda}{\sigma^2} \right)^2 p(x, \lambda) dx \tag{1.24}$$

By replacing the variable $\frac{x-\lambda}{\sigma} = t$, the integral (1.24) is reduced to the tabular

$$M\left[(\lambda^* - \lambda)^2\right] = D[\lambda^*] \geqslant \frac{\sigma^2}{n} \tag{1.25}$$

Here, σ^2 is the distribution variance of x. It is assumed that it is identical for all $x_i, i = \overline{1, n}$.

In this case, the evaluation of λ^* is not displaced. It is equal to the average arithmetic value of the sample function $(x_1, \ldots, x_n)$:

$$\lambda^* = \frac{1}{n}\sum_{i=1}^{n} x_i$$

For this reason $b = 0$. In practice, the variant analyzed can happen when evaluating the envelope of a useful signal received on a background of narrowband Gaussian noise with a spectral width F_n, represented by a fairly long sample duration function T_n. In this case $(x_1, \ldots, x_n)$ are the reference values, as determined using V. A. Kotelnikov's Theorem, $n = 2F_nT_n >> 1$. By increasing σ^2, which is the variance of the reference values, the jamming side can ensure a required variance of envelope determination errors for a useful signal λ.

In a similar fashion, it is possible, based on the Cramer–Rao inequality, to intentionally ensure the required variance values for target coordinate evaluation for any optimum reception (evaluation) techniques. Let us note that, in the case of white noise, the spectral noise density N_0 is used instead of reference value variances. The maximum of the probability function is one of the most widely used criteria for optimum evaluation in electronic systems. Evaluation on the basis of this criterion is equivalent to an effective one (i.e., conversion of the Cramer–Rao inequality into an equality when the parameter λ being evaluated enters into the probability function $p(x, \lambda) = p(x|\lambda)$ as an exponent [7]). In particular, this follows from (1.22).

1.2.4 Mathematical Models of Optimum Reception in Radio Communications and Radio Navigation Systems

As targets of EW, the greatest interest is represented by models of radio communications and radio navigation receiver systems used to control AAD forces and weapons. In the given case, procedures for distinguishing the two deterministic signals at the elementary transmission level are key to determining reception quality. The problem of optimally distinguishing the two deterministic signals can be reduced to a statistics problem of checking parametric hypotheses, which, in a well-known sense, is similar to the one that was analyzed in conjunction with the model of an optimum radar detector. Specifically, the problem is formulated in the following way [3, 12].

According to observation results in the interval $0 \leqslant t \leqslant T$ for wave $y(t)$, which is represented by the sum:

$$y(t) = \lambda s_1(t) + (1 - \lambda)s_2(t) + n(t) \quad 0 \leqslant t \leqslant T \tag{1.26}$$

where $n(t)$ is white noise, $s_1(t)$ and $s_2(t)$ are deterministic signals, and where it is necessary to find an optimum solution to the problem of an unknown parameter λ, which has two values 1 and 0. If $\lambda = 1$, then signal $s_1(t)$ is present in the observed wave $y(t)$. In the alternative case ($\lambda = 0$), signal $s_2(t)$ is present in the wave $y(t)$. The a priori probabilities that each of the signals is present are assumed to be known.

According to this statement, in the problem being analyzed, we are speaking of the verification of two alternative hypotheses H_0 and H_1: H_0 — the signal $s_2(t)$ is present in the sample function, $\lambda = 0$; and H_1 — the signal $s_1(t)$ is present in the sample function, $\lambda = 1$.

We denote the a priori probabilities of the corresponding hypotheses H_0 and H_1 by $P_{pr}(H_0)$ and $P_{pr}(H_1)$, $P_{pr}(H_0) + P_{pr}(H_1) = 1$. Accordingly, the conditional probabilities of errors of the first and second kind are designated by

$$P(H_1|H_0) = P(s_1|s_2); P(H_0|H_1) = P(s_2|s_1)$$

The total probability of a false decision about parameter λ is

$$P_e = P_{pr}(H_0)P(s_1|s_2) + P_{pr}(H_1)P(s_2|s_1) \tag{1.27}$$

According to the criterion of the ideal observer, a solution that minimizes the probability of P_e may be considered optimum in the given case. A receiver that implements this criterion is termed optimum from the standpoint of V. A. Kotelnikov's criterion, or, in other words, an ideal receiver where potential noise immunity has been implemented.

Let us examine the particular case of a symmetrical digital binary communications system for which the following equality holds true:

$$P_{pr}(H_1) = P_{pr}(H_0) = 0.5; \; E_1 = E_2 = E; \; P(s_1|s_2) = P(s_2|s_1) \tag{1.28}$$

In the conditions under consideration and according to the criterion of the ideal observer, the numerical value of the minimal error probability P_e for an optimum receiver can be found from the following expression [3, 9]:

$$P_e = \frac{1}{2} - \Phi_0\left(\sqrt{\frac{E}{N_0}(1 - r_s)}\right) \tag{1.29}$$

Here, r_s is the standard cross-correlation function between the signals s_1 and s_2; r_s varies within limits from $-1(s_1 = -s_2)$ to $+1(s_1 = s_2)$. The greatest noise immunity (the least P_e) is exhibited by a system with phase-modulated signals (PM), for which $r_s = -1$ and, correspondingly,

$$P_e = \frac{1}{2} - \Phi_0(q) \qquad q = \sqrt{\frac{2E}{N_0}} \tag{1.30}$$

In particular, for frequency modulation, $r_s = 0$ and P_e is at its maximum:

$$P_e = \frac{1}{2} - \Phi_0\left(\sqrt{\frac{E}{N_0}}\right) \tag{1.31}$$

Amplitude modulation occupies an intermediate position, where

$$P_e = \frac{1}{2} - \Phi_0\left(\frac{1}{2}\right) \tag{1.32}$$

The analysis done shows that even an ideal radio receiver can be attacked to the extent necessary (i.e., it stops distinguishing signals $(P_e = 0,5)$), given the signal is fully known and has a constant energy E, and provided the spectral noise density N_0 is great enough. In principle, the expressions given for P_e permit us to determine the threshold values for jamming signals.

1.2.5 Mathematical Models of Optimum Recognition Techniques in Electronic Systems

Recognition is one of the key problems of military radio electronics and becomes extremely acute in conditions of EW. All forms of military radio electronics involve this problem in one way or another and it is evident to an even greater degree in radar and radio ELINT.

In general, the problem of recognition may be formulated as follows [13, 14]. There exists a set of mutually exclusive classes $S = (S_1, \ldots, S_K)$. Each class is defined by a corresponding set of attributes. A certain object is analyzed based on the results of observing (measuring) its attributes. The goal of recognition is to establish that the object being analyzed is a member of one of these mutually exclusive classes $S_k, k = \overline{1, K}$.

The observation results for the object being analyzed can be represented as a matrix X:

$$X = \begin{pmatrix} x_{11}, x_{12}, \ldots, x_{1n} \\ x_{m1}, x_{m2}, \ldots, x_{mn} \end{pmatrix} = (x_{ij}) \qquad i = \overline{1, m};\ j = \overline{1, n} \tag{1.33}$$

Here, n is the quantity of observations for each of the attributes; and m is the quantity of attributes $x_i = (x_{i1}, \ldots, x_{in})$, $i = \overline{1, m}$ being observed for the given object.

Each jth column of the matrix (x_{ij}) is an m-dimensioned attribute vector for the jth observation (measurement). In this case, the attribute vector with the number j can be portrayed as a point in m-dimensioned space with the coordinates x_{ij}, $i = \overline{1, m}$. The set of attributes is assumed to be identical for all classes of targets being recognized. In the opposite case, the problem of recognition becomes trivial. In the conditions under consideration, the targets are distinguished only by the results of the observations (measurements) of attributes corresponding to x_{ij}. These differences can be quantitative (measurements) or qualitative. For any set of attributes $x_1, \ldots, x_m$, it is possible to define rules permitting the evaluation of the distance d_{lr} between targets of various classes S_l and S_r, which are represented by observation matrices $(x_{ij})_l$ and $(x_{ij})_r$. According to these rules, the targets S_l and S_r are made to correspond to a vector-column $d_{lr} = (d_1^{lr}, \ldots, d_q^{lr})^T$ (T is the transposition sign), comprising q scalars, which are termed interclass distances. They describe the degree of difference between the attributes x_i^l and x_i^r, $i = \overline{1, m}$ of the given classes. A mandatory prerequisite to any recognition process is the creation of an attribute space that includes the required set of attributes $x_1, \ldots, x_m$ reflecting the most typical ones needed to recognise a property of the targets in the given class.

In the case of radar signals [14, 15] reflected from aircraft, it is possible to use envelope parameters, spectral and polarization characteristics, as well as signal modifications caused by individual aircraft features, as attributes of $x_i, (i = \overline{1, m})$.

For ELINT systems, the specific signal parameters of the electronic systems subject to surveillance serve as the attributes of x_i, $(i = \overline{1, m})$.

As a rule, the attributes of x_i are multidimensional random values or sample functions of random processes. The analysis of recognition methods known to date shows that the only method providing a sufficiently complete description of targets being analyzed under real conditions is the statistical method [14, 15]. It takes into consideration destabilizing factors and permits us to quantitatively express the key factor, reliability of recognition, through the basic parameters of the recognition system. Under these conditions, the selection of a decision rule permitting us to establish that a specific sample function of observations (measurements) belongs to one of the mutually exclusive classes is made in accordance with the above-mentioned theory of statistical decisions. In many cases of interest, the dimensionality of the set S is equal to two $(K = 2)$. The observation results for the given incidence are compared with the model class description. As was noted before, in the theory of statistical hypotheses, an optimum decision is made on the basis of examining the likelihood ratio l:

$$l = \frac{p_n(x_1, \ldots, x_n|S_2)}{p_n(x_1, \ldots, x_n|S_1)} \lesseqgtr C \tag{1.34}$$

where $p_n(x_1, \ldots, x_n|S_j), j = 1, \bar{2}$, is the conditional combined probability density for selected (or reference) values, provided they are members of class S_j; and C is the threshold value.

The distinguishing feature is that, in the theory of statistical decisions, the probability densities mentioned are known, whereas, in the case of statistical recognition, they are, in principle, unknown and we substitute not the probability densities themselves in the decision rule (1.34), but their evaluations. It is not the likelihood ratio that is compared to the threshold, but its evaluation:

$$l^* = \frac{p_n^*(x_1, \ldots, x_n|S_2)}{p_n^*(x_1, \ldots, x_n|S_1)} \lesseqgtr C \tag{1.35}$$

This circumstance limits the possibilities of selecting quality criteria. The one that gives the best results is the criterion of maximum likelihood [13, 14, 16]. In specific conditions, the Neyman–Pearson criterion can also be used.

Statistical methods for parameter recognition, which assume a priori knowledge of the probability density $p_n(x_1, \ldots, x_n|S_j)$, are of practical interest in the domain of radio electronics. The task of recognition is reduced to evaluations of distribution parameters for random values of X_i, which form the attribute spaces of the targets being recognized. In particular, we check the statistical hypotheses for average values and variances of selected normal distributions of X_i attributes for the targets being recognized.

One of the basic results of the analysis performed is the conclusion that it is possible to deliberately simulate any of the targets of this class, provided the statistical characteristics of its X_i attributes are known. This, in particular, follows from the analysis of the likelihood ratios presented in (1.34) and (1.35). In the case of parameter recognition, the task of deliberate deception is simplified.

1.3 Mathematical Models of Automated Systems for the Control of AAD Forces as Targets of EW

1.3.1 Fundamental Principles

The simplest mathematical model reflecting, in the first approximation, the principal functional elements of an automated system for control of AAD forces would be a multichannel mass service system with rejects [17, 18]. A system of the type under consideration is designed to service, in a prescribed way, a random stream of requests arriving with a definite intensity. It includes a certain number of independent channels, each of which can service only one request, and a control system, which distributes the requests as they arrive among the free channels. The latter task is achieved through the use of appropriate information support, permitting the detection of both new requests and free channels. It is assumed that any request of a given class will be accepted for servicing, provided there is at least one free channel in the system. In the opposite case, a reject occurs and the request goes unserviced.

A full service cycle in an AAD system is performed in two stages. In the first stage (control of forces) we solve the problems of detection and recognition of airborne targets and their distribution among free fire channels. The second stage of servicing (individual guidance) is performed by military equipment (weapon) control systems.

In the conditions being analyzed, depending on the level of the groups in opposition (operational, tactical), the elements of the request stream and the channels of the service system are allocated in various ways. In operational level groupings, the initial unit of the request stream could be a regiment (squadron) and the service channel an AAD missile system battery (division) — a control and warning center (CWC) guiding AAD fighter planes. In tactical level groups (an airforce division (regiment), an antiaircraft missile system division (battery), or a tactical airforce CWC), the input stream is formed from individual airplanes and other aircraft that are overcoming (penetrating) the AAD system. The service channels are AAD missile system fire sections, fighter guidance channels, mobile AAD installations, and other control channels for AAD military equipment. Radar systems for detection, guidance, and target designation provide the technical base for information support systems for these groups on both the operational and tactical levels of antiaircraft defense. These radar systems, in turn, can also be viewed as targets of radio jamming by a group of active jammers, which may be represented by their own mathematical model as a multichannel mass service system with rejects. The examples given justify the requirement for examining the elements of the theory of mass service in greater detail than usual.

1.3.2 A Model of a Mass Service System with Rejects and Centralized Control

The mathematical model of a multichannel mass service system with rejects and centralized control is a system of differential equations for probabilities of state, which can be worked out based on the following assumptions: the random stream of input requests is Poissonian, the service time for a request is distributed in accordance with the exponential law, and all n channels are identical. Each of them may service any of the requests received. The system can, with a definite probability, be in any of $n+1$ states $X_0, X_1, \ldots, X_n$.

When performing an analysis of random processes in a system with discrete states, it is convenient to use oriented graphs. These graphs are sets of peaks and branches. Each peak corresponds to a certain system state and each of the branches, equipped with an arrow, indicates the direction of a possible transition to an adjacent state. The oriented graph shown in Figure 1.1 reflects the dynamics of changes in state of an n-channel mass service system with rejects, where at the input there is a constant influx of requests with intensity λ and each channel services the incoming requests with intensity μ. When all n channels are operating, the system services $n\mu$ requests per unit of time. If there are no requests, then the system is in state X_0. It is assumed that a request is in the service system for a far greater period than the average service time.

Before we begin deriving the differential equations for probabilities of

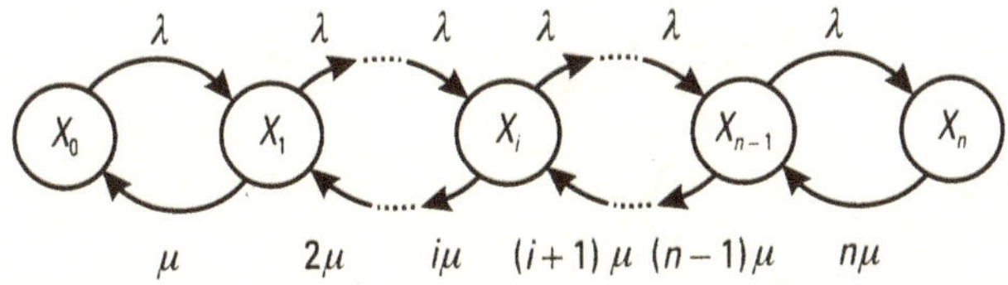

Figure 1.1 The dynamics of changes in state of an n-channel mass service system with rejects.

state based on the oriented graph we have mentioned, we would like to remind the reader of certain initial elements of the mass service theory [19, 20]. Real random processes can be regarded as Poissonian if two conditions are satisfied: uniformity and independence (absence of consequences). If a third condition is fulfilled, stability, then the stream is termed most elementary. Based on these three conditions, it is possible to derive an analytic expression for a one-dimensional Poisson distribution without using an asymptotic juncture from a binomial distribution, as is usually done in textbooks on probability theory. The condition of uniformity permits us to state with an accuracy to infinitely small values of a higher order that, during a sufficiently small time interval Δt, the probability that only one request will arrive is equal to $\lambda\Delta t$. As a result of the condition of independence (the absence of any consequence) the probability that a request will arrive during a certain time interval does not depend on the frequency and manner that requests arrived during previous moments of time. Assuming that all three conditions are satisfied, we can determine the probability $P_n(t + \Delta t)$ that exactly n requests will arrive within the time period $(t + \Delta t)$. Such an event can occur as the result of two uncombined events: during the time interval t exactly n requests were received and during the interval Δt there was not a single one; or during the time t there were $n - 1$ requests and over the period Δt there was one additional one. The probability of occurrence of the first event is $P_n(t)(1 - \lambda\Delta t)$, and the probability of the second event is $P_{n-1}(t)\lambda\Delta t$; therefore,

$$P_n(t + \Delta t) = P_n(t)(1 - \lambda\Delta t) + P_{n-1}(t)\lambda\Delta t$$

After self-evident transformations and converting to the limit as $\Delta t \to 0$, we obtain the initial differential equation for the solution of the problem posed:

$$\frac{\mathrm{d}P_n(t)}{\mathrm{d}t} = -\lambda P_n(t) + \lambda P_{n-1}(t) \tag{1.36}$$

In the case $n = 0$, (1.36) can be transformed to

$$\frac{\mathrm{d}P_0(t)}{\mathrm{d}t} = -\lambda P_0(t) \tag{1.37}$$

The solution of (1.37) with the initial condition $t = 0,\ P_0 = 1$ has the format

$$P_0(t) = \mathrm{e}^{-\lambda t} \tag{1.38}$$

The formula defines the probability that not a single request will be received over the time period t, when the average stream density is λ. Using (1.38), (1.36), and sequentially increasing n, we obtain the Poisson distribution formula we are looking for:

$$P_n(t) = \frac{(\lambda t)^n}{n!}\mathrm{e}^{-\lambda t} \tag{1.39}$$

For subsequent analysis, it is necessary to determine the average time $\bar{t}_\lambda$ between adjacent requests in the most elementary stream with an average intensity of λ. According to the definition of the mathematical expectation of a random value [3],

$$\bar{t}_\lambda = \int_0^\infty t p(t)\mathrm{d}t \tag{1.40}$$

where $p(t)$ is the probability density for the time intervals under consideration and the probability density $p(t)$, in turn, can be defined with the aid of the distribution function $F(t)$ for the time intervals shown. By definition,

$$F(t) = P(T < t)$$

Here, $P(T < t)$ is the probability that the time interval between adjacent requests will be less than t. This probability together with the probability of the opposite event (i.e., that not a single request will arrive during the time t), forms the entire group of events; that is,

$$P(T < t) + P(T \geqslant t) = 1$$

The probability $P(T < t)$ for the most elementary stream is determined using (1.38); thus,

$$F(t) = 1 - e^{-\lambda t}$$

The probability density $p(t)$ that we are looking for is equal to

$$p(t) = \frac{dF(t)}{dt} = \lambda e^{-\lambda t} \tag{1.41}$$

that is, the random intervals between adjacent requests in the most elementary stream are distributed according to the exponential law. It follows from (1.40) and (1.41) that

$$\bar{t}_\lambda = \frac{1}{\lambda} \tag{1.42}$$

It was noted earlier that the channel service time is also subject to the exponential law, so therefore the service time $\bar{t}_\mu$ and the average stream density for requests serviced μ are related in a ratio analogous to (1.42):

$$\bar{t}_\mu = \frac{1}{\mu} \tag{1.43}$$

Let us move on to the derivation of the differential equations for the probabilities of state for the mass service system shown in Figure 1.1. We will limit ourselves in the beginning to a single-channel system ($n = 1$) in which there can be only two states, X_0 and X_1, with corresponding probabilities P_0 and P_1:

$$P_0 + P_1 = 1 \tag{1.44}$$

Let us determine the probability $P_0(t + \Delta t)$ that, at the moment of time $(t + \Delta t)$, the single-channel mass service system with rejects will be in the state X_0. In the conditions being considered, this can happen as a result of the occurrence of two uncombined events: at the moment in time t, the system was free and, during the time Δt, not a single request arrived; or at the moment in time the system was occupied providing service (was in state

X_1) and during the time Δt the channel became free. According to what was said,

$$P_0(t+\Delta t) = P_0(t)(1-\lambda\Delta t) + P_1(t)\mu\Delta t$$

After algebraic transformations and conversion to the limit as $\Delta t \to 0$, we obtain the differential equation for the probability of $P_0(t)$:

$$\frac{dP_0(t)}{dt} = -\lambda P_0(t) + \mu P_1(t) \tag{1.45}$$

Reasoning in a similar fashion, it is possible to obtain a differential equation for $P_1(t)$:

$$P_1(t+\Delta t) = P_1(t)(1-\mu\Delta t) + P_0(t)\lambda\Delta t$$

Converting to the limit, we obtain

$$\frac{dP_1(t)}{dt} = -\mu P_1(t) + \lambda P_0(t) \tag{1.46}$$

Equations (1.45) and (1.46) illustrate the well-known mnemonic rule [18, 20] for creating a differential equation for probabilities of state for an n-channel system. According to this rule, the time derivative for the probability of the given state is equal to the algebraic sum of the probabilities, where the number of terms is equal the quantity of branches entering and exiting the peak of the graph being analyzed. The terms corresponding to the outgoing branches have a negative sign and are equal to the product of the probabilities of state multiplied by the stream intensities required to place the system in an adjacent state. The incoming branches correspond to positive addends, which are the products of the probabilities of the adjacent states multiplied by the stream intensities causing the transition to an adjacent state. The number of incoming and outgoing branches is identical.

Based on the mnemonic rule and the original oriented graph, we can write down the system of differential equations we are seeking for the probabilities of state of an n-channel mass service system with rejects (Figure 1.1):

$$\frac{dP_0(t)}{dt} = -\lambda P_0(t) + \mu P_1(t)$$

$$\frac{dP_i(t)}{dt} = -(\lambda + i\mu)P_i(t) + \lambda P_{i-1}(t) + (i-1)\mu P_{i+1}(t) \qquad (1.47)$$

$$\frac{dP_n(t)}{dt} = -n\mu P_n(t) + \lambda P_{n-1}(t)$$

The system of equations (1.47) can be solved given the initial conditions:

$$P_0(0) = 1 \quad P_i(0) = 0, (i = \overline{0, n}) \qquad (1.48)$$

Besides this, normalization conditions must be satisfied:

$$\sum_{i=0}^{n} P_i(t) = 1 \qquad (1.49)$$

We remind our readers that, in the given case, we are considering a mathematical model for mass service reflecting, in the first approximation, the operation of an automated system for distributing targets among fire channels. The parameter λ is understood to be the average intensity of the stream of airplanes (helicopters, winged missiles) entering the zone of responsibility. The parameter μ determines the average stream density of aircraft fired on in the case where centralized control of the firing system has not been disrupted. The combat potential of the automated control system (ACS) is determined by its throughput, the probability of servicing each of the requests in the input stream and the average number of service channels. Let us determine these indices for an automated control system working in a steady state. In this case, the derivatives from the probabilities of state assume a zero value and the system of equations assumes the format:

$$\begin{aligned}
0 &= -\lambda P_0 + \mu P_1 \\
0 &= -(\lambda + \mu)P_1 + \lambda P_0 + 2\mu P_2 \\
0 &= -(\lambda + i\mu)P_i + \lambda P_{i-1} + (i+1)\mu P_{i+1} \qquad (1.50) \\
0 &= -n\mu P_n + \lambda P_{n-1}
\end{aligned}$$

With the assistance of the system of algebraic equations in (1.50), it is possible to express the probabilities of states P_i using P_0:

$$P_1 = \frac{\lambda}{\mu} P_0 \quad P_2 = \frac{\left(\frac{\lambda}{\mu}\right)^2}{2!} P_0$$
$$P_3 = \frac{\alpha^3}{3!} P_0 \quad P_i = \frac{\alpha^i}{i!} P_0,\ i = \overline{0, n} \tag{1.51}$$

where

$$\alpha = \frac{\lambda}{\mu} \tag{1.52}$$

The probability P_0 can be determined using the normalization condition (1.49):

$$P_0 \sum_{i=0}^{n} \frac{\alpha^i}{i!} = 1$$
$$P_0 = \frac{1}{\sum_{i=0}^{n} \frac{\alpha^i}{i!}} \tag{1.53}$$

Therefore,

$$P_i = \frac{\alpha_i / i!}{\sum_{i=0}^{n} \frac{\alpha^i}{i!}} \quad (i = \overline{0, n}) \tag{1.54}$$

An aircraft is serviced when, during its flight within the zone of responsibility of the automated control system, at least one of the fire channels is free. The respective service probability P_{serv} is equal to

$$P_{serv} = 1 - P_n \tag{1.55}$$

where

$$P_n = \frac{\alpha^n / n!}{\sum_{i=0}^{n} \frac{\alpha^i}{i!}}$$

According to (1.55), when control is centralized,

$$P_{\text{serv}} = \frac{\sum_{i=0}^{n-1} \alpha^i / i!}{\sum_{i=0}^{n} \alpha^i / i!} \tag{1.56}$$

The service probability P_{serv} can also be determined directly as the ratio of the stream density of serviced requests λ_0 to the density of the input stream λ. In turn,

$$\lambda_0 = \mu \bar{K} \tag{1.57}$$

where $\bar{K}$ is the average number of channels in service. It is necessary to know $\bar{K}$ when evaluating the operational and tactical indices for effectiveness of a mass service system:

$$\bar{K} = \sum_{i=0}^{n} iP_i$$

By definition, and also taking into consideration (1.57),

$$P_{\text{serv}} = \frac{\lambda_0}{\lambda} = \frac{\mu \bar{K}}{\lambda} = \frac{\bar{K}}{\alpha} \tag{1.58}$$

$$\bar{K} = \alpha P_{\text{serv}} \tag{1.59}$$

In order to link the operation of an automated control system with the electronic systems analyzed earlier, which are direct targets of EW, it is necessary to clarify what is meant by service at the given stage of operation of an AAD system. As was noted earlier, when considering the stage of controlling forces, we are solving the problem of target detection (surveillance), target distribution among fire channels, and the monitoring of the trajectories of those targets which have been selected for tracking. In airforce fighter control systems, this stage is called long-range guidance. The quality (accuracy) of service is not evaluated in its entirety at this stage. Radio jamming of radar for detection, guidance, and target designation forming part of an automated control system results only in varying degrees of disruption of centralized control.

1.3.3 A Variant of a Multichannel Mass Service System with Decentralized Control

Let us consider the functional variant of a mass service system with disrupted centralized control. Practically speaking, here there is no automated control system and each fire channel targets its weapons at airborne targets only within the bounds of its fairly limited zone of responsibility. Service requests (airborne targets) are distributed among fire channels with equal probability, independent of whether the channel is servicing a request (aiming the attack system) or not [18]. The corresponding oriented graph is given in Figure 1.2. If i of n available channels are occupied providing services, then, given an overall average target input stream density of λ, the average target stream density arriving at the remaining free $n-i$ channels is equal to $\lambda \frac{n-1}{n}$. Targets arriving in the zone of responsibility of the occupied channels remain unserviced. Assuming a steady service process as shown in the graph in Figure 1.2 and the rule stated earlier, we can immediately write down a system of algebraic equations for the probabilities of states P_i, $(i = \overline{0, n})$:

$$
\begin{aligned}
0 &= -\lambda P_0 + \mu P_1 \\
0 &= \lambda P_0 - \left(\lambda \frac{n-1}{n} + \mu\right) P_1 + 2\mu P_2 \\
0 &= \lambda \frac{n-i+1}{n} P_{i-1} - \left(\lambda \frac{n-1}{n} + i\mu\right) P_i + (i+1)\mu P_{i+1} \\
0 &= \frac{\lambda}{n} P_{n-1} - n\mu P_n
\end{aligned}
\tag{1.60}
$$

The initial conditions are: $P_0(0) = 1$; $P_i(0) = 0, (i = \overline{1, n})$.

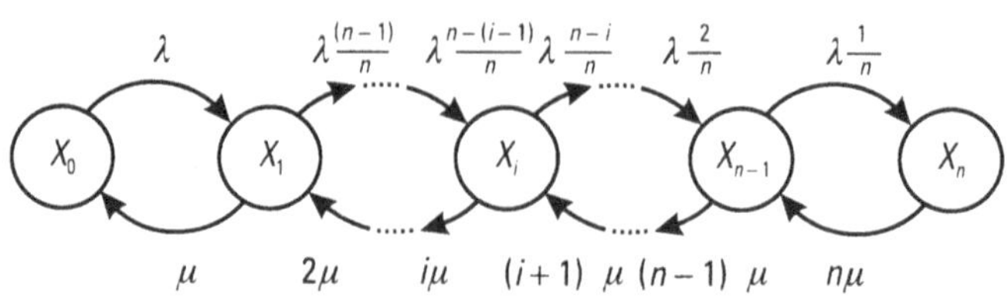

Figure 1.2 A mass service system with disrupted centralized control.

With the help of the system of equations in (1.60), it is possible to express the probabilities of states P_i, $(i = \overline{1, n})$ using P_0 [18]:

$$P_i = \left(\frac{\lambda}{n\mu}\right)^i \frac{n(n-1)\dots(n-(i+1))}{i!} P_0 \quad (i = \overline{0, n})$$

or

$$P_i = \chi^i C_n^i P_0 \quad C_n^i = \frac{n!}{i!(n-1)!} \tag{1.61}$$

Here, P_0 is defined based on normalized conditions:

$$\sum_{i=0}^{n} P_i = 1 \quad \sum_{i=0}^{n} P_i = P_0 \quad \sum_{i=0}^{n} \chi^i C_n^i = P_0(1+\chi)^n$$

$$P_0 = \frac{1}{(1+\chi)^n} \tag{1.62}$$

$$\chi - \frac{\lambda}{n\mu} = \frac{\alpha}{n} \tag{1.63}$$

In the given case, the average number of channels providing service is equal to

$$\bar{K} = \sum_{i=0}^{n} iP_i = P_0 \sum_{i=0}^{n} i\chi^i C_n^i$$

Taking into consideration (1.62), we obtain

$$\bar{K} = \frac{\chi}{(1+\chi)^n} \frac{\partial}{\partial\chi} \sum_{i=0}^{n} \chi^i C_n^i = \frac{\chi}{(1+\chi)^n} \frac{\partial}{\partial\chi} (1+\chi)^n$$

After transformations we get

$$\bar{K} = \frac{n\chi}{1+\chi} \tag{1.64}$$

In analogy to (1.58), the probability of a request being serviced P'_{serv} is equal to

$$P'_{serv} = \frac{\lambda_0}{\lambda} = \frac{\mu\bar{K}}{\lambda} = \frac{1}{1+\chi} = \frac{1}{1+\dfrac{\lambda}{n\mu}} \tag{1.65}$$

In Figures 1.3 and 1.4, we show the dependency of the service probabilities P_{serv} and P'_{serv} on the number of fire channels n and the value of α, which is the ratio of the average input stream densities λ and requests serviced (airborne targets) μ. Calculations were performed using (1.56) and (1.65). Decentralization of control results in the greatest decrease in service probability when the average parameter values are equal to $\alpha = \frac{\lambda}{\mu} (\alpha = 8, \ldots, 20)$ and the number of channels is equal to $n = 6, \ldots, 24$. Depending on specific conditions, this decrease comprises approximately 15, . . . , 40%. In both cases, service is assumed to be accurate.

Let us pause and examine in greater detail the question of service accuracy. The AAD system being analyzed performs its services under conditions of conflict. As a result of airforce counteraction, this service is not accurate. It is possible to speak of successful servicing, resulting in the destruction of a target, only with a certain probability. It would seem to be expedient to take this circumstance into consideration quantitatively by introducing the concept of an average stream density for successfully serviced airplanes μ^*. This should be understood to mean the average stream density of serviced airplanes without consideration to counteractions μ, multiplied by the probability of successful servicing P_{ss}:

$$\mu^* = \mu P_{ss} \tag{1.66}$$

In this case, the analogy with the well-known concept of an average stream density for successful shots Λ is evident, which is understood to be the product of the average density of the stream of shots λ multiplied by the average probability a target will be destroyed by each shot $P_{destruct}$ [17]:

$$\Lambda = \lambda P_{destruct} \tag{1.67}$$

To a significant extent, the probability of successful servicing P_{ss} depends on the quantitative and qualitative effectiveness indices of the EW measures taken by the airforce. In particular, when an automated control system for AAD forces is jammed by electronic systems, decentralization of control may occur. One of the variants was considered above (the breakdown of automatic target tracking in a target designation system).

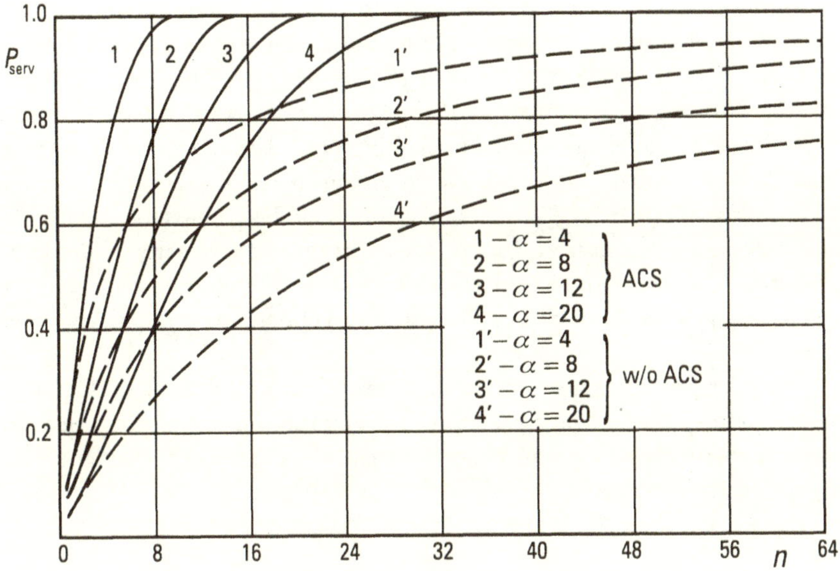

Figure 1.3 Dependency of the service probabilities P_{serv} and P'_{serv}.

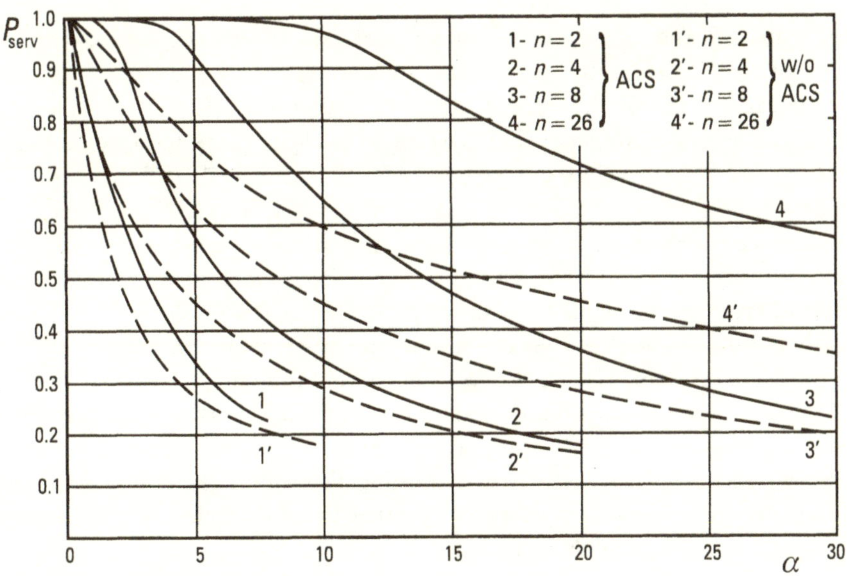

Figure 1.4 Dependency of the service probabilities P_{serv} and P'_{serv}.

As in the case considered (an automated control system), jamming that affects control systems for electronic weapons can lead to a breakdown in tracking or to a significant increase in the error variance for the determination of target coordinates by trackers. In the end, all these factors determine the successful service probability P_{ss}. The considerations given show that the probabilities of servicing an automated system for control of AAD forces P_{serv} and P'_{serv}, in the dynamics of EW, must be determined giving due consideration to P_{ss} (i.e., the average stream density of successful servicings μ^*). The latter requires evaluations of the operational quality of the electronic system for direct control of AAD defense combat equipment (weapons).

Similar problems also occur when examining models of multichannel mass service systems, which reflects the dynamics of using a group of active jammers to attack a group of radar systems of the detection, guidance, and target designation type.

1.4 Mathematical Models of Automated Systems for the Control of AAD Weapons as Targets of Electronic Warfare

1.4.1 Fundamental Principles

In principle, the operational dynamics of an automated system for the control of AAD combat equipment (weapons) can be described using a canonical system of differential stochastic equations (DSE) of state and observation (measurement). This system is similar to that encountered in the case of electronic trackers, more precisely, (1.9) and (1.10). The respective conclusions drawn for trackers as EW targets are also valid in this case. Below, we consider only certain individual issues associated, above all, with radio-controlled AAD strike systems. Generalized block diagrams of missile control systems, guidance laws and approximate methods for evaluating the quality and combat effectiveness of linearized systems for missile control all comprise parts. A knowledge of the issues enumerated is required to perform, in the first approximation, analytical evaluations of the capabilities of control systems as EW targets, as influenced by limited intensity jamming signals, and, on their basis, to determine threshold values for jamming parameters. High level jamming signals that cause a breakdown in tracking are considered in the chapters that follow.

Differential stochastic equations of the type (1.9) and (1.10) are used in the case where in-depth examination is required [21].

In Figures 1.5, 1.6, and 1.7, we give examples of generalized block diagrams for missile homing and command guidance systems [21–23].

Missile homing systems are automatic. The relative parameters of missile and target movement are measured by radar or optoelectronic systems located on the missile itself termed, in this case, homing heads. Depending on the principles of their construction and their frequency bands, they can be subdivided into radar, television, thermal, and laser

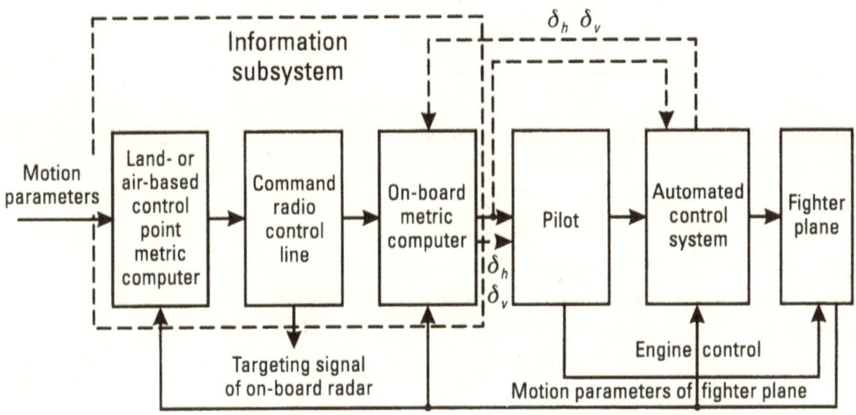

Figure 1.5 Generalized block diagram for a missile homing system.

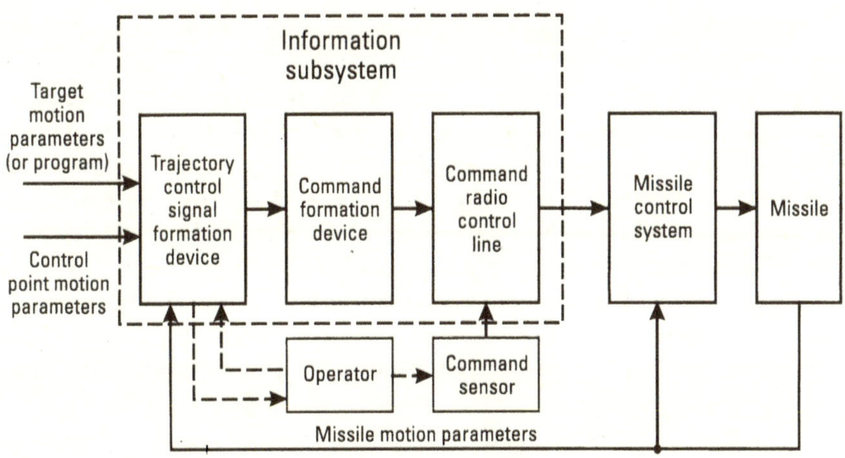

Figure 1.6 Generalized block diagram for a missile command guidance system (type 1).

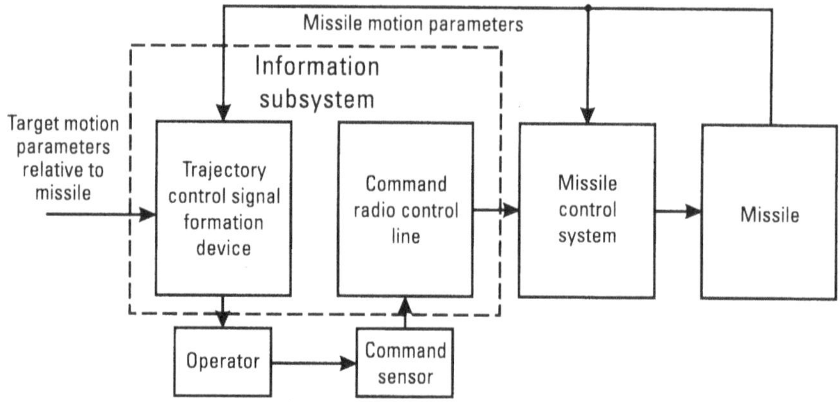

Figure 1.7 Generalized block diagram for a missile command guidance system (type 2).

homing heads. Signals are fed from the output of the homing head into the missile control system. This system includes sensors providing information about the parameters of the movement of the missile per se, as well as computer and control devices for the generation of signals to control the trajectory of the missile and to ensure its stability.

Missile command control systems can be divided into two types. In systems of the first type (Figure 1.6), the trackers measuring the relative motion parameters of the target and missile are located at the control point (mobile or stationary). The signal generation device controlling the trajectory evaluates the movement parameters of the missile and target relative to the control point and produces coordination parameters for all control channels. The command generation device performs the transformation of input signals into required control signals. Command radio control lines then transmit them to the missile control system. The latter performs the same functions as a homing system. There is a possible variant where specific missile movement parameters are generated at a command point and are then compared to actual movement parameters in the control system onboard the missile, resulting in the determination of coordination signals. In the case of nonautomatic control, the information subsystem contains an additional display (for example, television or thermovision), permitting an operator to make corrections to the trajectory of the missile with the help of a command transmitter.

The block diagram of a radio control command system of the second type is shown in Figure 1.7. Its distinguishing feature is that the initial information sources for relative target movement parameters are located on the missile and the output signals are transmitted to a control point, where they are used by an operator to generate control commands. Radio control command systems of the second type are always nonautomatic.

In the general case, the movement of the missile through space is described by a fairly complex system of nonlinear differential equations with variable coefficients. It is possible to research them fully only with the help of modern computer systems. The problem becomes significantly simpler if the movement of the missile through space is represented as the sum of two independent planar movements — longitudinal and lateral. Such a representation is based on actually implemented systems, where missile movement is controlled independently by yaw and pitch. Thus, two independent channels are considered separately for the control of the longitudinal and lateral movements of the missile [22, 23].

As a rule, missiles used in AAD systems are axisymmetric (i.e., they have axial aerodynamic symmetry). As a result, the equations for the longitudinal and lateral movement of such missiles coincide in form. This latter circumstance permits us to limit ourselves to considering movement in one plane only, with a subsequent recalculation of results obtained in the orthogonal plane.

1.4.2 Guidance Laws and Effectiveness Indicators for Control of AAD Missiles

To control the trajectories of homing AAD missiles ("ground-to-air" and "air-to-ground") by yaw and pitch, the proportional navigation (PN) method is used. Command guidance of the first type (Figure 1.6) is often performed using the combined method, also called the three-point method. In command control systems of the second type (Figure 1.7), the same PN law is implemented as with homing. The PN method can also be implemented in a three-point variant [23, 24].

In missile control systems using PN, the normal acceleration component j_N of the missile being guided is proportional to the line of sight rate ω of the target. Accordingly, the coordination equation Δ, which determines the trajectory control algorithm for the missile, can be written in the two following equivalent variants [23]:

$$\Delta = N\omega - j_N \tag{1.71*}$$

$$\Delta = N_0 V_{ap}\omega - j_N \tag{1.72}$$

*Formula numbers (1.68–1.70) are unused and intentionally left blank.

where N and N_0 are proportionality coefficients; and V_{ap} is the closing velocity of the missile on the target (aircraft).

Figure 1.8 shows a kinematic diagram of the movement in one plane of a homing missile m relative to a target (airplane) a in a system of coordinates with an origin at point A. The coordinates of the missile and target are fixed at a certain moment in time t, corresponding to: the closing velocity vector V_{ap}; the velocity vectors of the target and missile V_a and V_m; the distance between them D; the angles ε and ψ reckoned from a certain fixed (gyrostabilized) direction AA and determining the orientation of the target line of sight ma and the velocity vector V_m; and q, which is the yaw angle of the missile. The missile's relative velocity vector V_{rel} in a system of coordinates with its origin at point C is equal to the difference of the velocity vectors V_m and V_a of the missile:

$$V_{rel} = V_m - V_a$$

The minimal guidance error h occurring at a given moment of time and normally termed miss distance (instantaneous miss) is used as the effectiveness indicator for the guidance system (and other systems for the destruction of airborne targets). The miss distance is defined in the presentation plane (i.e., the plane that is perpendicular to the target line of sight). The miss distance is understood to be the minimum distance $h(t)$, where the missile will fly past the target provided that, beginning at a given moment of time t, the movement of the missile and the target are uniform

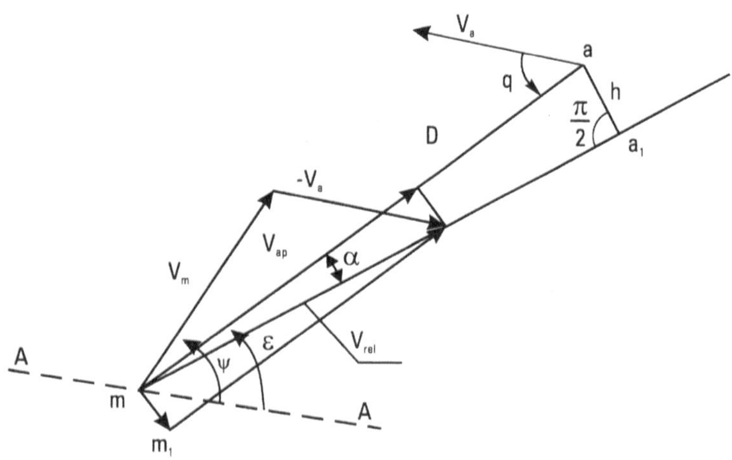

Figure 1.8 Kinematic diagram of the movement in one plane of a homing missile *m*.

and rectilinear. The components of the miss distance h_x and h_y are considered in the presentation plane along the axes of coordinates CX and CY (by yaw and pitch), with the origin of the coordinates at point C:

$$h^2 = h_x^2 + h_y^2 \tag{1.73}$$

The value of miss distance $h(t)$ for a fixed moment of time t according to Figure 1.8 is equal to

$$h = \frac{D^2 \dot{\varepsilon}}{V_{rel}} \tag{1.74}$$

where $\dot{\varepsilon} = \omega$ is the line of sight rate of the target.

Actually,

$$h = D \sin \alpha \tag{1.75}$$

$$\sin \alpha = \frac{mm_1}{V_{rel}} \tag{1.76}$$

The vector mm_1 is the transversal (lateral) V_T component of the relative velocity V_{rel}. By definition,

$$V_T = D\omega = D\dot{\varepsilon} \tag{1.77}$$

where $\dot{\varepsilon} = \omega$ is the line of sight rate. Equation (1.74) is a direct consequence of (1.75), (1.76), and (1.77).

Equations (1.71) and (1.74) indicate that the operating quality of the angle measuring channel of a homing head determines to a great extent the military capabilities of a missile as an AAD fire system. This permits us to view the angle measuring channel of a homing head as a major and direct object of jamming when performed with the help of appropriate jamming signals. In those cases where an optimum algorithm to control the trajectory of a missile is being implemented, it should be kept in mind that, along with measurement of the angles (angular velocities) of the target, it is also necessary to know the distance to the target D, as well as the approach velocity to the target V_{ap} [22, 23].

As will subsequently be shown, the jamming of automatic range and velocity tracking channels is much less difficult than the jamming of an angle measuring channel. Even when automatic range and velocity tracking channels are attacked, the missile can complete its homing process as long as the automatic direction tracking channel is functional. This circumstance once again confirms the thesis stated above that jamming of angle measuring channels in modern AAD missile control systems is of primary importance.

Fairly complete quantitative evaluations of jamming effectiveness can be obtained by examining its affect on closed-loop tracking systems for controlling a missile, taking into consideration its kinematic and dynamic elements.

1.4.3 A Model of a Kinematic Element in the Homing Circuit of a Missile

A component that reflects the kinematics of the relative movement of a missile when it is homing in on (being guided towards) a target (airplane, helicopter) is normally termed a kinematic element [23, 24]. In the case of homing, a kinematic element reflects the change over time of the missile–target range vector. As an example, let us define the parameters of a kinematic element when the relative movement of a missile occurs in a single plane (Figure 1.9). In relative motion, the position of the range vector *ma* is characterized by two equations that reflect the forward and rotational movements of the centers of mass of the missile and target. Both equations

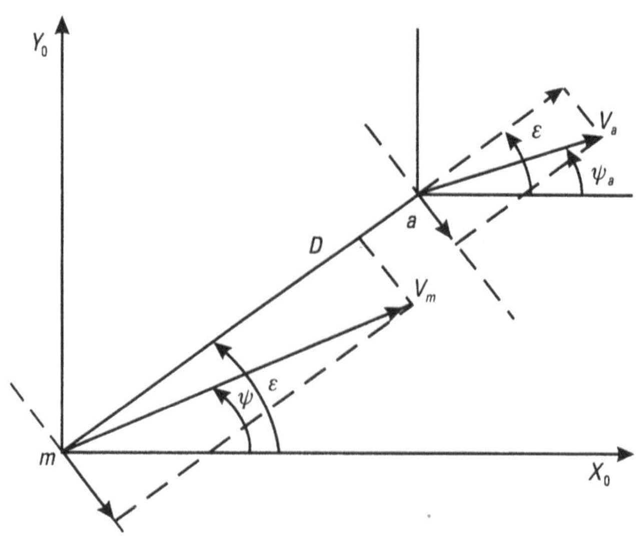

Figure 1.9 Parameters of a kinematic element in a single plane.

follow directly from Figure 1.9. The forward relative motion corresponds to the equation defining the approach velocity $V_{ap} = -\dot{D}$:

$$\dot{D} = V_a \cos(\varepsilon - \psi_a) - V_m \cos(\varepsilon - \psi) \tag{1.78}$$

The rotational relative motion corresponds to the equation defining the transversal velocity component:

$$D\dot{\varepsilon} = V_m \sin(\varepsilon - \psi) - V_a \sin(\varepsilon - \psi_a) \tag{1.79}$$

Kinematic equations (1.78) and (1.79) are nonlinear. It is possible to linearize them for courses close to those for pursuit and encounter. In these cases, the differences of angles $(\varepsilon - \psi)$ and $(\varepsilon - \psi_a)$ turn out to be quite small and (1.78) and (1.79) can be transformed into the following format:

$$\begin{aligned} \dot{D} &= V_a - V_m \\ D\dot{\varepsilon} + \dot{D}\varepsilon &= V_a\psi_a - V_m\psi \end{aligned} \tag{1.80}$$

Let us assume that the velocities V_m and V_a are constant and differentiate them (1.80) with respect to t. We obtain:

$$-2V_{ap}\,\dot{\varepsilon} + D\ddot{\varepsilon} = V_a\dot{\psi}_a - V_m\dot{\psi} \tag{1.81}$$

In the case of the linearization being analyzed, the transversal and normal components for the acceleration of the missile and the airplane are quite close, permitting us to write it down in the following format:

$$\dot{\omega} - \frac{2V_{ap}}{D}\omega = \frac{1}{D}(j_{Na} - j_N) \tag{1.82}$$

where j_{Na} and j_N are the normal components for the acceleration of the target (airplane) and the missile; $\omega = \dot{\varepsilon}$. Let us write down the equation (1.82) in operator format:

$$\left(p - \frac{2V_{ap}}{D}\right)\omega = \frac{1}{D}(j_{Na} - j_N) \tag{1.83}$$

where p is the differentiation symbol. By definition [23, 25], the transfer function of the kinematic element $W_{KE}(p, t)$ can be defined as the ratio of the line of sight rate $\omega(t)$ to the difference of the normal components for the acceleration of the target and missile $\Delta j(t) = j_{Na} - j_N$. Therefore,

$$W_{KE}(p, t) = \frac{\omega(t)}{\Delta j(t)} \tag{1.84}$$

Taking into consideration (1.82), we obtain

$$W_{KE}(p, t) = \frac{1}{Dp - 2V_{ap}} = \frac{K_{KE}}{T_{KE}p - 1} \tag{1.85}$$

Here, $K_{KE} = \frac{1}{2V_{ap}}$ is the transmission coefficient of the kinematic element; and $T_{KE} = \frac{D}{2V_{ap}}$ is the time constant of the kinematic element.

In principle, a kinematic element is not stationary due to the change in the dynamics of range guidance D. The element shown in (1.85) can be considered to be quasi-static and represented as an integrator, to which positive feedback has been applied (Figure 1.10). In reality, according to the transformation rules for dynamic elements, the transfer function for the closed-loop element in the form given with positive feedback is equal to

$$W_{KE} = \frac{\frac{1}{D}\frac{1}{p}}{1 - \frac{2V_{ap}}{D}\frac{1}{p}} = \frac{1/2V_{ap}}{T_{KE}p - 1}$$

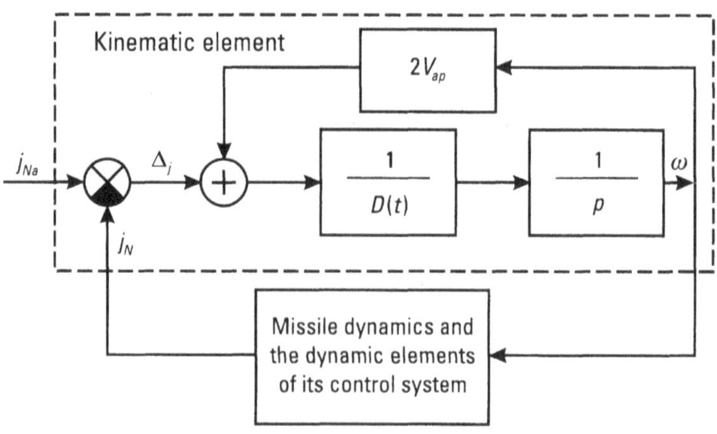

Figure 1.10 Integrator with positive feedback.

which corresponds to (1.85). Due to the positive feedback shown, the kinematic element is not stable. By introducing a circuit with negative feedback into the model, the positive feedback can be neutralized. The negative feedback circuit contains the dynamic elements of the missile and its control systems. In the case of PN homing, when a signal is generated that linearly depends on V_{ap}, the system remains stable for all angles of attack [23].

1.4.4 Mathematical Models of Angle Measuring Devices for Onboard Radar and Radar Homing Heads

As was noted before, in the conditions of EW, devices for measuring angles and line of sight rate comprise the dynamic elements in a system for controlling the trajectory of a homing missile, determining the magnitude of its miss distance at all stages of its guidance. As a rule, they consist of: a tracking system, including a range finder; a power amplifier; a drive; and transmitters of correction signals. Range finders can be radar, thermal, television, and laser in nature. In the material that follows, greater attention will be paid to radar range finders and, accordingly, to radar angle measuring devices. Gyroscopic devices, which provide needed correction to the measuring devices and ensure their stabilization in space, are a mandatory component of measuring devices for angles and angular velocities in a missile (airplane, helicopter). It is possible to differentiate position and velocity correction. Position correction is associated with measuring indicators for angles, and velocity correction with measuring devices (sensors) for angular velocities [23].

Approximate methods for evaluating the quality of control systems can be derived from linearized models of angle measuring devices and other trackers. As examples, we analyze below several actual variants of dynamic elements of this kind that permit us to quantitatively evaluate the dynamic and random errors of angle measuring channels.

In Figure 1.11, we show the geometrical interrelationships occurring when a fighter plane homes in on a predetermined intercept point [23]. Here, Fa is the target line of sight, ε is the required intercept angle, ε_F is the value of the intercept angle as measured at the given moment, ESL is the instantaneous position of the equal signal line of the antenna of the fighter's onboard radar, and Fx is the direction of the longitudinal axis of the fighter plane. A simplified block diagram of an automatic angle measuring device with velocity correction and indicator stabilization, which provides the homing specified to keep the fighter on course, is given in Figure 1.12. Here, $W_1(p)$, $W_2(p)$, and $W_3(p)$ are the transfer functions of: the range

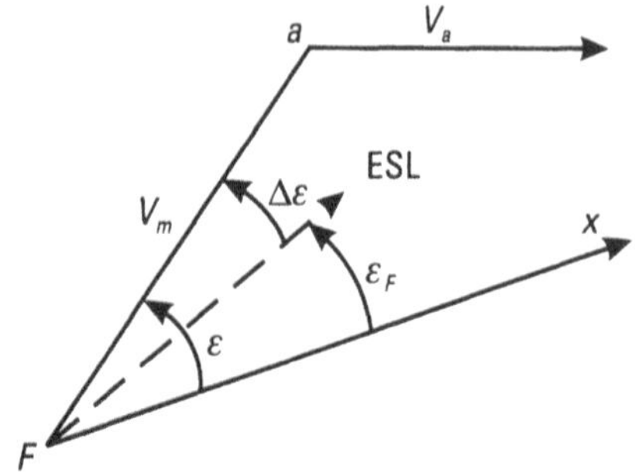

Figure 1.11 The geometrical interrelationships occurring when a fighter plane homes in on a predetermined intercept point.

finder; the onboard radar antenna drive together with its power amplifier; and the angular velocity sensor. It is assumed that the stabilization circuit provides for a high degree of coupling for radar antenna motion and of fighter yawing, which permits us to eliminate the yawing angle from our considerations.

The instantaneous dynamic error is

$$\Delta\varepsilon = \varepsilon - \varepsilon_F \tag{1.86}$$

where

$$\varepsilon_F = \Phi_1(p)\Delta\varepsilon$$

$\Phi_1(p)$ is the transmission function of the closed range-finder circuit, including $W_1(p)$ and $\Phi_{ST}(p)$.

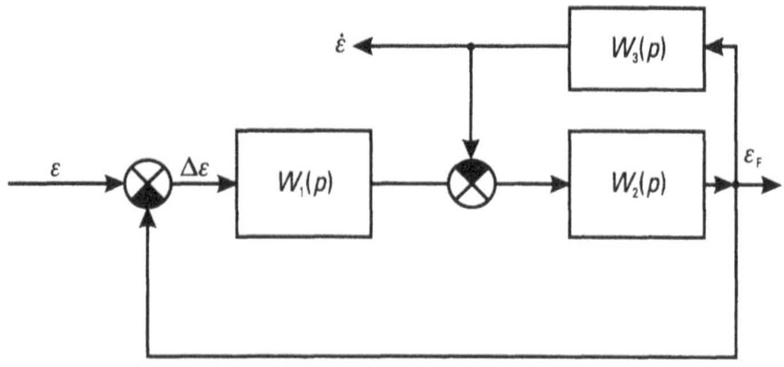

Figure 1.12 A simplified block diagram of an automatic angle measuring device with velocity correction and indicator stabilization.

According to Figure 1.12,

$$\Phi_1(p) = \frac{W_1(p)\Phi_{ST}(p)}{1 + W_1(p)\Phi_{ST}(p)} \tag{1.87}$$

Here,

$$\Phi_{ST}(p) = \frac{W_2(p)}{1 + W_2(p)W_3(p)} \tag{1.88}$$

is the transfer function of the closed-loop stabilization circuit of the radar antenna.

We can assume [23]

$$W_1(p) = \frac{K_{DF}}{T_{DF}\,p + 1} \quad W_2(p) = \frac{K_{DR}}{p(T_{DR}\,p + 1)} \quad W_3(p) = K_{AV}\,p$$

where T_{DF} and T_{DR} are the time constants for the range finder and antenna drive, K_{DF}, K_{DR}, K_{AV} are the gains of: the range-finder device, the output of which is an inertia amplifier for direct current; the radar antenna drive together with its power amplifier; and the angular velocity sensor, which, in the given case, is a differentiation circuit. Consequently, using the values taken for the transfer functions and assuming $T_{DF} >> \frac{T_{DR}}{1+K_{DR}\,K_{AV}}$, with the help of (1.88) and (1.87), we obtain

$$\Phi_{ST}(p) = \frac{K_{DR}}{p(1 + K_{DR}\,K_{AV})\left(\dfrac{T_{DR}}{1 + K_{DR}\,K_{AV}}p + 1\right)}$$

$$\Phi_1(p) = \frac{K_V}{T_{DF}\,p^2 + p + K_V} \tag{1.89}$$

where

$$K_V = \frac{K_{DF}\,K_{DR}}{1 + K_{DR}\,K_{AV}} \tag{1.90}$$

is the gain for the angular velocity of the angle measuring channel under consideration.

It should be noted that expression (1.89) for the transfer function of the closed-loop circuit for the range finder $\Phi_1(p)$ is common and is also valid for other types of angle measuring devices. The formula (1.90), which defines

K_{V}, is for a range finder of the type being considered. For example, if position correction and power stabilization are performed by the angle measuring device, then [23],

$$K_{V} = K_{DF} K_{A} K_{GD} \tag{1.91}$$

Here, K_{DF} is the gain of the range finder, which is assumed to be unsmoothed; K_{A} is the gain of the power amplifier; and K_{GD} is the gain of the gyro drive.

As an object of EW, the amplitude and frequency responses of the range finder, as well as its passband $\Delta F_{\ni\varepsilon}$, are important parameters. The latter can be determined if we know the amplitude and frequency response $\Phi_1(j\omega)$. In order to find it, it is necessary to substitute $j\omega$ for p in (1.89):

$$|\Phi_1(j\omega)|^2 = \frac{K_{V}^2}{(K_{V} - T_{DF}\omega^2)^2 + \omega^2} \tag{1.92}$$

Accordingly,

$$\Delta F_{\ni\varepsilon} = \frac{1}{2\pi}\int_0^{\infty} |\Phi_1(j\omega)|^2 d\omega \tag{1.93}$$

From (1.92) and (1.93) we obtain

$$\Delta F_{\ni\varepsilon} = \frac{K_{V}}{4} \tag{1.94}$$

In Figure 1.13 we show the amplitude and frequency responses of a tracking angle measuring device, constructed using (1.92), for various values of K_{V} and T_{DF}. The dependency of $|\Phi_1(j\omega)|^2$ on the parameters shown is noteworthy. In particular, this presents a possibility of decreasing angular tracking quality by deliberately changing the gain of the angle measuring channel for line of sight rate K_{V} using jamming signals generated by emission sources distributed in space.

1.4.5 Dynamic Errors in an Angle Measuring Device

According to (1.85), (1.86), and (1.89), the value of the instantaneous dynamic error $\Delta\varepsilon$, in operator format, is determined by the formula:

$$\Delta\varepsilon = (1 - \Phi_1(p))\varepsilon = \frac{p(T_{DF}p + 1)}{T_{DF}p^2 + p + K_{V}}\varepsilon \tag{1.95}$$

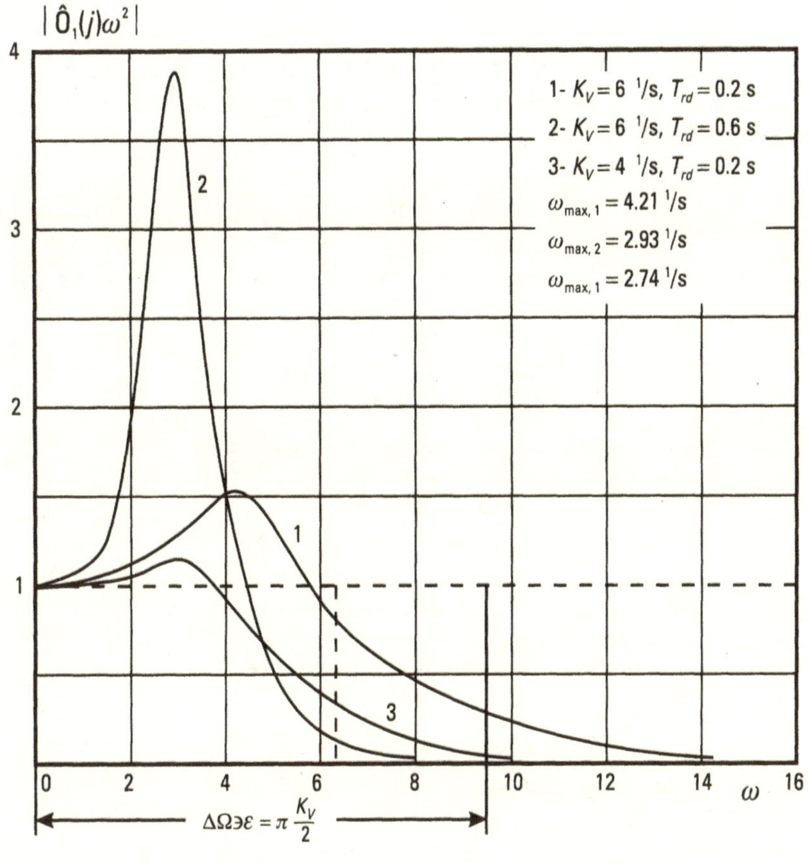

Figure 1.13 The amplitude and frequency responses of a tracking angle measuring device.

The steady state value of the dynamic error $\Delta\varepsilon_S$ is determined by the transfer function:

$$W_0(p) = \frac{p(T_{DF}p + 1)}{T_{DF}p^2 + p + K_V}$$

and the input effect ε by presenting both in the form of corresponding series. In the case of the automatic angle estimator under consideration [23],

$$\Delta\varepsilon_S = \frac{1}{K_V}\left(\dot{\varepsilon}_0 - \frac{\ddot{\varepsilon}_0}{K_V} + T_{DF}\,\ddot{\varepsilon}_0\right) + \frac{\ddot{\varepsilon}_0}{K_V}t \tag{1.96}$$

Here, $\dot{\varepsilon}_0$ and $\ddot{\varepsilon}_0$ are the angular velocity and the line of sight rate at $t = 0$.

According to Figure 1.11 and the block diagram in Figure 1.12, the instantaneous dynamic error $\Delta\omega$ for measuring the line of sight rate is determined using the equation:

$$\Delta\omega = \omega - \omega_{I'}$$

where $\omega = \dot{\varepsilon}$ and $\omega_{I'} = \ddot{\varepsilon}_{I'}$ are the input and measured values of the line of sight rate:

$$\dot{\varepsilon}_{I'} = \Phi_1(p)p\varepsilon = \Phi_1(p)\dot{\varepsilon}$$

$$\Delta\omega = (1 - \Phi_1(p))\omega = \frac{p(T_{DF}p + 1)}{T_{DF}p^2 + p + K_{I'}}\omega \tag{1.97}$$

The steady state value of the dynamic error $\Delta\omega_S = \Delta\dot{\varepsilon}_S$ is determined by analogy to $\Delta\varepsilon_S$. This follows directly from the comparison of (1.95) and (1.97):

$$\Delta\omega_S = \frac{\omega_0}{K_{I'}} \tag{1.98}$$

where $\dot{\omega}_0$ is the steady state value of the line of sight rate.

It follows from (1.96) and (1.98) that the steady state value of the dynamic errors $\Delta\varepsilon_S$ and $\Delta\omega_S$ are inversely proportional to $K_{I'}$.

The effective passband $\Delta F_{\ni\varepsilon}$ of the tracking system, for which the input signal is the angle ε and the output signal is the angular velocity $\omega_P = \dot{\varepsilon}_P$, is determined from (1.97) using the transfer function $\Phi_1(p)p = \Phi_2(p)$, after conversion to the complex gain $|\Phi_2(j\omega)|^2$. Unlike the case of the amplitude and frequency response $|\Phi_1(j\omega)|^2$ considered before, the equivalent rectangular amplitude and frequency response with band $\Delta F_{\ni\varepsilon}$ is normalized not to one, but to the maximum value $|\Phi_2(j\omega)|^2$, which is equal to $K_{I'}^2$ [23]:

$$\Delta F_{\ni\varepsilon} = \frac{1}{2\pi K_{I'}^2}\int_0^\infty |\Phi_2(j\omega)|^2\, d\omega = \frac{1}{4T_{DF}} \tag{1.99}$$

The maximum value $|\Phi_2(j\omega)|^2$ is shifted along the frequency axis relative to the origin of the coordinates by a value of

$$f_M = \frac{1}{2\pi}\sqrt{\frac{K_V}{T_{DF}}} \tag{1.100}$$

$$|\Phi_2(j\omega)|^2 = \frac{K_V^2\omega^2}{(K_V - T_{DF}\,\omega^2)^2 + \omega^2} \tag{1.101}$$

According to (1.101), we have calculated and show in Figure 1.14 the amplitude and frequency responses of a tracker for line of sight rate $|\Phi_2(j\omega)|^2$ for various values of K_V. As in the case of an angle measuring

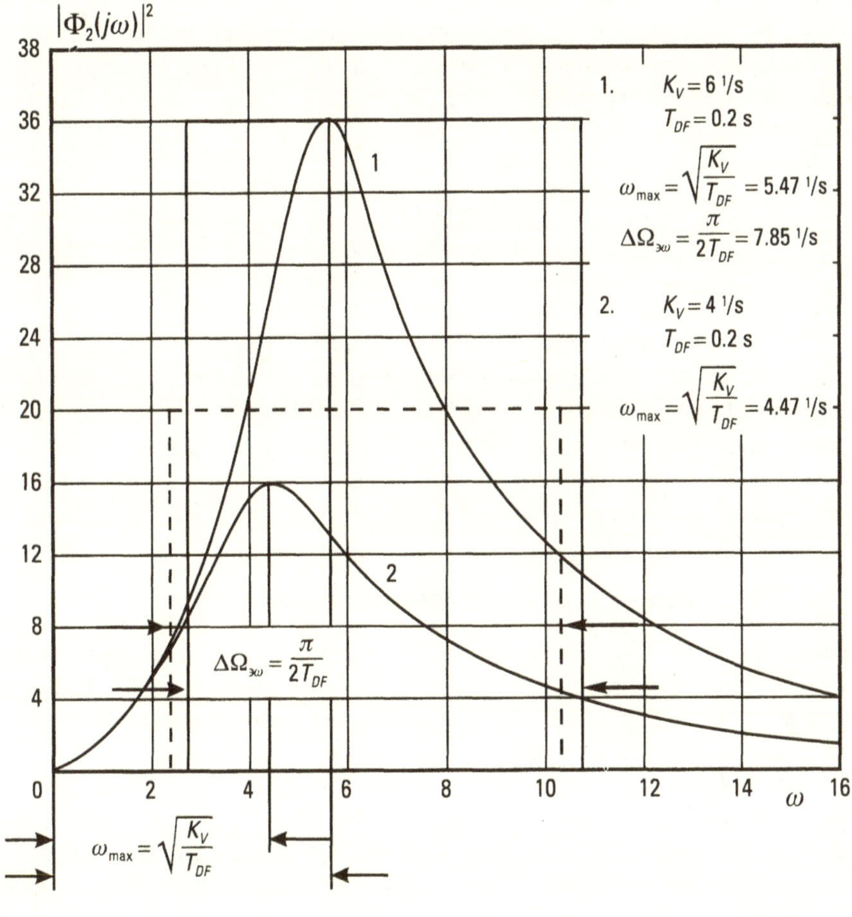

Figure 1.14 The amplitude and frequency responses of a tracker for line of sight rate $|\Phi_2(j\omega)|^2$.

device, the measurements of K_V lead to the deformation of the amplitude and frequency response, which can reduce the quality of measurement for the line of sight rate of the target.

In principle, the possibility of deliberately changing K_V results from its dependence on the gain of the range finder K_{DF}, which in turn is directly associated with the slope of the angle response of the angle measuring device K_α:

$$K_{DF} = K_\alpha K'_{DF} \tag{1.102}$$

where K'_{DF} takes into consideration the gains of the other elements of the automatic range tracking system [25].

The slope of the angle response is determined by the pattern of the antenna of the range finder $F(\Phi, \Theta)$, and Φ and Θ are the angles in a spherical system of coordinates [25, 26]. If range finding is performed in a single plane, then,

$$K_\alpha = \left| \frac{\dfrac{dF(\Theta)}{d\Theta}}{F(\Theta)} \right| \quad \Theta = \Theta_0 \tag{1.103}$$

where Θ_0 is the angle which corresponds to the equal-signal line.

The formula (1.103) usually assumes that the pattern of the antenna is determined using a single point-source. If there is more than one source of emission, or if the source is not a point, then the pattern of the antenna and, accordingly, the direction finding characteristics are deformed, which, in the end, leads to changes in K_V.

1.4.6 Random Errors in an Angle Measuring Device

The examination of angle measuring channels in onboard radar systems and radar homing heads makes it possible to evaluate deterministic effects. Let us review in linear approximation some of the methods for determining random errors in tracking angle measuring devices.

Random errors can occur due to an influence on the range finder of a group of random emission sources distributed in time and space. Their effect on the range finder can be considered to be space and time noise. In the material that follows, the term angle noise will be used more frequently.

Under the influence of angle noise, the angle ε (Figure 1.11) will change randomly. As follows from the block diagram shown in Figure 1.12, the instantaneous angle error $\Delta\varepsilon$ will also change randomly determined by the transfer function of the closed loop circuit of the range finder $\Phi_1(p)$:

$$\Delta\varepsilon(t) = \Phi_1(p)\varepsilon(t) \tag{1.104}$$

If $\varepsilon(t)$ is a random process with spectral density $S_{\varepsilon n}(\omega)$, the variance D_ε of angle $\Delta\varepsilon$ in linear approximation can be found from the equation:

$$D_\varepsilon = \frac{1}{2\pi}\int_0^\infty |\Phi_1(j\omega)|^2 S_{\varepsilon n}(\omega)d\omega \tag{1.105}$$

where $|\Phi_1(j\omega)|^2$ is determined in the given case using (1.93).

If $S_{\varepsilon n}(\omega) = const$, then it follows from (1.105) and (1.92) that

$$D_\varepsilon = S_{\varepsilon n}(\omega)\frac{K_V}{4} \tag{1.106}$$

The fluctuation variance D_ω of the line of sight rate $\Delta\omega$ can also be determined on the basis of the block diagram shown in Figure 1.12. In particular, it follows from the figure that

$$\Delta\omega = \Phi_1(p)W_3(p)\varepsilon = p\Phi_1(p)\varepsilon \tag{1.107}$$

Fluctuations in ε cause fluctuations in $\Delta\omega$ also. If $S_\varepsilon(\omega)$ is the spectral density of the input angle noise, then the variance D_ω in linear approximation will be equal to

$$D_\omega = \frac{1}{2\pi}\int_0^\infty |\Phi_2(j\omega)|^2 S_\varepsilon(\omega)d\omega \tag{1.108}$$

where $|\Phi_2(j\omega)|^2$ is determined from (1.101).

If $S_\varepsilon(\omega) = const$, then, taking into consideration (1.99), we obtain

$$D_\omega = \frac{K_V^2}{4T_{DF}}S_\varepsilon(\omega) \tag{1.109}$$

For the dynamic structure under consideration, the optimum transition process occurs when

$$T_{DF} K_V = 1 \tag{1.110}$$

Taking into consideration (1.110), it follows from (1.106) and (1.109) that the mean square errors σ_ω and σ_ε are linearly dependent on each other:

$$\sigma_\omega = K_V \sigma_\varepsilon \tag{1.111}$$

where

$$\sigma_\omega = \sqrt{D_\omega} \qquad \sigma_\varepsilon = \sqrt{D_\varepsilon} \tag{1.112}$$

There exists an optimum value K_V, which corresponds to the minimum total error value and is equal to the sum of the squares of the dynamic and random errors [23].

1.4.7 A Linearized Mathematical Model of a Closed Loop Circuit in a Radar Homing Head (under the Influence of Angle Noise)

In order to quantitatively determine the miss value of the missile $h(t)$, it is necessary to analyze a closed loop homing circuit, in which, besides the kinematic element and the angle measuring device, we must take into consideration other dynamic elements (i.e., the control system of the missile and the dynamics of the missile itself). In the case of PN homing, the input and output influences in the closed loop circuit are the normal acceleration of the target (airplane, helicopter) j_{Na} and the current missile miss distance $h(t)$. First approximation estimates of the dynamic and random errors in a linearized closed loop homing system can be obtained using the example of a missile moving in a single plain.

In Figure 1.15 we show a simplified block diagram permitting an approximate evaluation of the dynamic homing error, which is understood in the given case to be the missile miss distance h_K taking place at the end of the homing process [23]. It is assumed that the variable coefficients of the homing circuit are "frozen." The transfer function $W_{KE}(p)$ is determined using (1.84) (Figure 1.10) and reflects the properties of the kinematic element. Accordingly, the transfer function $W_{j\omega}(p)$ reflects the tracking angle measuring device of the missile control system and the missile itself as an object of control. The output value $h(t)$ and the input value j_{Na} (Figure 1.15) are interrelated by the following formula, resulting from the definition of a closed loop tracking system:

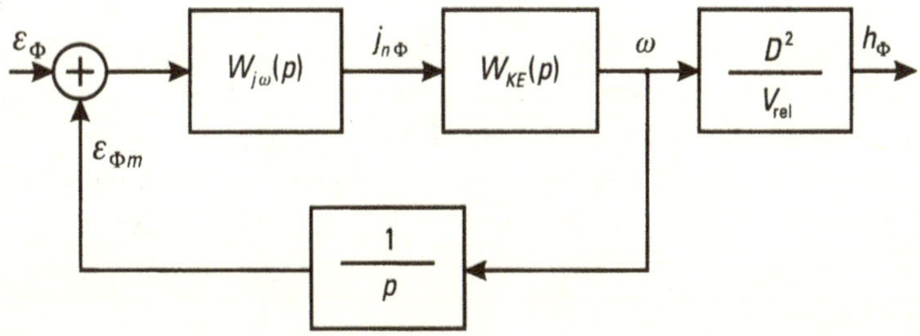

Figure 1.15 A simplified block diagram permitting an approximate evaluation of the dynamic homing error.

$$h(t) = \frac{D^2}{V_{rel}} \Phi_{HC}(p) j_{Na} \tag{1.113}$$

Here,

$$\Phi_{HC}(p) = \frac{W_{KE}(p)}{1 + W_{KE}(p) W_{j\omega}(p)} \tag{1.114}$$

is the transfer function of the closed loop homing circuit.

According to methods well known in the theory of automatic control and assuming an input value $j_{Na} = j_{NO} = const$, the steady state miss distance h_S, which in the given case represents the dynamic error, will be equal to

$$h_S = \frac{D^2}{V_{rel}} [\Phi_{HC}(p)]_{p=0}\, j_{NO} \tag{1.115}$$

It is assumed that the speed of the missile is sufficiently high.

To have the possibility of quantitatively evaluating the stability of a missile in the homing process, as well as its dynamic and random errors, it is necessary to determine, in first approximation, the equivalent structure and parameters of the transfer function $W_{j\omega}(p)$ (Figure 1.15). In the case where PN is used, $W_{j\omega}(p)$ can be represented as the equivalent resonant element, which has the following format [23]:

$$W_{j\omega}(p) = \frac{K_{H\omega} K_{HR}}{T_{\ni\omega} T_{HR}\, p^2 + (T_{\ni\omega} + T_{HR})\, p + 1} \tag{1.116}$$

where: K_{HR} and T_{HR} are the gain and time constant of the equivalent element under consideration; and $K_{H\omega} = N_0 V_{ap} = N$ is the gain of the homing head along the signal generation channel of the control system for the missile according to the PN law (1.71).

If the velocity of a missile in the homing process is sufficiently high (no less than several hundred meters a second), then it is possible to assume that $K_{HR} = 1$ and $T_{HR} = 0$. Then the transfer function $\Phi_{HC}(p)$ can be represented using the expression:

$$\Phi_{HC}(p) = \frac{K_{KE}(T_{\ni\omega} p + 1)}{T_{KE} T_{\ni\omega} p^2 + (T_{KE} - T_{\ni\omega}) p + K_{H\omega} K_{KE}^{-1}} \tag{1.117}$$

where $T_{\ni\omega} = \frac{1}{K_V}$ is the equivalent time constant of the measuring instrument for the line of sight rate; $T_{KE} = \frac{D}{2V_{ap}}$ is the time constant of the kinematic element (1.84); and $K_{KE} = \frac{1}{2V_{ap}}$ is the gain of the kinematic element.

Taking into consideration (1.117), the expression (1.115) can be converted to the following form:

$$h_S = \frac{4T_{KE}^2 V_{ap} j_{NO}}{V_{rel}(N_0 - 2)} \tag{1.118}$$

Designating the range to the target at the final moment of the homing process of the missile by D_K and the time of guided flight of the missile by $t_K = 2T_{KE} = \frac{D_K}{V_{ap}}$, we can convert (1.117) to the following format:

$$h_S = \frac{t_K^2 j_{NO}}{N_0 - 2}$$

Here it was assumed that $V_{ap} = V_{rel}$. The quadratic dependency of the final (steady state) miss distance on the unguided flight time t_K is worthy of note.

The transfer function $\Phi_{HC}(p)$, written in the format of (1.117), permits us to determine the stability conditions of the missile in the homing process and to establish their relationship to the parameters of the basic dynamic elements. In the case of the system under consideration, which is described by a second-order differential equation, it is expedient to use the algebraic criterion of stability. Accordingly, it is necessary to make a characteristic equation for the system and define the conditions for which its coefficients are positive. In the given problem, we obtain the characteristic equation by replacing p by λ in the denominator of (1.117):

$$T_{KE} T_{Э\omega} \lambda^2 + (T_{KE} - T_{Э\omega})\lambda + K_{H\omega} K_{KE} - 1 = 0 \tag{1.119}$$

Since $T_{KE} \geqslant 0$, $T_{Э\omega} > 0$, stability of the homing circuit occurs if

$$T_{Э\omega} < T_{KE} = \frac{D}{2V_{ap}} \tag{1.120}$$

$$K_{H\omega} > \frac{1}{K_{KE}} = 2V_{ap} \tag{1.121}$$

If $K_{H\omega} = N_0 V_{ap}$, then it follows from (1.121) that the navigation constant must be greater than 2:

$$N_0 > 2 \tag{1.122}$$

In as much as $T_{Э\omega} = \frac{1}{K_{V}}$, where K_V is the gain for line of sight rate, it follows from (1.120) that K_V has the limitation:

$$K_V > \frac{2V_{ap}}{D} \tag{1.123}$$

The last inequality indicates that, through the use of jamming, there is a potential possibility to deliberately change the parameters of one of the dynamic elements (of the range finder) of the homing circuit of the missile so as to disrupt its stability.

The miss variance, which occurs under the influence of angle noise on the homing circuit of the missile, can be evaluated with the help of (1.74). If $\dot{\varepsilon}(t) = \omega(t)$ is the sample function of a random process with a zero mathematical expectation and a variance of D_ω, then the miss variance D_h is determined according to the well-known rules of the theory of probability [3, 20]. In as much as

$$h = \frac{D^2 \omega}{V_{rel}}$$

then

$$D_h = \left(\frac{D^2}{V_{rel}}\right)^2 D_\omega \tag{1.124}$$

The value of the miss variance at the end of the homing process (at the moment when unguided flight begins) D_{hk} can be accordingly written, by analogy with (1.124), simply by adding the additional index K:

$$D_{hK} = \left(\frac{D_K^2}{V_{rel}}\right)^2 D_{\omega K} \tag{1.125}$$

In order to determine the variance $D_{\omega K}$, it is necessary to convert the block diagram in Figure 1.15 to the form shown in Figure 1.16 [22]. Here the input to the dynamic block diagram is the random angle ε_Φ (ref. Figure 1.8) generated by angle noise or other jamming radio signals influencing the angle measuring device. At the input to the homing circuit, we have the action of the random line of sight rate $\dot{\varepsilon}_\Phi$. The dynamic element with the transfer function $W_{j\omega}(p)$ is equivalent to a group of dynamic elements representing a radar homing head, a flight control system or a missile as an object of control. The transfer function of the kinematic element $W_{KE}(p)$ is determined using (1.81) and (1.84), where $j_{Na} = 0$ and $j_{N\varphi}$ is substituted for j_N. The random normal acceleration $j_{N\varphi}$ is generated by the random interference angle ε_Φ. The mathematical expectations ε_Φ and $\dot{\varepsilon}_\Phi$ are assumed to be equal to zero. The noise ε_Φ is assumed to be white, with a constant one-sided spectral density of $S_{\varepsilon\Sigma}(\omega) = S_{\varepsilon\Sigma}$. In view of the nonstationary nature of the tracking system under consideration, resulting basically from the kinematic element, the solution of this problem in statistics is difficult in its generalized form. Below we give an approximate variant of a solution, based on the principle of "freezing" the variable coefficients, according to which, at definite segments of the trajectory and during a limited time, the parameters of the dynamic elements are assumed to be constant and the closed loop homing circuit to be stationary.

The variance D_ω in the stationary linearized homing circuit, as represented in Figure 1.16, will be equal to

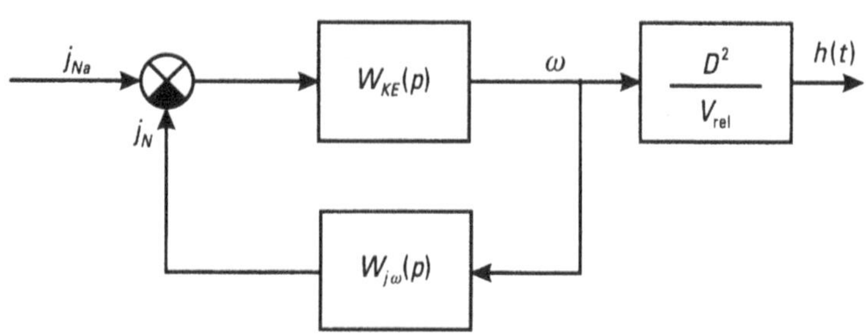

Figure 1.16 A dynamic block diagram to determine the variance $D_{\omega K}$.

$$D_\omega = \frac{S_{\varepsilon\Sigma}}{2\pi} \int_0^\infty |\Phi_{HC}(j\omega)|^2 d\omega \tag{1.126}$$

where $\Phi_{HC}(j\omega)$ is the complex gain of the closed homing loop, in which ε_Φ is active at the input, and $\dot{\varepsilon}_\Phi$ at the output.

As a specific variant of a solution to the problem, let us consider the case where the radar homing head of the missile is equivalent to a differentiating element with a transfer function $K_{H\omega}p$. The differentiating element provides for transformation of the fluctuations of angle ε_Φ into fluctuations of angular velocity $\dot{\varepsilon}_\Phi$. The missile control system and the missile as an object of control are represented as an inertial element $1/[(T_{\ni\omega} + T_{HR})p + 1]$ with a gain of K_{HR}, which is equal to one. This is valid if in (1.116) the following condition is satisfied:

$$T_{\ni\omega} T_{HR} p^2 << (T_{\ni\omega} + T_{HR})p + 1$$

According to what as been said,

$$W_{j\omega}(p) = \frac{K_{H\omega}p}{(T_{\ni\omega} + T_{HR})p + 1}$$

Sequentially performing the required transformations, we obtain the following expression, which defines the square of the absolute value of the complex gain:

$$|\Phi_{HC}(j\omega)|^2 = \frac{N_0^2 V_{ap}^2 K_{KE}^2 \omega^2}{|-(T_{\ni\omega} + T_{HR})T_{KE}\,\omega^2 + (T_{KE} - T_{\ni\omega} - T_{HR})j\omega + K_{\ni\omega}K_{KE} - 1|} \tag{1.127}$$

The integral (1.126) is determined using the well-known method [22, 27]. Accordingly,

$$D_\omega = \frac{N_0^2 S_{\varepsilon\Sigma}}{16 T_{KE}(T_{\ni\omega} + T_{HR})(T_{KE} - T_{\ni\omega} - T_{HR})} \tag{1.128}$$

If the speed of the missile is sufficiently high, then T_{HR} can be assumed to be equal to zero. Then,

$$D_\omega = \frac{N_0 S_{\varepsilon\Sigma}}{16 T_{KE} T_{Э\omega}(T_{KE} - T_{Э\omega})} \tag{1.129}$$

After replacing T_{KE} and $T_{Э\omega}$ by their values $T_{KE} = \frac{D}{2V_{ap}}$, $T_{Э\omega} = \frac{1}{K_V}$, the formula for D_ω is converted to the format:

$$D_\omega = \frac{N_0^2 V_{ap}^2 K_V^2 S_{\varepsilon\Sigma}}{4D(DK_V - 2V_{ap})} \tag{1.130}$$

According to (1.125) and (1.130), the miss variance $D_{h\text{K}}$ being sought, which occurs at the end of the homing process (i.e., at a distance of D_K from the airplane (target)), will be equal to

$$D_{h\text{K}} = \frac{N_0^2 D_\text{K}^3 V_{ap}^2 K_V^2 S_{\varepsilon\Sigma}}{4V_{rel}^2(D_\text{K} K_V - 2V_{ap})} \tag{1.131}$$

The variance value of the resulting miss distance permits us to determine the probability that an airplane (target) will be hit by a single missile. As an example, let us consider the case when the hit probability can be determined using the diagram proposed by A. N. Kolmogorov. According to this diagram, the single-shot hit probability $P_{hit}(1)$ is determined using the formula:

$$P_{hit}(1) = \int_{-\infty}^{+\infty}\int P(h_x, h_y)\, p(h_x, h_y) \mathrm{d}h_x\, \mathrm{d}h_y \tag{1.132}$$

Here, $P(h_x, h_y)$ is the conditional probability that the target will be hit by a single missile, provided its resulting miss distances, defined in the presentation plane, are equal to h_x and h_y. $p(h_x, h_y)$ is the probability density of miss distances h_x and h_y, defined in the same presentation plane. We understand the presentation plane to be a plane perpendicular to the line of sight of the target and passing through its center of mass (or other conditional center). The conditional probability $P(h_x, h_y)$ is also called the hit distribution. In the specific case where the hit distribution can be approximated by a Gaussian curve:

$$P(h_x, h_y) = \exp\left(-\frac{h_x^2 + h_y^2}{2R_{ЭФ}^2}\right) \tag{1.133}$$

where $R_{\ni\Phi}$ is the effective target hit radius and the missile miss distances h_x and h_y are uncorrelated Gaussian random values with a zero mean and identical variances $D_{hx} = D_{hy}$; that is,

$$p(h_x, h_y) = \frac{1}{2\pi D_{hx}} \exp\left(-\frac{h_x^2 + h_y^2}{2D_{hx}}\right) \tag{1.134}$$

the single-shot hit probability $P_{hit}(1)$, defined by (1.132), is equal to [23, 24]

$$P_{hit}(1) = \frac{1}{1 + \dfrac{D_{hx}}{R_{\ni\Phi}^2}} \tag{1.135}$$

This concludes our analysis of mathematical models of angle measuring channels and closed loop radar homing head circuits, which comprise one of the basic targets of EW in an AAD system. Using analogous diagrams, it is possible to define linearized models of AAD missile command guidance systems and other electronic systems for the control of weapons. Some of the results obtained can be used directly. In particular, this is valid for the linearized angle measuring tool model. It can be used when examining automatic tracking channels for range and velocity. The combined analysis of tracking radar sensors for direction, range, and velocity permits us to define mathematical models of radar sights for antiaircraft artillery complexes, mobile antiaircraft installations, and also radar sights for AAD fighter planes viewed as EW targets.

In conclusion, we would like to formulate some general considerations showing the desirability of using linearized models of trackers for radio electronic systems that control weapons (military equipment) viewed as targets of radio jamming. With the exception of certain specific cases where nonlinearity must be considered as a matter of principle (for example, when evaluating conditions of tracking failure), it can be stated that the degree of jamming needed to attack the tracking system of a linearized tracking system, at the input to an initial tracking system with nonlinearities, is sufficient to attack it, provided that the intensity of jamming at the input is maintained within the limits of the linear part. In other words, if the jamming effect needed at the input to a tracking system to provide the required degree of jamming has been defined within the limits of its linearized part, then the consideration of effects generated by additional jamming influences in the area of nonlinearities will not, as a rule, lead to a reduction in the overall jamming effect.

References

[1] Simonyan, R. G., *O nastupatel'nykh operatsiyakh v armii SSHA* (*About Offensive Operations in the US Army*). Moscow: *Voennaya mysl'* (*Military Thought*), Vol. 1, 1989, pp. 69–77.

[2] Brusnitsin, N. A., *Global'naya tekhnicheskaya razvedka SSHA* (*US Global Technical Intelligence*). Moscow: *Voennaya mysl'* (*Military Thought*), Vol. 10, 1990, pp. 57–65.

[3] Tikhonov, V. I., and Yu. N. Bakaev, *Statisticheskaya teoriya radiotekhnicheskikh ustroystv* (*The Statistical Theory of Radio Technical Devices*). Moscow: N.E. Zhukovsky Airforce Academy, 1976, pp. 5–119, 180–220, 263–416.

[4] Helstrom, C. W., *Statisticheskaya teoriya obnaruzheniya signalov* (*Statistical Theory of Signal Detection*). Moscow: NIL, 1963, pp. 106–158.

[5] Wilks, S., *Matematicheskaya statistika* (*Mathematical Statistics*). Moscow: *Nauka* (*Science*), 1967, pp. 353–390, 402–426.

[6] Bronshteyn, I., and K. Semendyaev, *Spravochnik po matematike* (*Mathematics Handbook*). Moscow: *Nauka* (*Science*), 1986, pp. 72–73.

[7] Tikhonov, V. I., and V. N. Kharisov, *Statisticheskiy analiz i sintez radiotekhnicheskikh ustroystv i sistem* (*Statistical Analysis and the Synthesis of Radio Technical Devices and Systems*). Moscow: *Radio i svyaz'* (*Radio and Communications*), 1991, pp. 221–309.

[8] Kazakov, I. E., and V. M. Artem'ev, *Optimizatsiya dinamicheskikh sistem sluchaynoy struktury* (*Optimization of Dynamic Systems of Random Structure*). Moscow: *Nauka* (*Science*), 1983, pp. 13–22, 178–204.

[9] Tikhonov, V. I., *Optimal'nyy priem signalov* (*Optimal Signal Reception*). Moscow: *Radio i svyaz'* (*Radio and Communications*), 1983, pp. 68–156.

[10] Cramer, H. P., *Matematicheskaya statistika* (*Mathematical Methods of Statistics*). Moscow: IIL, 1948, pp. 277–287, 513–540.

[11] Tikhonov, V. I., *Statisticheskaya radiotekhnika* (*Statistical Radio Technology*). Moscow: *Radio i svyaz'* (*Radio and Communications*), 1982, pp. 427–477.

[12] Velichkin, A. I., G. S. Azarov, and Yu. V. Sayutin, *Sredstva svyazi i sistemy peredachi dannykh VVS* (*Airforce Communications Equipment and Data Transmission Systems*). Moscow: N. E. Zhukovsky Airforce Academy, 1985, pp. 69–103.

[13] Fomin, Ya. A., and G. P. Tarlovskiy, *Statisticheskaya teoriya raspoznavaniya obrazov* (*The Statistical Theory of Image Recognition*). Moscow: *Radio i svyaz'* (*Radio and Communications*), 1986, pp. 30–56.

[14] Fomin, Ya. A., and A. V. Savich, *Optimizatsiya raspoznayushchikh sistem* (*Optimization of Recognition Systems*). Moscow: *Mashinostroenie* (*Machine Building*), 1993, pp. 3–37.

[15] Nebabin, V. G., and V. V. Sergeev, *Metody i tekhnika radiolokatsionnogo raspoznavaniya* (*Methods and Technology of Radar Recognition.*). Moscow: *Radio i svyaz'* (*Radio and Communications*), 1984, pp. 7–28, 135–144. English Translation: Artech House, 1995.

[16] Duda, R., and P. Hart, *Raspoznavanie obrazov i analiz stsen* (*Image Recognition and Scene Analysis*). Moscow: *Mir* (Peace), 1976, pp. 20–51.

[17] Venttsel', E. S., *Issledovanie operatsiy* (*Operations Research*). Moscow: *Sov. radio* (*Soviet Radio*), 1972, pp. 238–281, 291–357, 446–509.

[18] Ovcharov, L. A., *Prikladnye zadachi teorii massovogo obsluzhivaniya* (*Applied Problems in Mass Service Theory*). Moscow: *Mashinostroenie* (*Machine Building*), 1968, pp. 92–204.

[19] Feller, W., *Vvedenie v teoriyu veroyatnostey i ee prilozheniya* (*An Introduction to Probability Theory and its Applications*). Moscow: IIL, 1952, pp. 374–410.

[20] Venttsel', E. S., *Teoriya veroyatnostey* (*The Theory of Probability*). Moscow: *Fiz.-mat.literat.* (*Physics and Mathematics Literature*), 1962, pp. 504–547.

[21] Maksimov, M. V., and V. I. Merkulov, *Radioe'lektronnye sledyashchie sistemy* (*Radio Electronic Tracking Systems*). Moscow: *Radio i svyaz'* (*Radio and Communications*), 1990, pp. 10–42.

[22] Maksimov, M. V., and G. I. Gorgonov, *Radioe'lektronnye sistemy samonavedeniya* (*Radio Electronic Homing Systems*). Moscow: *Radio i svyaz'* (*Radio and Communications*), 1982, pp. 6–46. English Translation: Artech House, 1988.

[23] Maksimov, M. V., G. I. Gorgonov, and V. S. Chernov, *Aviatsionnye sistemy radio-upravleniya* (*Aircraft Radio Control Systems*). Moscow: N. E. Zhukovsky Airforce Academy, 1984, pp. 11–46, 50–197.

[24] Gutkin, L. S. et al., *Radioupravlenie reaktivnymi snaryadami i kosmicheskimi apparatami* (*Radio Control of Jet Propelled Projectiles and Space Equipment*). Moscow: *Radio i svyaz'* (*Radio and Communications*), 1968, pp. 225–281.

[25] Vagapov, V. B., and B. Kh. Krivitskiy, *Osnovy avtomatiki radioe'lektronnogo oborudovaniya* (*The Basics of Automation in Radio Electronic Equipment*). Kiev: Kiev School of Higher Aircraft Engineering, 1983, pp. 10–110.

[26] Vakin, S. A., and L. N. Shustov, *Osnovy radioprotivodeystviya i radiotekhnicheskoy razvedki* (*The Basics of Radio Countermeasures and Radio Technical Intelligence*). Moscow: *Sov. radio* (*Soviet Radio*), 1968, pp. 9–122, 259–342.

[27] James, H. M., N. B. Nichols, and R. S. Phillips, *Teoriya sledyashchikh sistem* (*Theory of Servomechanisms*). Moscow: IIL, 1953, pp. 165–288.

2

Mathematical Models of Signals, Systems and Techniques for Electronic Jamming

2.1 A General Description of the Basic Elements of Electronic Jamming

In principle, as has already been noted, electronic attack can be implemented by jamming, changing the electrical and magnetic characteristics of the environment, or by reducing the radar and thermal detectability of aircraft. At the present time, the basic means of electronic attack in the radio frequency band is the use of various jamming signals that directly affect electronic systems. These signals are deliberate electromagnetic emissions with the appropriate amplitude, frequency, phase, polarization, space and time characteristics. Practically speaking, jamming signals are produced by jammers (electronic attack systems).

In a given electronic environment, jammers are used in particular ways depending on the specifics of the targets being jammed and the capabilities of the jammers.

In the general case, electronic attack systems comprise an information support and control system, a subsystem for producing jamming signals, high-frequency amplifiers and generators with modulators, and antenna devices. ELINT systems, knowledge bases and computer data that enter into the EW complex serve as the basis for the information support and control system. The ELINT subsystem seeks out, receives and processes radio signals from jamming targets. By comparing signals coming in from ELINT and a priori information contained in the knowledge base and computer data:

- The electronic environment is evaluated;
- The jamming target is determined, as well as the type and parameters of the jamming signal needed;
- A means of attacking the jamming target is selected and executed.

In specific cases, there can be other variants for the structure and functioning of jamming.

Jamming methods are determined by the characteristics of the jamming targets and the abilities of jammers to cause information damage to the victim, given specific conditions of the operational and tactical environment. As was noted earlier, by information damage we mean the quantity of information that the attacked side loses during a specific time interval as a result of the effect of jamming or other EW measures. In the specific case where radar is being attacked by jamming, the magnitude of the information damage corresponds to the size of the corridor jammed.

Thus, jamming signals, as well as systems and techniques of jamming, can be considered to be the basic elements of electronic attack. Mathematical models serve as the point of departure for quantitative evaluation of jamming effects and the definition of criteria for jamming.

2.2 Mathematical Models of Jamming Signals

2.2.1 Fundamental Principles

Three types of jamming signals have been defined to date: destructive, masking and deception. They also occur in combinations. As a rule, masking and deception jamming signals are additive (i.e., they combine with the useful signal in the victim receiver).

Destructive jamming signals are implemented using deliberate high-energy electromagnetic radiation [1, 2]. The effect of destructive jamming signals is to cause irreversible damage to input components in the receivers of the targets being jammed. This damage remains even after the jamming operation has finished. In order to restore the receiver or other device, it is necessary to replace (or repair) the damaged component.

Masking jamming signals, acting on the receiver in combination with the useful signal, exclude or, to a significant extent, hinder the decision about detection and recognition (classification) of useful signals at the input to the receiver. In this case, the effects of jamming are reversible. After jamming operations end, the properties of the receiver are restored.

The basic parameters of deception jamming signals are intentionally made to appear like the signal parameters of the targets being simulated, which can result, for example, in the victim systems for control of forces and weapons being redirected from the real targets to false ones. Just as with masking jamming, deception jamming signals donot directly cause irreversible changes in the targets being jammed. The latter circumstance permits the side being jammed to exclude or attenuate the jamming effect by using the differences between the jamming and useful signals that result from specific circuits in the jamming devices generating them, and also the way they are used in the dynamics of EW. For this reason it is necessary to analyze the question of radio signal parameters in greater detail.

When it is far enough from the emission source, any radio signal can be considered locally to be a plane wave with lateral fluctuations in the electromagnetic field with respect to their direction of propagation (the Umov–Poynting vector), as represented by their electrical and magnetic vectors.

A monochromatic linearly polarized plane wave at a sufficiently great distance D from the emission source can be represented analytically in the following way (Figure 2.1):

$$\begin{aligned} E &= E_0 \cos(\omega t - kD - \varphi) \\ H &= H_0 \cos(\omega t - kD - \varphi) \end{aligned} \tag{2.1}$$

Correspondingly, here E and H are the electrical and magnetic vectors; and E_0 and H_0 are the complex amplitudes (envelopes) E and H in the section being analyzed, which decrease proportionately to $\frac{1}{D}$ as the distance between them and the emission source increases; $k = \frac{2\pi}{\lambda}$ is the wave number, λ is the wavelength, and φ is the initial phase; p is the Umov–Poynting vector, $p = \frac{1}{2}\mathrm{Re}(EH)$;

$$\frac{E}{H} = Z_r \tag{2.2}$$

Z_r is the wave impedance of the environment:

$$Z_r = \sqrt{\frac{\mu_a}{\varepsilon_a}} \tag{2.3}$$

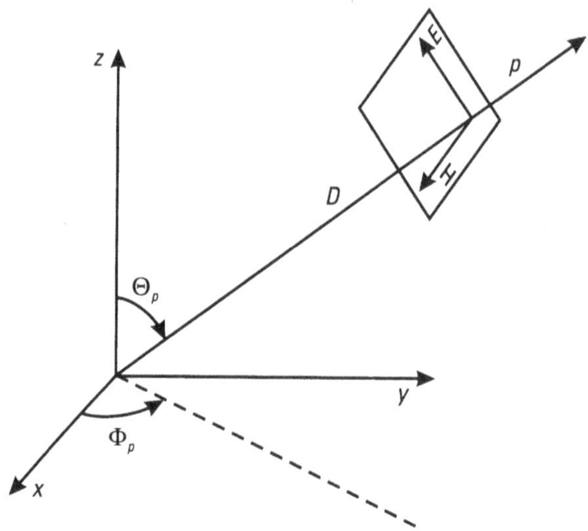

Figure 2.1 Analytical representation of a monochromatic linearly polarized plane wave.

μ_a is the absolute magnetic permeability; and ε_a is the absolute dielectric constant.

For free space: $\varepsilon_a = \varepsilon_0 = \frac{1}{36\pi} 10^{-9} F/m$

$$\mu_a = \mu_0 = 4\pi 10^{-7} \, {}^{\Gamma h}\!/_{m} \quad Z_r = Z_0 = 120\pi \text{ ohm}$$

In a number of cases, the polarization of a plane sinusoidal electromagnetic wave can be elliptical. In particular, this is characteristic of jamming signals meant for the jamming of electronic systems with random polarization. Elliptical polarization occurs when the electrical (or magnetic) vector is the sum of two orthogonal sinusoidal components shifted in phase by an angle δ. Accordingly, for a locally plane wave in the section being analyzed, the electrical vector E can be represented as the sum of two orthogonal components E_Φ and E_Θ with amplitudes $E_{\Phi 0}$ and $E_{\Theta 0}$, where

$$\begin{aligned} E_\Phi &= E_{\Phi 0} \cos(\omega t - kD) \\ E_\Theta &= E_{\Theta 0} \cos(\omega t - kD - \delta) \end{aligned} \tag{2.4}$$

The equations (2.4) can be solved relative to $\cos(\omega t - kD)$ and $\sin(\omega t - kD)$:

$$\frac{E_\Phi}{E_{\Phi 0}} = \cos(\omega t - kD)$$

$$\left(\frac{E_\Theta}{E_{\Theta 0}} - \frac{E_\Phi}{E_{\Phi 0}}\cos\delta\right)\frac{1}{\sin\delta} = \sin(\omega t - kD)$$

Raising both of the last two equations to the second power and adding them together, we obtain the equation sought for a polarized ellipse:

$$\frac{E_\Phi^2}{E_{\Phi 0}^2} + \frac{E_\Theta^2}{E_{\Theta 0}^2} - 2\cos\delta\frac{E_\Phi}{E_{\Phi 0}}\frac{E_\Theta}{E_{\Theta 0}} = \sin^2\delta \tag{2.5}$$

which describes the electrical vector $E = E_\Phi + E_\Theta$.

The polarization ellipse is defined by the ratio of the semiaxes $m = \frac{b}{a}$, the phase shift δ and the angle ψ between the main axis of the ellipse and the axis of the abscissa (0_Φ). The symbol δ defines the direction of rotation of the end of the electrical vector along its elliptical orbit (Figure 2.2). The direction of rotation of the electrical vector in the plane, perpendicular to the Umov–Poynting vector, depends on the observation point. If the observer is located at the emission point and looks out along the outgoing wave, then rotation for the symbol δ adopted in Figure 2.2 occurs to the right of the electrical vector (clockwise). At the receiver end, which is the orientation of the Umov–Poynting vector, rotation to the left (counterclockwise) is observed. As a rule, in the material that follows, the direction of rotation will be defined relative to an observer at the receiving end.

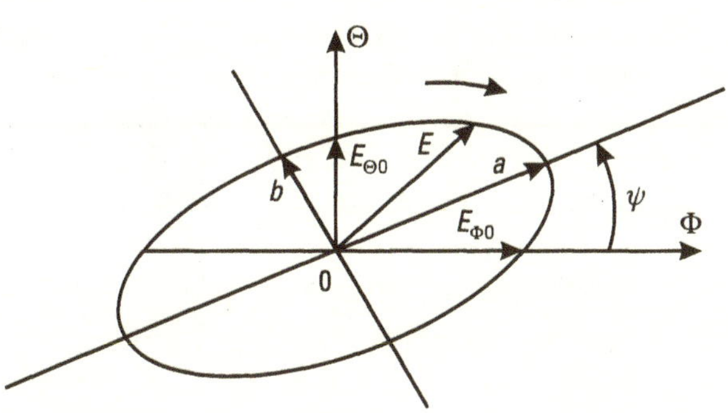

Figure 2.2 Direction of rotation of the end of the electrical vector along its elliptical orbit.

Practically speaking, the parameters of the polarization ellipse are determined using the well-known method with the assistance of a polarization diagram [3].

At the input to the receiver, after transformation in the antenna device, the sinusoidal signal $U(t)$ can be represented in the following format:

$$U(t) = U(t)\cos(\omega t - \varphi(t)) \tag{2.6}$$

where $U(t)$ is the amplitude (envelope), ω is the frequency, and $\varphi(t)$ is the initial phase.

Thus, the sinusoidal radio signal is defined by the envelope $U(t)$, the carrier frequency ω, the initial phase $\varphi(t)$ and the polarization of the electrical vector. In the general case for a locally plane wave, the latter is elliptical and is characterized by the three parameters of a polarization ellipse (m, ψ and the symbol δ). In turn, the parameters of the radio signal depend on the spatial coordinates of the emission sources or the receivers and their derivatives, which justifies speaking of the phase space of radio signals and its phase coordinates.

Depending on the modulation (keying) methods used, the parameters of the radio signal can be divided into informational and subordinate. Modulated parameters are called informational and unmodulated (unkeyed) are called subordinate. Modulation of the signal parameter permits the transmission or receipt of information using a radio signal. In many cases, modulation (keying) is performed on one or two of the radio signal parameters. The remaining parameters, which are of necessity present in the signal, are subordinate. For example, in the case of amplitude modulation, the envelope is the information parameter and the carrier frequency, the initial phase and the polarization are subordinate.

In principle, sinusoidal radio signals can be distinguished from one another, to a certain degree, if at least one of their parameters does not correspond. Keeping additive jamming signals in mind, the statement made above can be formulated in the following way: if at least one of the information or subordinate parameters of the jamming signal remains unchanged (deterministic) for a sufficiently long time, then at least one way can be proposed to null the jamming effect under consideration. Let us clarify this statement using the polarization parameter of a jamming signal with circular polarization as an example. The latter occurs if $\delta = \frac{\pi}{2}$ and $E_{\Phi 0} = E_0 = E_0$. Equation (2.5) is transformed into the equation for a circle:

$$E_\Phi^2 + E_\Theta^2 = E_0^2 \tag{2.7}$$

For the sake of clearness, let us assume that

$$E_\Phi = E_0 \cos \omega t \qquad E_\Theta = E_0 \cos\left(\omega t - \frac{\pi}{2}\right)$$

That is, the component E_Φ in the signal being studied leads E_Θ in phase by $\frac{\pi}{2}$. For an observer at the transmitting end, the rotation of the electrical vector in the signal being studied is counterclockwise (at the receiving end, a clockwise rotation is observed). In order to reduce the jamming effect, it is sufficient to adjust the electrical length of the feeder system in the receiving antenna on the side being jammed so as to provide for a supplementary shift in phases E_Θ by $\frac{\pi}{2}$ (or an uneven number of times by $\frac{\pi}{2}$). In this case, the difference in phases between E_Φ and E_Θ at the input to the receiver being jammed will be equal to π, and the amplitude of the total jamming field will be $E = E_\Theta + E_\Phi = 0$.

If an elliptical polarization occurs in the jamming signal, then it is possible to obtain the same result as in the case of circular polarization by varying the electrical length of the feeders for the corresponding components E_Φ and E_Θ and by leveling off their amplitudes at the input to the receiver. Analogous conclusions can be drawn both for the other parameters of the radio signal and for its phase coordinates (the angle of incidence, intervals and the distances between the emission sources or receivers etc.).

In order to exclude or significantly hinder the possibility of reducing the effectiveness of jamming using simple nulling circuit algorithms, well-known since the 1950s, it is necessary to deliberately introduce elements of a priori uncertainty into jamming signal parameters. Practically speaking, this is achieved by introducing elements of randomness (randomization) into all parameters of a radio signal and its phase coordinates. In other words, the changes in the parameters of a signal and its phase coordinates in time and space must represent a random process. This relates both to the information and subordinate parameters of signals. Destructive jamming signals are the only exception. The degree of a priori uncertainty in these and other parameters of jamming signals, including their phase coordinates, can vary. This is determined both by the assumed conditions of military use and the technological capabilities of the conflicting sides to develop and manufacture appropriate electronic technology.

In the material that follows, stability with respect to countermeasures against jamming signals, systems and techniques of jamming will be called information stability. Correspondingly, indices for information stability will also be considered.

2.2.2 Destructive Jamming Signals

Destructive jamming effects, independent of the frequency band of radio emissions, can be implemented using radiation sources of sufficiently high energy. In the optical band of emissions, destructive effects can be attained with the assistance of high-energy lasers [2]. In principle, in the radio band, it is possible to employ relativistic and other generators of UHF radiation [1] for this purpose. Above all, the direct target of such radiation is the semiconductor devices comprising the input components of receivers (mixers, detectors, parametric amplifiers). The heat effect or the rupturing of the field of a p-n junction results in irreversible changes in the electrical properties of the input components, due to which mixers, envelope detectors and other modern electronic equipment can no longer function. This circumstance permits us to consider these radiation sources to be electronic destruction systems and not jamming systems. For this reason, sometimes we do not speak of the destruction effect, but of functional destruction [1]. In the material that follows, we will use the terms destruction effect and destructive jamming signal. Let us evaluate the required energy indices for sources of destructive jamming effects.

At the present time, the input signal energy required to cause irreversible damage to p-n junctions in receiver input components is characterized by the following values [1, 4]. In the band from 1 to 10 Hz, the p-n junction in mixers and detectors burns out when acted upon by a pulse of up to 10 nsec in length with an energy of 0.1–1 mJ. The effectiveness of pulses that are longer than 10 nsec depends on their power. Single pulses at a frequency of less than 10 GHz put input semiconductor devices out of service when the power of the pulse is greater than 5W. At higher frequencies (greater than 10 GHz), irreversible damage occurs when the pulse power of the input signal is 0.5W. If we are dealing with the effect of a series of pulses, then the criterion for destructive radiation can be reduced by 20 dB.

In order to evaluate the required effective radiated power for a source of destructive jamming signals in the radio band, it is necessary to determine the power of the jamming emission P_{rec} that is acting on the semiconductor component and compare it with its threshold value P_{thresh}. Assuming that the radiation is moving in free space and the maximums for the antenna radiation patterns of the target and the jamming system correspond, we obtain

$$P_{\text{rec}} = \frac{P_j G_j}{4\pi D_j^2} A_{\text{rec}} \gamma_j \eta_j \tag{2.8}$$

Here, P_jG_j is the effective radiated power of the jamming system, the value to be determined; P_j is its power and G_j is the gain of the antenna; A_{rec} is the effective area of the receiving antenna of the target being jammed; γ_j is a coefficient allowing for the difference in polarization of the antennas of the target and the jamming system; and η_j is the attenuation coefficient for the jamming emission in the waveguide transmission line (or feeder). This number also takes into consideration the attenuation in safeguard devices against destructive effects; and D_j is the distance from the source of destructive radiation to the target being jammed.

The minimum required effective radiated power $(P_jG_j)_{min}$ occurs when $\eta_j = \gamma_j = 1$. When solved with respect to $(P_jG_j)_{min}$ and represented in logarithmic form (in decibels), (2.8) assumes the following form:

$$(P_jG_j)_{min}\ \text{dBW} = 11 + 2D_j\ \ \text{dBm} + P_{thresh}\ \text{dBW} - A_{rec}\ \text{dBm}^2 \qquad (2.9)$$

Let us make an approximate estimation of the order of the required effective radiated power $(P_jG_j)_{min}$ for two cases:

- The target being jammed is a radar of the detection, guidance and target designation type, where jamming is being performed at a range of $D_j = 100$ km (50 dBm), $A_{rec} = 10\ \text{m}^2$ (10 dBm^2);
- The target being jammed is the radar homing head of a surface-to-air missile, where jamming is being performed at a range of 4 km (36 dBm), reducing by approximately 20 times the probability of a hit by a single missile; $A_{rec} = 0.07\ \text{m}^2$ (-11.5 dBm^2).

In both cases, the threshold value is assumed to be identical and equal to 5W (7 dBW).

In the first case, the required effective radiated power is $(P_jG_j)_{min} = 108$ dBW ($10^{10,8}$W), and, in the second case, $(P_jG_j)_{min} =$ 101.5 dBW ($10^{10,15}$W) (i.e., the order of the values is about the same). As already has been noted above, when the destructive action is repeated multiple times, the required effective radiated power is reduced only by 20 dB and remains quite high. In the second instance, we should note that multiple repetitions of the jamming effect may not occur due to time limitations. In the examples analyzed, we have not taken into consideration the various reductions of the jamming emission that occur in practice, particularly due to the use of high-frequency switches acting on the sidelobes of the antenna radiation pattern, as well as other reasons.

The attenuation of the jamming effect due to the use of high-frequency switching diodes (p-i-n diodes, etc.) and polarization discrimination can comprise approximately 30–40 dB. In conclusion, the required effective radiated power of the destructive jamming source must be 130–140 dBW. The practical implementation of such radiation sources is problematic, even more so since the electronic efficiency of super high energy generators in the UHF band does not exceed 0.1–0.15.

We should note that, in the optical emission band, destructive jamming effects implemented using high energy lasers require significant expenditures of energy. Thus, for the destruction of input components in thermal homing heads, it is necessary to have an emission energy of 10 J/cm^2.

Bearing in mind the problems mentioned above, as well as others associated with the development and use of super high energy emission sources, destructive jamming signals will not be considered further in this work.

Let us take a brief look at jamming actions that cause reversible destructive effects and result basically from the limited dynamic range of receivers K_{drr}. The latter may be defined as the ratio of the maximum and minimum values of the input signal powers, where information losses in the receiver during processing do not exceed a certain permissible value

$$K_{drr} = \left(\frac{P_{rec\ max}}{P_{rec\ min}}\right) \tag{2.10}$$

Knowing the dynamic range of a receiver K_{drr} and the limit of its sensitivity $P_{rec\ min}$, it is possible to determine the lower power limit of the input jamming signal $P_{rec\ max}$. Under the influence of a high-power signal, the losses of information in the receiver during processing exceeds a permissible level.

Depending on the sensitivity limit $P_{rec\ min}$, the dynamic range of the receiver K_{drr} and the distance D_j from the jammer to the target being jammed, the value of the required effective radiated power causing a dynamic overload in the receiver varies over a wide range of values. In the case of radar for detection, guidance and target designation with an effective antenna area $A_{rec} = 10$ dBm^2, when it is jammed along the main lobe of its antenna radiation pattern with a range of $D_j = 50$ dBm (100 km) and where the dynamic range of the receiver is $K_{drr} = 60$–70 dB and the sensitivity limit is -150 to -170 dBW, the minimum effective radiated power $(P_jG_j)_{min}$ required for a reversible destructive effect varies in a range of -20 to $+20$ dBW. When the sensitivity of the receiver is degraded, the minimum effective radiated power increases accordingly.

Thus, for a sensitivity of −100 dBW, we obtain: $(P_jG_j)_{\min} = 70$ dBW (10^7W). Accordingly, in order to ensure the jamming of a radar along the sidelobes of its antenna radiation pattern at a level of −30 dB, the effective radiated power must be increased to 10 and 50 dBW. If we are jamming a radar homing head with $K_{\text{drr}} = 90$–100 dB and $P_{\text{recmin}} = -150$ dBW with a range of $D_j = 36$ dBm (4 km) along the main lobe of the antenna radiation pattern, then the required effective radiated power P_jG_j must not be less than 24–34 dBW (250–2500W). In the latter case (self-protection), it is necessary to keep in mind the possibility that the missile will home in on the jamming emission source.

Based on what has been said, it is possible to come to a conclusion about the limited potential of reversible destructive jamming actions of the type considered. As will be shown in the material that follows, situations can arise in the dynamics of EW where reversible destructive effects can significantly increase the information damage, and this circumstance should be kept in mind. In the material that follows, our basic attention is devoted to the analysis of deception and masking jamming signals.

2.2.3 Mathematical Models of Masking and Deception Jamming Signals

Depending on technological capabilities, the specifics of the countermeasures implemented in jamming targets, and the techniques of jamming selected, jamming signals can be continuous in time or pulse, distributed in space or concentrated, and modulated or unmodulated. According to what was said earlier, in all the mentioned variants of jamming signals, as well as in others, there must be a suitable degree of randomization both for information and subordinate parameters of jamming signals and for their phase coordinates. Thus, as the basis for models of jamming signals, be they masking or deception, it is necessary to use mathematical models representing sample functions in time and space of random processes with structures that vary to reflect the specifics of the parameters of the jamming targets being presented.

2.2.4 Unmodulated Frequency and Time Masking Jamming Signals

In many cases, unmodulated forward noise masking jamming signals can be represented in time using a model of white noise. In particular, this occurs when the spectrum width of the jamming signal Δf_j significantly exceeds that of the useful signal Δf_s.

A mathematical model of white noise can be derived by converting limits from Gaussian noise with a uniform spectrum, defined in the domain of positive frequencies $(0, f_j)$, to Gaussian noise with a uniform spectrum, defined for the entire frequency band $(-f_j, +f_j)$, with subsequent unlimited extension of the band. If the spectrum of the original noise, generated by a specific source, is defined in the band $(0, f_j)$ and its spectral density N_0 is constant, then, when converting to white noise, its spectrum is defined for the band $(-f_j, +f_j)$ and the spectral density will also be constant, but equal to $\frac{N_0}{2}$ (Figure 2.3).

As an initial sample function, let us analyze a Gaussian noise of duration $T(t_0 \leqslant t \leqslant t_0 + T)$ defined in the frequency band $(-f_j, +f_j)$ with a spectral density $\frac{N_0}{2}$. According to the Sampling Theorem (V. A. Kotelnikov's Theorem), if $f_j T >> 1$, then the sample function of the random process being analyzed is defined to a sufficient degree of accuracy by $n = 2f_j T$ noncorrelated and, for a Gaussian process, independent one-dimensioned random values (voltage and current) with a mathematical expectation equal to zero and identical variances $D_j = \sigma^2$. In this case the n-dimensioned probability density $p_n(u_1, \ldots, u_n)$ can be represented in the following manner:

$$p_n(u_1, \ldots, u_n) = (2\pi\sigma^2)^{-\frac{n}{2}} \exp\left(-\sum_{i=1}^{n} \frac{u_i^2}{2\sigma^2}\right) \tag{2.11}$$

The variance σ^2 is defined by the value of the correlation function $R(\tau)$ at $\tau = 0$; that is,

$$\sigma^2 = R(0) \tag{2.12}$$

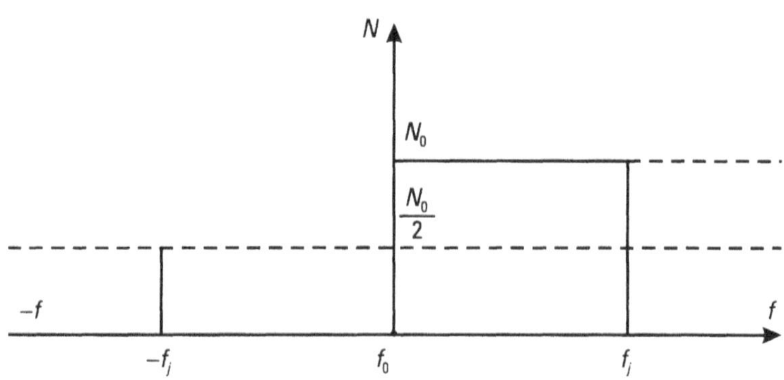

Figure 2.3 The spectrum of the original noise when converting to white noise.

According to the Khinchin–Wiener formula, in the given case:

$$R(\tau) = \int_{-f_n}^{+f_n} \frac{N_0}{2} e^{i2\pi f\tau} df \tag{2.13}$$

After well-known transformations, we obtain

$$R(\tau) = \frac{N_0}{2} 2f_n \frac{\sin 2\pi f_n \tau}{2\pi f_n \tau} \tag{2.14}$$

In particular, it follows from (2.14) that the time interval τ_K between adjacent independent (noncorrelated) one-dimensioned random values (correlation time) is determined by the upper frequency f_n:

$$\tau_K = \frac{1}{2f_n} \tag{2.15}$$

which corresponds to the Sampling Theorem.

As $f_n \to \infty$, $\tau_K = \Delta t \to 0$ and then (2.16)

The variance σ^2 being sought is equal to

$$\sigma^2 = R(0) = \frac{N_0}{2} 2f = \frac{N_0}{2} \frac{1}{\Delta t} \tag{2.17}$$

Accordingly,

$$p_n(u_1, \ldots, u_n) = \left(\frac{\pi N_0}{\Delta t}\right)^{-\frac{n}{2}} \exp\left(-\frac{\Delta t}{N_0} \sum_{i=1}^{n} u_i^2\right) \tag{2.18}$$

It is possible to determine the correlation function $R(\tau)$ for white noise from (2.13) by assuming the integration limits to be infinite:

$$R(\tau) = \int_{-\infty}^{+\infty} \frac{N_0}{2} e^{i2\pi f\tau} df = \frac{N_0}{2} \delta(\tau) \tag{2.19}$$

where

$$\delta(\tau) = \int_{-\infty}^{+\infty} e^{j2\pi f\tau} df \tag{2.20}$$

is the δ function.

White noise is δ-correlated.

Using expressions (2.11), (2.14), (2.18) and (2.19), we can define a model for white noise with the required degree of accuracy, provided $n = 2f_n T >> 1$. This condition is not always fulfilled in real masking jamming signals. In this case, the formula for the sample function of a jamming signal of duration T, in the form of a sum of noncorrelated random values, can be obtained with the help of the Karhunen–Loeve Expansion (canonical expansion), which is a Fourier series of a special type [5–7].

The initial element in the operation mentioned is the expansion of the function $u(t)$ into a series of orthonormal components $f_i(t)$. In this case,

$$u(t) = \sum u_i f_i(t) \tag{2.21}$$

where u_i is the expansion coefficient and $f_i(t)$ is a system of real orthonormal functions in the interval $0 < t < T$. The functions f_i satisfy the conditions:

$$\int_0^T f_i(t) f_j(t)\, dt = \delta_{ij} = \begin{cases} 1, i = j \\ 0, i \neq j \end{cases} \tag{2.22}$$

The coefficients u_i of series (2.21), according to (2.22), are determined by the formula:

$$u_i = \int_0^T f_i(t)\, u(t)\, dt \quad 0 < t < T \tag{2.23}$$

If $u(t)$ is a sample function for a random masking signal defined in the interval $0 < t < T$, then u_i in (2.21) consists of random values. The correlation function of random values u_i and u_j for a stationary process is defined as the average value of their product in the interval $(0, T)$:

$$\overline{u_i u_j} = \frac{1}{T}\int_0^T u_i(t_1)\, u_j(t_2)\, \mathrm{d}t \tag{2.24}$$

Let us determine u_i and u_j using (2.23) and substitute them into (2.24), obtaining

$$\overline{u_i u_j} = \int_0^T\int_0^T f_i(t_1) f_j(t_2)\, \overline{u(t_1)\, u(t_2)}\, \mathrm{d}t_1\, \mathrm{d}t_2 \tag{2.25}$$

$$\overline{u(t_1)\, u(t_2)} = \mathrm{R}(t_2 - t_1) \tag{2.26}$$

where $\mathrm{R}(t_2 - t_1)$ is the correlation function of the masking jamming signal. Accordingly, we obtain

$$\overline{u_i u_j} = \int_0^T\int_0^T f_i(t_1) f_j(t_2)\, \mathrm{R}(t_2 - t_1)\, \mathrm{d}t_1\, \mathrm{d}t_2 \tag{2.27}$$

It is necessary to find a system of orthonormal functions $\{f_i(t)\}$ that convert (2.27) to zero when i and j are different. In principle, this occurs if

$$\int_0^T \mathrm{R}(t - \tau) f_j(\tau)\mathrm{d}\tau = \lambda_j' f_j(t) \quad 0 < t < T \tag{2.28}$$

Then, taking into consideration (2.27),

$$\overline{u_i u_j} = \lambda_j' \delta_{ij} \tag{2.29}$$

The latter means that u_i and u_j are noncorrelated and their variances are equal to the numbers λ_j', which are eigenvalues of the homogeneous integral equation (2.28). The functions $f_j(\tau)$ are eigenfunctions of the integral equation cited. The task of expanding the random process represented by the correlation function $R(t)$ into noncorrelated summands of a sample function amounts to solving the integral equation (2.28), the kernel of which comprises the correlation function $\mathrm{R}(\tau)$ of the original random process (in the given case, the sample function of the masking signal).

If the spectrum width of the random process is limited by a range

$(-f_j, +f_i)$; its spectral density $N/2$ within the range is constant; and the sample function $u(t)$ is considered in the interval $(0, T)$, then the correlation function $R(\tau)$ is determined by (2.14), which, in the given case, can be represented in the following way:

$$R(t, \tau) = P_j \frac{\sin \omega(t - \tau)}{\omega(t - \tau)} \tag{2.30}$$

where $P_j = \frac{N}{2} 2f_j = Nf_j$ is the effective power of the jamming signal:

$$\omega = 2\pi f_j \tag{2.31}$$

The integral equation (2.28) is converted to the form:

$$\int_{-\frac{T}{2}}^{+\frac{T}{2}} P_j \frac{\sin \omega(t - \tau)}{\omega(t - \tau)} f(\tau) d\tau = \lambda'_n f(t) \tag{2.32}$$

The latter equation has been quite well studied [6, 8]. Its eigenfunctions are termed circular spherical functions and its eigenvalues are expressed in radial spherical functions. Both have been tabulated [8]. One should bear in mind that the data provided in the table has been obtained for an integral equation represented in the following format:

$$\lambda_n f_n(t) = \int_{-\frac{T}{2}}^{+\frac{T}{2}} \frac{\sin \omega(t - \tau)}{\omega(t - \tau)} f_n(\tau) d\tau \qquad n = 0, 1, 2, \ldots \tag{2.33}$$

Besides this, the parameter $c = \pi f_n T$ also figures in the tables.

In order to find the eigenvalues of λ'_n, as determined from (2.32), it is necessary to multiply the tabular value of λ_n by P_j. To a great extent, its eigenvalues are determined by the number of sample values $(2f_n T + 1)$. In the case where $n > 2f_n T + 1$, the eigenvalues of λ_n decrease rapidly. If the number of sample values increases, then $\lambda_n \to 1$. As an example, let us look at four instances, more specifically, where $c = 4$ and, accordingly, $2f_n T = 2.55$; $c = 8$, $2f_n T = 5.1$; $c = 16$, $2f_n T = 10.2$; and, finally, $c = 30$, $2f_n T = 19.1$. If we accept T to mean the duration of the signal being concealed by jamming, then, in the first case, the values of the spectrum widths of the jamming and useful signals are commensurable; in subsequent cases the spectrum width of the jamming signal is approximately

Table 2.1
Eigenvalues of λ_n

$c = 4$ $2f_nT = 2.55$	$\lambda_0 = 0.996$; $\lambda_1 = 0.912$; $\lambda_2 = 0.519$; $\lambda_3 = 0.110$; $\lambda_4 = 0.009$; $\lambda_5 = 0.0004$
$c = 8$ $2f_nT = 5.1$	$\lambda_0 = \lambda_2 = 0.9999$; $\lambda_2 = 0.997$; $\lambda_3 = 0.961$; $\lambda_4 = 0.748$; $\lambda_5 = 0.320$; $\lambda_6 = 0.061$; $\lambda_7 = 0.0061$
$c = 16$ $2f_nT = 10.2$	$\lambda_0 = \lambda_1 = \lambda_2 = 1$; $\lambda_3 = \lambda_4 = \lambda_5 = \lambda_6 = 0.999$; $\lambda_7 = 0.993$; $\lambda_8 = 0.949$; $\lambda_9 = 0.754$; $\lambda_{10} = 0.375$; $\lambda_{11} = 0.098$; $\lambda_{12} = 0.015$; $\lambda_{13} = 0.0017$
$c = 30$ $2f_nT = 19.1$	$\lambda_0 = \lambda_1 = \ldots = \lambda_9 = 1$; $\lambda_{10} = \ldots = \lambda_{14} = 0.999$; $\lambda_{15} = \lambda_{16} = 0.994$; $\lambda_{17} = 0.921$; $\lambda_{18} = 0.707$; $\lambda_{19} = 0.356$; $\lambda_{20} = 0.106$; $\lambda_{25} = 4.74 \times 10^{-6}$

two, five and ten times greater than the frequency band of the useful signal (which is being concealed). The corresponding values of λ_n are shown in Table 2.1 [8].

An analysis of Table 2.1 shows that, beginning when $\Delta f_n = f_n$ is five times greater than $\Delta f_i = \frac{1}{T}$, a model of a masking jamming signal with a homogeneous spectrum can be built based on the Sampling Theorem, and the one-sided spectral density of the jamming signal can be determined with sufficient precision in accordance with Figure 2.3:

$$N_0 = \frac{P_n}{\Delta f_n} \tag{2.34}$$

If the spectrum of the jamming signal is not flat, then it is possible to determine the equivalent spectrum width $\Delta f_{j\ni}$ and spectral density $N_{0\ni}$:

$$N_{0\ni} = \frac{P_{n\ni}}{\Delta f_{n\ni}} \tag{2.35}$$

where

$$\Delta f_{n\ni} = \frac{1}{S_{0f}(f_0)} \int_0^{\infty} S_{0f}(f)\mathrm{d}f \tag{2.36}$$

$S_{0f}(f)$ is the one-sided spectral density, defined in the range $(0, f)$; and $S_{0f}(f_0)$ is the value of the spectral density at frequency f_0, corresponding to the maximum value $S_{0f}(f)$:

$$P_{j\ni} = \int_0^\infty S_{0f}(f)\mathrm{d}f \tag{2.37}$$

In many cases, $\Delta f_{n\ni}$ is taken to be equal to the spectrum width at the level the half-value of the spectral density.

As was noted earlier, the probability density of the instantaneous values for the masking jamming signal is assumed to be Gaussian. In view of the linearity of the operation of canonical expansion, the noncorrelated (in the given case, independent) random values u_i and the sampling values corresponding to them, as determined form (2.32), are also distributed according to a normal law with a mathematical expectation equal to zero and variances:

$$\sigma_i^2 = \lambda_n' = \lambda_n P_n \tag{2.38}$$

If the eigenvalues of λ_n are defined by (2.33), then

$$\lambda_n' = \lambda_n \frac{P_n}{2\Delta f_n}$$

Taking what has been said into consideration, (2.18), which defines the multidimensional probability density of the jamming signal, can be represented in the following manner:

$$p_n(u_0, \ldots, u_n) = \prod_{i=0}^{n} (2\pi p_n \lambda_i)^{-\frac{1}{2}} \exp\left(-\frac{1}{p_n}\sum_{i=0}^{n}\frac{u_i^2}{2\lambda_i}\right) \tag{2.39}$$

In a number of problems involving electronic jamming, it is convenient to represent a masking signal $u_j(t)$ as a superposition of quasi-sinusoidal oscillations of a special type, or, to be more precise [9]:

$$u_j(t) = \sum_{n=1}^{N} c_n \cos(\omega_n t - \varphi_n) \tag{2.40}$$

where $\varphi_1, \ldots, \varphi_n$ are random phases, evenly distributed in the interval $[0, 2\pi]$:

$$c_n = \sqrt{2S_{0,f}(f_n)\Delta f} \quad \omega_n = 2\pi f_n, f_n = n\Delta f \tag{2.41}$$

$S_{0,f}(f_n)$ is the one-sided spectral density of the jamming signal.

The signal $u_j(t)$ is represented as a sum of sinusoidal components with a constant amplitude, but with a random phase.

2.2.5 Time and Space Masking Jamming Signals (Angle Noise)

Up to this point, we have been examining masking jamming signals defined in a frequency and space domain. As a result of their operation, they can (potentially) exclude the possibility of receiving information for a certain period of time in a particular frequency band. One of the most important tasks of jamming is the creation of conditions where the side being jammed is deprived of the possibility of, or significantly impaired from, receiving information from a certain area in space or in a given range of solid angles for a definite period time. This task can be solved using space and time jamming signals.

For example, practically speaking, space and time masking jamming signals can be obtained by using point radiation sources of the required density located within the limits of a given solid angle. If the energy intensity index for frequency and time jamming is their spectral density with a dimension of *Watt/Hertz*, then the corresponding index for space and time jamming is their angular (space) density with a dimension of *Watt/steradian.* Depending on the specific problem, sources of jamming emissions distributed within the limits of a given area in space (or solid angle) can be of various types. In the case where a radar system is being jammed from sufficiently distant positions, these can be, for example, stationary sources of frequency and time masking jamming. In the general case, the randomization of both jamming emissions and the coordinates of their sources must be provided for in such a way as to ensure the required stability with respect to countermeasures by the side being jammed. Let us clarify what has been said based on one of the simplest examples: the generation of space and time jamming using emission sources. Let us assume that n quasisinusoidal jamming sources with identical amplitudes E_0 and random phases ψ_j have been placed in a range of angles $(-\Phi_j/2, +\Phi_j/2)$ at equal angular intervals $\Delta\Phi_j$ and at the same distance, distributed uniformly with an interval of $(-\pi, +\pi)$. Furthermore, let us assume that the angular interval $\Delta\Phi_j$ does not exceed the width of the main lobe $(\Phi_{0.5})$ of the antenna radiation pattern of the receiver being jammed, at a level of -3 dB. Then the voltage $u(\Phi_i, t)$ at the input to the receiver in the case where the maximum antenna radiation pattern lies within the limits of angles $(-\Phi_j/2, +\Phi_j/2)$, Figure 2.4, is defined by the following expression:

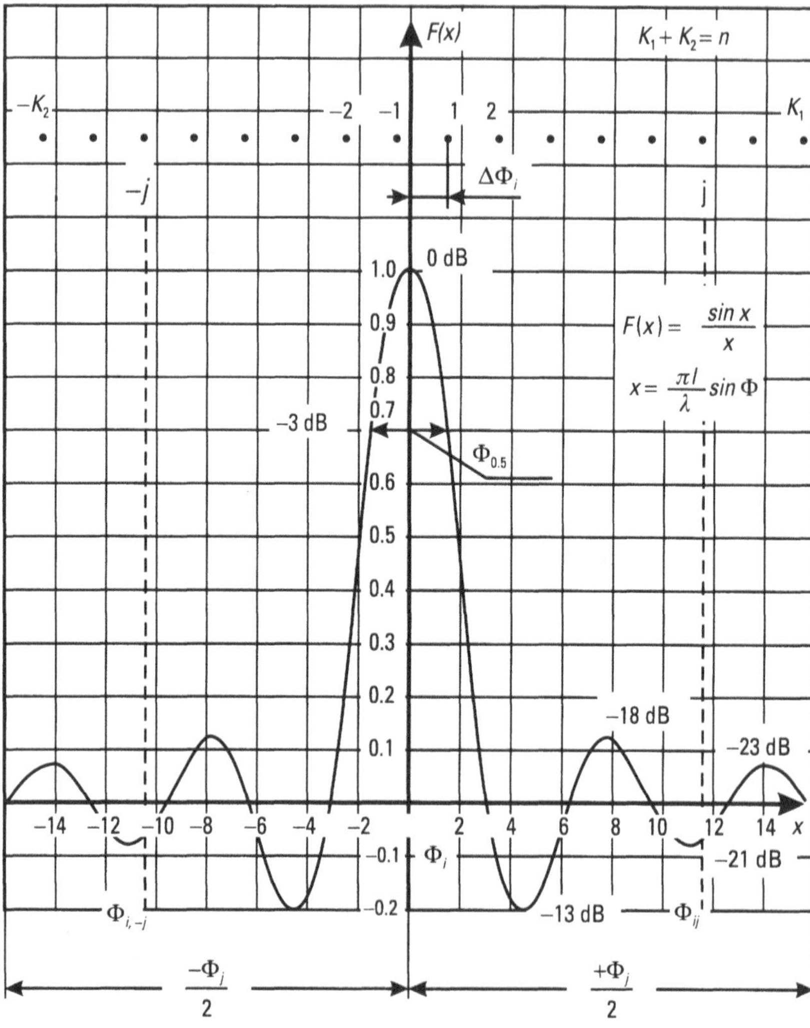

Figure 2.4 The voltage $u(\Phi_i, t)$ at the input to the receiver in the case where the maximum antenna radiation pattern lies within the limits of angles $(-\Phi_j/2, +\Phi_j/2)$.

$$u(\Phi_i, t) = BE_0 \left(\sum_{j=1}^{K_1} u_{i,j} + \sum_{|j|=1}^{K_2} u_{i,-j} \right) \tag{2.42}$$

where $u_{i,j} = F(\Phi_{i,j}) \cos(\omega t - \psi_j)$:

$$u_{i,-j} = F(\Phi_{i,-j}) \cos(\omega t - \psi_{-j})$$

$$\Phi_{i,j} = \Delta\Phi_i + (j-1)\Delta\Phi_j \quad j = \overline{1, K_1}$$

$$\Phi_{i,-j} = \Delta\Phi_j - \Delta\Phi_i + (|j|-1)\Delta\Phi_j \quad |j| = \overline{1, K_2}$$

$F(\Phi)$ is the normalized antenna radiation pattern in the field; B is the coefficient, taking into consideration the specific conditions and limitations that accompany the emission of radiation; $\Phi_{i,j(-j)}$ is the angle between the directions of the $j(-j)$ jamming source and the maximum antenna radiation pattern; $\Delta\Phi_i$ is the angle between the maximum antenna radiation pattern and source 1 (Figure 2.4); and K_1 and K_2 are the number of emission sources taken into consideration when determining $u(\Phi_i, t)$ $(K_1 + K_2 = n)$.

If $\Phi_{i,j} = 0$, then $F(\Phi_{i,j}) = 1$. Normally, when $\Delta\Phi_j \approx \Phi_{0.5}$, besides the basic source, for which $F(\Phi_i)$ is close to one, two or three additional sources to the left or right of its maximum antenna radiation pattern are taken into consideration depending on the level of attenuation of its side lobes.

It follows from (2.42) and Figure 2.4 that, when the angle Φ_i changes in the interval $(-\Phi_j/2, +\Phi_j/2)$ under examination, the value of $u(\Phi_i, t)$ remains without significant changes. This, in principle, excludes the possibility of the side being jammed obtaining information about the angular coordinates of the enemy target in the interval of angles Φ_i being analyzed. The special feature of the scheme being considered is its deterministic nature when the number of jamming sources n is small. The angular coordinates of each of them can be determined. This permits the side being jammed, for example, to reduce the effectiveness of the jamming operation by using an adaptive phased antenna array. It is possible to increase or maintain the degree of the jamming effect either by increasing the angular density of jamming sources $n \to \infty$, or, in accordance with the principle stated above, by randomizing their angular coordinates. The latter, in the given conditions, amounts to a random (more precisely pseudo-random) dislocation of sources within the limits of the given range of angles and the turning of each of them on at random moments of time and with random time intervals of operation. This randomness in the operating time of jamming sources causes random modification in time and angles of the jamming radiation arriving at the antenna of the victim receiver. Time and space jamming signals of this type are called angle noise. The spectral density of angle noise has a dimensionality of rad^2/Hertz (grad^2/Hertz). The simplest example of angle noise is a jamming signal generated by two jamming sources, switched according to a random law, symmetrically distributed along the axis of the ordinates at angles Φ_j and $-\Phi_j$ (Figure 2.5). If the distribution law for the number of intersections with the axis of the abscissae is random, then such signals are normally termed telegraph signals. The simplest model of a telegraph signal occurs when the number of zero-level intersections with the signal being observed is distributed according to a Poissonian distribution. In this case, according to (1.39), the probability that, within time t, exactly n zero-level intersections (i.e., n emission source switches) should occur, $P_n(t)$ is equal to:

$$P_n(t) = \frac{(\nu t)^n}{n!}\exp(-\nu t) \qquad (2.43)$$

where ν is the average number of emission source switches over the unit of time $(\nu, \frac{1}{c})$.

The correlation function $R_\Phi(\tau)$ and the spectral density $S_\Phi(\omega)$ of a telegraph signal of the type being analyzed are known [9–11]. In the case where it is symmetrical (Figure 2.5), they are defined by the expression:

$$R_\Phi(\tau) = \Phi_j^2 \exp(-2\nu|\tau|) \text{ rad}^2 \quad S_\Phi(f) = \frac{8\Phi_j^2\nu}{4\nu^2 + \omega^2} \text{ rad}^2/\text{Hertz} \qquad (2.44)$$

Here, $S_\Phi(\omega)$ is the spectral density of stationary angle noise, generated by two switched sources with an average switching frequency of ν and distributed symmetrically relative to a certain fixed direction.

In the dynamics of jamming generation, the case of an asymmetrical distribution of switched sources (Figure 2.6) is of special interest. Following [9, 11], we determine the spectral density for the random ξ, $(0 \leqslant \xi \leqslant 1)$. We

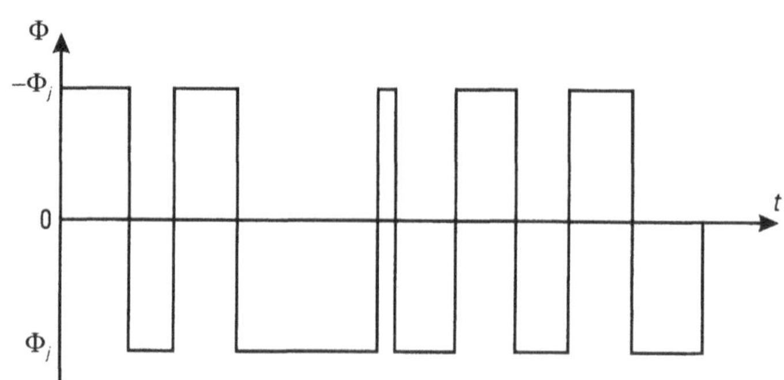

Figure 2.5 The simplest example of angle noise.

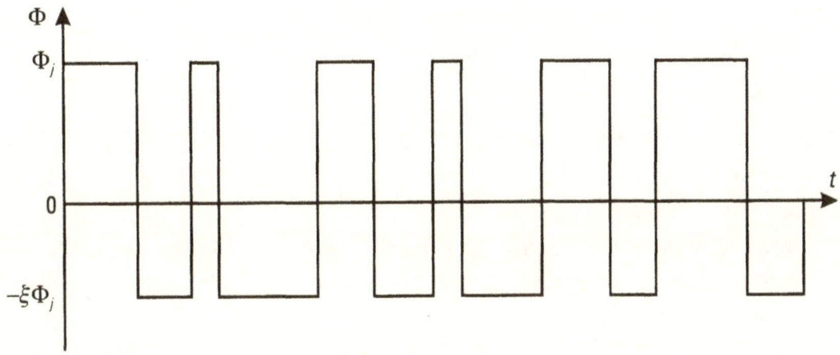

Figure 2.6 An asymmetrical distribution of switched sources.

determine the spectral density in two steps. In step one, the correlation function $R_{\Phi\xi}(\tau)$ of the random process being analyzed is determined. In step two, the spectral density $S_{\Phi\xi}(\omega)$ we are seeking is determined using of the Khinchin–Wiener transformation.

Assuming the random process to be stationary, we use the known definition of the correlation function [9, 11, 12]:

$$R_{\Phi\xi}(\tau) = \overline{\Phi_j(t)\,\Phi_j(t+\tau)} - M^2_{\Phi\xi} \tag{2.45}$$

Here, the overstrike designates smoothing with respect to τ, and M_Φ is the mathematical expectation $\Phi_j(t)$.

Depending on τ, as a result of multiplying together and smoothing of $\Phi_j(t)$ and $\Phi_j(t+\tau)$, the following three outcomes can occur:

$\overline{\Phi_j(t)\Phi_j(t+\tau)} = \Phi_j^2$, if the positive values of the signal are multiplied together;

$\overline{\Phi_j(t)\Phi_j(t+\tau)} = \xi\Phi_j^2$, when the negative values are multiplied together;

$\overline{\Phi_j(t)\Phi_j(t+\tau)} = -\xi\Phi_j^2$, when the positive and negative components of the signal are multiplied together.

The occurrence of the first and second outcomes is equally likely. The respective probabilities P_1 and P_2 are equal to each other and their sum is equal to 1. Therefore, $P_1 = P_2 = 0.5$. In both cases, the outcomes indicated occur if, during the smoothing interval τ being analyzed, there is an even number of source switches. The third outcome occurs when the number of switches is odd. Keeping in mind the Poissonian distribution of the number of intersections and taking into account what has been said, we can write

$$\overline{\Phi_j(t)\,\Phi_j(t+\tau)} = \Phi_j^2 P_1(P_0(\tau) + P_2(\tau) + \cdots)$$
$$+ \xi^2\Phi_j^2 P_2(P_0(\tau) + P_2(\tau) + \cdots) - \xi\Phi_j^2(P_1(\tau) + P_3(\tau) + \cdots) \qquad (2.46)$$

where $P_i(\tau)$ is determined from (2.43).

Sequentially transforming (2.46) and keeping in mind the well-known expansions of an exponential function into a series, we obtain

$$\overline{\Phi_j(t)\,\Phi_j(t+\tau)} = \Phi_j^2 \exp(-\nu|\tau|)\left(\frac{1}{2}\left(1 + \frac{(\nu|\tau|)^2}{2!} + \frac{(\nu|\tau|)^4}{4!} + \cdots\right)\right.$$
$$+ \frac{1}{2}\xi^2\left(1 + \frac{(\nu|\tau|)^2}{2!} + \frac{(\nu|\tau|)^4}{4!} + \cdots\right)$$
$$\left. - \xi\left(\nu|\tau| + \frac{(\nu|\tau|)^3}{3!} + \frac{(\nu|\tau|)^5}{5!} + \cdots\right)\right) \qquad (2.47)$$

$$\overline{\Phi_j(t)\,\Phi_j(t+\tau)} = \Phi_j^2\left(\frac{1}{4}(1 + \exp(-2\nu|\tau|)) + \frac{1}{4}\xi^2(1 + \exp(-2\nu|\tau|))\right.$$
$$\left. - \frac{1}{2}\xi(1 - \exp(-2\nu|\tau|))\right) \qquad (2.48)$$

Let us determine the mathematical expectation $\Phi_j(t)$:

$$M_{\Phi\xi} = P_1\Phi_j + P_2(-\xi\Phi_\Phi) = \frac{1}{2}\Phi_j(1 - \xi) \qquad (2.49)$$

Taking into consideration (2.45), (2.48) and (2.49), we obtain

$$R_{\Phi\xi}(\tau) = \frac{\Phi_j^2}{4}(1 + \xi)^2 \exp(-2\nu|\tau|) \qquad (2.50)$$

The spectral density being sought $S_{\Phi\xi}(\omega)$ is found with the assistance of the Fourier transformation (Khinchin–Wiener), which, in this case, we express in the following manner:

$$S_{\Phi\xi} = 4\int_0^\infty R_{\Phi\xi}(\tau)\cos\omega\tau d\tau \qquad (2.51)$$

or taking into consideration (2.50), and also [13]:

$$S_{\Phi\xi}(\omega) = \frac{\Phi_j^2(1+\xi)^2 2\nu}{4\nu^2 + \omega^2} \tag{2.52}$$

Accordingly,

$$S_{\Phi\xi}(f) = \frac{2(1+\xi)^2 \Phi_j^2 \nu}{4\nu^2 + (2\pi f)^2} \tag{2.53}$$

When $\xi = 1$, which is a symmetrical telegraph signal, we receive the previously known result (2.44). In this case, $S_{\Phi\xi}(f)$ attains the maximum value. If the condition is not stationary (i.e., Φ_j changes with time), then the approximate values of the spectral density can be obtained from the smoothed values $\overline{\Phi_j}$ and $\overline{\xi}$. In the general case, the spectral density of angle noise can be found by simulation modeling methods using a fast Fourier transform.

2.2.6 Modulated Frequency and Time Masking Jamming Signals

If we ignore the polarization, which can be considered separately, then the model of a quasisinusoidal jamming wave $U_n(t)$ can serve as the initial concept for building models of frequency and time masking jamming:

$$U_n(t) = U_j(t)\cos(\omega t + \psi_j(t)) \tag{2.54}$$

The parameters $U_n(t)$ and $\psi_j(t)$ change slowly when compared to $\cos\omega t$ of the time function. Depending on which parameter of the signal is modulated (keyed), we can distinguish amplitude, phase and frequency noise modulation, as well as phase modulation of a random telegraph signal. There exist various combined types of modulation (keying).

In the case of jamming signals obtained with the help of amplitude noise modulation [10, 14],

$$u_n(t) = U_0(1 + \overline{m}_{\mathrm{AM}}\xi(t))\cos(\omega t + \psi_j(t)) \tag{2.55}$$

where $U_0 = const$ is the amplitude of the modulated carrier wave; $\xi(t)$ is the random modulation process (modulating noise); and $\overline{m}_{\mathrm{AM}}$ is the effective coefficient of the amplitude noise modulation:

$$\overline{m}_{\mathrm{AM}} = \frac{\sigma_\xi}{h} \tag{2.56}$$

σ_ξ is the effective (mean square) value of the modulating noise; and h is the limiting level for the maximum values of the modulating noise.

The modulating noise is limited to provide a sufficient (on the average) depth of modulation in the carrier wave, keeping in mind the conditions of generation actually implemented in the transmitter being modulated. If the modulating noise is not limited, then the modulation of the carrier wave will not be deep (on the average) and the majority of the power will be concentrated on the carrier frequency. The form of the spectrum of the modulated wave, when the values of $\overline{m}_{\mathrm{AM}}$ are suitable, reproduces the form of the spectrum of the modulating wave. Besides the sidetones, the spectrum of the modulated wave also contains a discrete component on the carrier wave.

The width of the spectrum of the modulated wave ΔF_j is equal to double the value of the width of the spectrum of the modulating noise ΔF_{JM}, if the latter occupies a frequency band from 0 to ΔF_{JM}:

$$\Delta f_j = 2\Delta F_{\mathrm{JM}} \tag{2.57}$$

The modulated parameter $\psi_j(t)$ of the carrier wave, in the case of phase modulation of the noise and linearity of the modulation characteristic, can be represented in the following way [10]:

$$\psi(t) = m_{\mathrm{PM}}\xi(t) \tag{2.58}$$

where m_{PM} is the slope of the modulation characteristic of the phase modulator; and $\xi(t)$ is the modulating noise with a variance σ_Φ^2 and a zero mathematical expectation. Correspondingly, the sinusoidal wave, phase modulated by noise, has the following appearance:

$$u_j(t) = U_0 \cos(\omega_0 t + m_{PM}\xi(t)) \tag{2.59}$$

If the modulating noise $\xi(t)$ is a stationary Gaussian random process, then the correlation function of the phase-modulated sinusoidal wave will have the form [10]:

$$K(\tau) = \frac{U_j^2}{2}\exp\left(-m_{PM}^2\sigma_\Phi^2(1 - r_\Phi(\tau))\right) \times \cos\omega_0\tau \tag{2.60}$$

where $r_\Phi(\tau)$ is the standard correlation function of the modulating process.

The spectral density of the phase-modulated masking jamming signal using noise can be determined with the assistance of the Khinchin–Wiener transformation. However, in the general case, the expression for the spectral density $S_{PM}(\omega)$ turns out to be rather unwieldy. In practice, frequently only two extreme cases are of interest: to be more precise, when $m_{PM}\sigma_\Phi << 1$ and $m_{PM}\sigma_\Phi >> 1$.

In the first case [10],

$$S_{PM}(\omega) \approx \pi U_j^2 \delta(\omega - \omega_0) + \frac{U_j^2 m_{PM}^2}{2} S_\Phi(\omega - \omega_0) \tag{2.61}$$

Here, $S_\Phi(\omega - \omega_0)$ is the spectral density of the modulating process $\xi(t)$, and $\delta(\omega - \omega_0)$ is the δ-function.

The distribution of the spectral density in the given case is near to that which occurs in the case of amplitude modulation by noise.

In the second case ($m_{PM}\sigma_\Phi >> 1$), the distribution of the spectral density also contains a discrete component (a sinusoidal wave on the carrier). However, it is greatly attenuated in comparison to the first instance. The spectral density of the continuous part of the spectrum in the first approximation is described by the Gaussian curve:

$$S_{PM}(\omega) \approx \frac{U_j^2}{2} \frac{\sqrt{2\pi}}{m_{PM}\sigma_\Phi \Delta\Omega_\Phi} \exp\left(-\frac{3(\omega - \omega_0)^2}{2m_{PM}^2 \sigma_\Phi^2 \Delta\Omega_\Phi^2}\right) \tag{2.62}$$

where $\Delta\Omega_\Phi$ is the width of the spectrum of the modulating noise.

It is assumed that the spectral density $S_\Phi(\omega)$ of the modulating noise is distributed evenly in the frequency band $0, S_\Phi(\omega)$:

$$S_\Phi(\omega) = \begin{cases} \sigma_\Phi^2 \dfrac{2\pi}{\Delta\Omega_\Phi}, 04\omega 4\Delta\Omega_\Phi \\ 0, \omega > \Delta\Omega_\Phi \end{cases}$$

The spectrum width of phase-modulated noise (ΔF_{JPM}), in the case being analyzed, can be approximately evaluated in the following way [10]:

$$\Delta f_{JPM} \approx m_{PM}\sigma_\Phi \Delta\Omega_\Phi \sqrt{\frac{2\pi}{3}} \tag{2.63}$$

The discrete component in the spectrum of the phase-modulated wave is, in essence, a subordinate feature of the given type of jamming signal that can be used by the side being jammed to attenuate the jamming effect.

The spectrum of a carrier wave, phase manipulated by a random telegraph signal, is free from the discrete component. The jamming signal in this case can be expressed in the following way [15]:

$$u_j(t) = U_j l(t) \cos(\omega_0 t + \psi_0) \tag{2.64}$$

where $l(t)$ is a single function that determines the sign of the envelope of the modulated wave and is equal to +1 or −1, depending on which phase of the telegraph signal is occurring at the given moment of time; and ψ_0 is the fixed initial phase. The telegraph signal, providing phase modulation $U_{pmd}(t)$, and the phase-modulated sinusoidal wave $u_n(t)$, are shown in Figure 2.7. The number of alternations of sign (the modulation frequency) is a random value subject to a Poissonian distribution (2.43).

2.2.7 The Spectral Density of a Phase-Modulated Signal

The spectral density $S_{pmd}(f)$ is determined using the known spectral density of the envelope $S_{env}(\omega)$ [15]:

$$S_{pmd}(\omega) = \frac{1}{4} S_{env}(\omega) \tag{2.65}$$

The correlation function $R_{env}(\tau)$ and the spectral density $S_{env}(\tau)$,

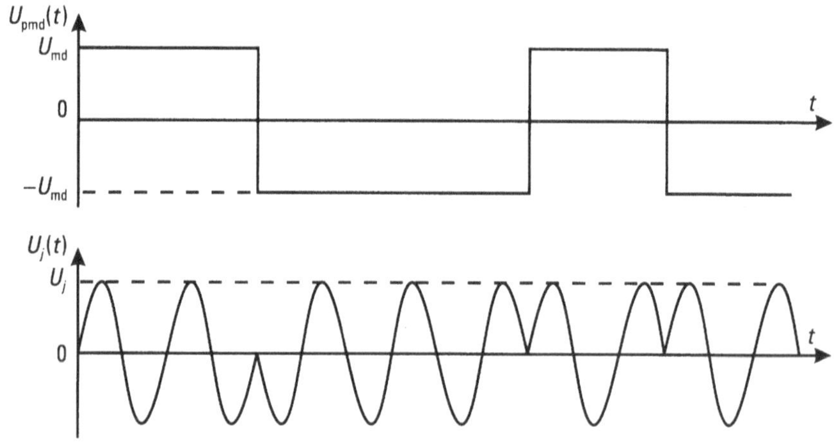

Figure 2.7 A telegraph signal providing phase modulation $U_{pmd}(t)$ and a phase-modulated sinusoidal wave $u_n(t)$.

according to (2.44), are represented by the expressions:

$$R_{env}(\tau) = U_j^2 \exp(-2\nu|\tau|) \tag{2.66}$$

$$S_{env}(f) = \frac{8U_j^2\nu}{\omega^2 + 4\nu^2} \tag{2.67}$$

The spectral density of the phase-modulated wave is equal to

$$S_{pmd}(\omega) = \frac{2U_j^2\nu}{(\Delta\omega)^2 + 4\nu^2} \tag{2.68}$$

Here,

$$\Delta\omega = \begin{cases} \omega_0 - \omega, \omega < \omega_0 \\ \omega - \omega_0, \omega > \omega_0 \end{cases}$$

The spectrum width of the masking jamming signal Δf_{cpmd}, defined at the level of the half-value for the maximum of the spectral density, is equal to

$$\Delta f_{cpmd} = \frac{2\nu}{\pi} \tag{2.69}$$

The instantaneous value of the masking jamming signal, formed by means of frequency modulation of the carrier wave using noise $\xi(t)$, is determined in the following way:

$$u_j(t) = U_j \cos\left(\omega_0 t + K_{FM} \int_0^1 \xi(t)\mathrm{d}t\right) \tag{2.70}$$

where K_{FM} is the proportionality coefficient.

A carrier wave frequency modulated using noise is characterized by the following indices:

- The effective frequency deviation σ_f, which is equal to the mean square value of the instantaneous frequency of the modulated wave;
- The effective modulation index m_{FM}, which is equal to the ratio of σ_f to the maximum frequency of the spectrum of the modulating noise $F_{j\max}$:

$$\overline{m}_{FM} = \frac{\sigma_f}{F_{j\max}} \tag{2.71}$$

If the spectral density of the modulating noise is constant in the frequency band, then,

$$\overline{m}_{FM} = \frac{\sigma_f}{\Delta F_{\text{JM}}} \tag{2.72}$$

The spectral density of the frequency-modulated noise of the wave $(N_{FM}(\omega))$ can be determined with the help of well-known procedures [10]. Simple analytical expressions are obtained in two extreme cases: $\overline{m}_{FM} >> 1$ and $\overline{M}_{FM} << 1$. The spectral density of the modulating noise is assumed to be uniform in the band $(0, \Delta F_{\text{JM}})$.

If $\overline{m}_{FM} >> 1$ is broadband frequency-modulated noise, then,

$$S_{FM}(\omega) = \frac{U_{\Pi}^2}{2\sqrt{2\pi}} \frac{1}{\overline{m}_{FM}\Delta F_{\text{JM}}} \exp\left(-\frac{(\omega - \omega_0)^2}{8\pi^2 \Delta F_{\text{JM}}^2 \overline{m}_{FM}^2}\right) \tag{2.73}$$

or

$$S_{FM}(f) = \frac{U_j\sqrt{2\pi}}{2\sigma_f} \exp\left(-\frac{(f - f_0)^2}{2\sigma_f^2}\right) \tag{2.74}$$

The spectrum width of the frequency-modulated noise of the wave in the given case is determined using the expression:

$$\Delta F_{JFM} = \sqrt{2\pi}\sigma_f = \sqrt{2\pi}\overline{m}_{FM}\Delta F_{\text{JM}} \tag{2.75}$$

If $\overline{m}_{FM} << 1$, then narrowband modulation occurs:

$$S_{FM}(\omega) = U_j^2 \frac{\pi^2 \sigma_f \overline{m}_{FM}}{(\pi^2 \sigma_f \overline{m}_{FM})^2 + (\omega - \omega_0)^2} \tag{2.76}$$

Accordingly, the spectrum width of frequency-modulated nose is equal to

$$\Delta F_{JFM} = \frac{\pi^2}{2} \overline{m}_{FM}\sigma_f = \frac{\pi^2}{2} m_{FM}^2 \Delta F_{\text{JM}} \tag{2.77}$$

As follows from their name, deception jamming signals must be represented, according to their information and subordinate parameters, using the same mathematical models as the signals being simulated (i.e., the useful signals). In the given case, the specifics of a disturbance become clear when analyzing the effect of the combination of jamming (disturbance) and useful signals on specific electronic systems, taking into consideration the noise-protection circuits they contain. This, in particular, relates to the jamming of radar systems working in tracking mode. Due to the reason indicated, the mathematical models of deception jamming signals will be analyzed below in the appropriate sections, where we present techniques for the jamming of specific electronic systems. Examples of deception jamming are, for example, the signals of a radar system reflected from a discrete chaff cloud, a false target, a radar decoy and a significant part of those response disturbances implemented using the retransmission of signals received, having imparted a jamming modulation to them.

2.3 Mathematical Models of Systems and Techniques for Jamming

Jamming signals are produced by systems especially designed for electronic attack. In the dynamics of EW, jamming systems are used in various ways according to the specific conditions of the electronic environment.

Jamming systems are quite varied. They basically comprise active jammers with an appropriate information support system. The basic targets for jamming using airplane- and helicopter-based jamming systems are AAD radar systems of various types (for control of AAD forces and weapons etc.). Besides these jammers, passive jamming is widely used, as well as optoelectronic jamming systems.

The determining element in a modern airborne jamming system is automatic active jammers with an information support system permitting the control both of active and passive jamming systems and optoelectronic jamming systems in the dynamics of EW.

In Figure 2.8, we give a variant of a block diagram for an automatic jammer, which is the logical result of the analysis of a smoothed radar environment, occurring at the current time in the dynamics of EW. The variant given for the structure of a jammer permits, in principle, the formation of jamming signals both for the jamming of radar systems for control of AAD forces and weapons [16, 17]. Block diagrams of onboard jamming systems, presented in materials from the avionics conference [17], served as the initial prototype for the variant given.

In the variant being analyzed, the information support system of the

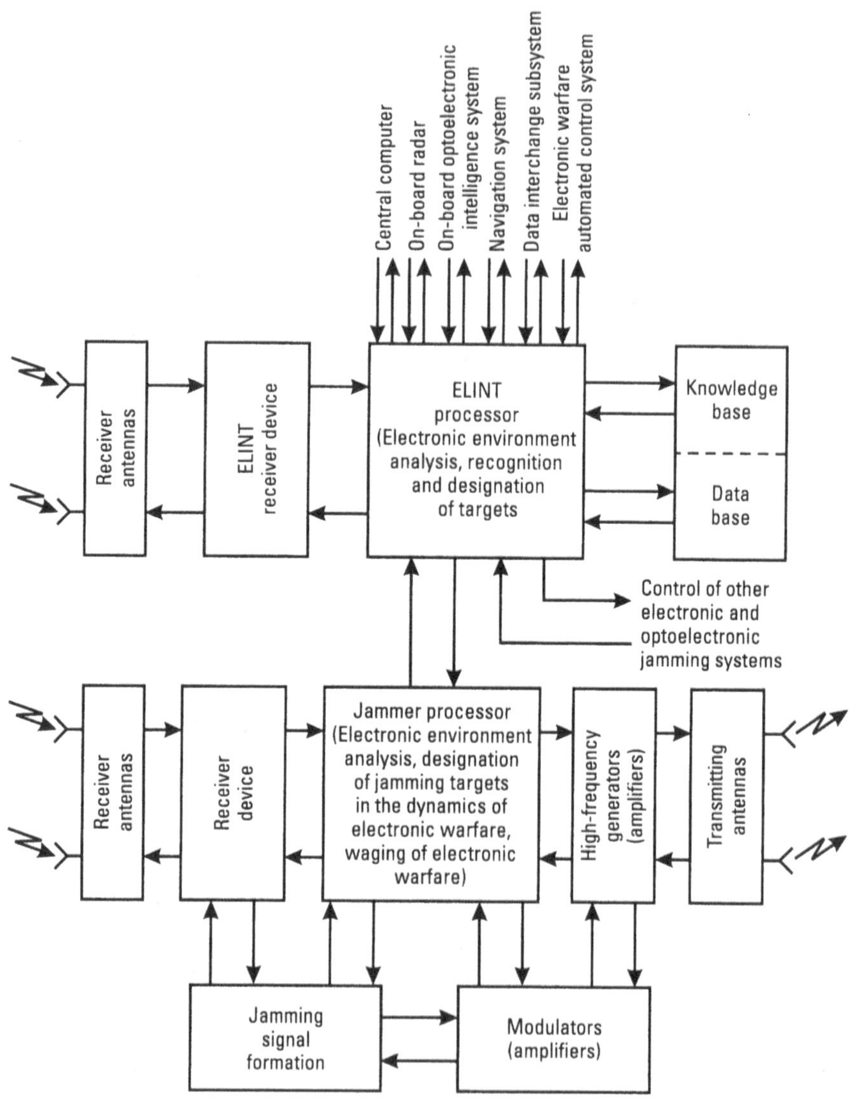

Figure 2.8 A block diagram for an automatic jammer.

jammer and the jamming system as a whole are represented as an ELINT station, with its processor, knowledge- and databases, a receiver, and the processor of the jammer itself. Besides this, the information support system also contains onboard information systems: the central onboard computer with its knowledge- and databases, the onboard radar system, the onboard optoelectronic intelligence system, the navigation and data interchange (radio communications) systems, and, possibly, an automated control system for EW systems. Future EW systems could include an information control system, an executive device system (active jammers, optoelectronic jamming systems, systems for controlling detectability, antiradiation missiles, etc.), an automated control system, a system to monitor the functional effectiveness of the EW complex, and an EW operator training system.

A stream of signals from radar stations, which, in the given case, are the operations targets, arrive at the receiving antennas of the ELINT devices and jammers. For example, the number of radar stations in an automated AAD force control system is quite large and their modes of operation are, to a great extent, independent, which permits us to view the arriving stream of signals from radar working in detect mode as being Poissonian. The onboard automatic jamming system of an airplane, taken separately, performing jamming of the radar system of an automated system for control of AAD forces, can be considered to be a single channel mass service system with rejects. In the given case it represents a single channel basically because of the limited possibilities of isolating transmitter and receiver devices on a single aircraft. As a rule, while transmitting, ELINT receivers are not able to detect signals from a radar irradiating the aircraft in the corresponding frequency band. At the same time, the entire set of jammers located on the various airplanes of a battle formation, when employed in specific ways, can be viewed as a multichannel service system with rejects.

After the radar signal has been detected, an analysis of the electronic environment is conducted and the operations target identified. Then, a jamming signal is generated. Depending on its type, either direct radio-frequency (RF) jamming (direct noise interference) is generated or jamming modulation of RF radiation is performed. Jamming signals are emitted in the appropriate direction by the jammer transmitter antennas. The average emission time for a jamming signal $\bar{t}_j$ is limited. This is determined by a number of conditions both technical and methodological in nature. The time $\bar{t}_j$ can be considered to be the average service time for the operations target of one jammer.

According to what has been mentioned, the mathematical model of the jammers on a single aircraft, reflecting the dynamics of changes in its states, can be represented in the form of differential Kolmogorov–Chapman equations (1.45) and (1.46), which determine the likelihood of states $P_0(t)$ and $P_1(t)$ of a single channel mass service system with rejects. Here, $P_0(t)$ is the probability that the jammer is free for servicing and $P_1(t)$ is the probability that the jammer is occupied servicing a request already received. The results of solving the equations for stationary conditions are given in (1.51). Accordingly,

$$P_0 = \frac{\mu}{\lambda + \mu} \tag{2.78}$$

$$P_1 = \frac{\lambda}{\lambda + \mu} \tag{2.79}$$

Here, λ is the average density of the signal stream arriving at the jammer from a radar system working in detect mode:

$$\mu = \frac{1}{\bar{t}_j} \tag{2.80}$$

is the average stream density of jamming signals emitted by the onboard jammer; and $\bar{t}_j$ is their average duration.

It is convenient to consider the mathematical models of other jamming systems, including the variant considered for an automatic jammer working in the mode of jamming radar for weapons control, relative to the target being jammed. We will do this in the material that follows.

Let us note that the practical implementation of onboard airborne electronic jamming systems and ELINT represents a complex technical and technological problem. It is necessary to provide for the construction of sufficiently precise, high-speed automated electronic equipment functioning in a wide band of frequencies from tenths to several tens of GHz and satisfying the general technical requirements imposed on airforce equipment.

2.3.1 Mathematical Models of Electronic Jamming Techniques

Jamming techniques are highly varied. They are, to a large extent, determined by the current electronic environment and the problem at hand. Among the most characteristic, we can include methods of concealing airplanes (helicopters) and other targets from stationary (fixed) areas (Figure 2.9) using jamming; methods of screening airplanes (helicopters) in battle formation (Figure 2.10) using jamming; the mutual screening of airplanes using electronic jamming systems; the self-screening of airplanes (helicopters) using electronic jamming systems; and the combination of electronic jamming with the destruction of electronic targets using firepower.

As a rule, mathematical models reflecting the characteristic special

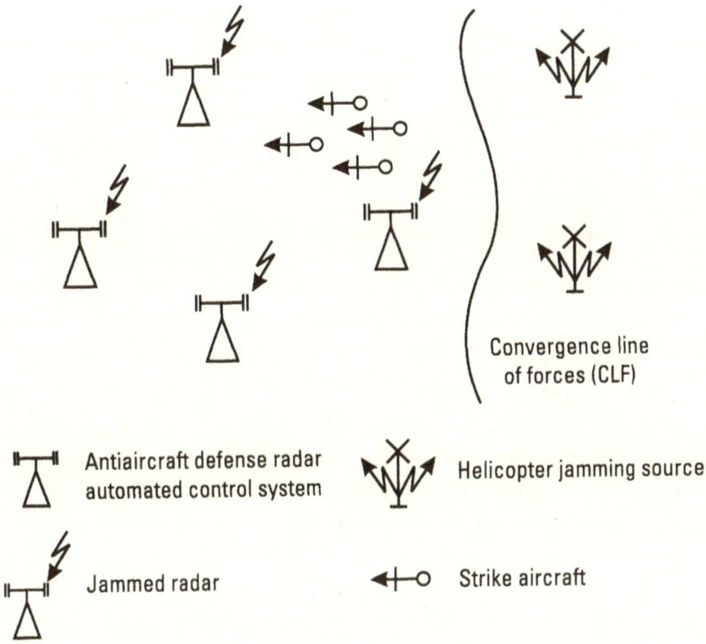

Figure 2.9 Methods of concealing airplanes (helicopters) and other targets from stationary (fixed) areas using jamming.

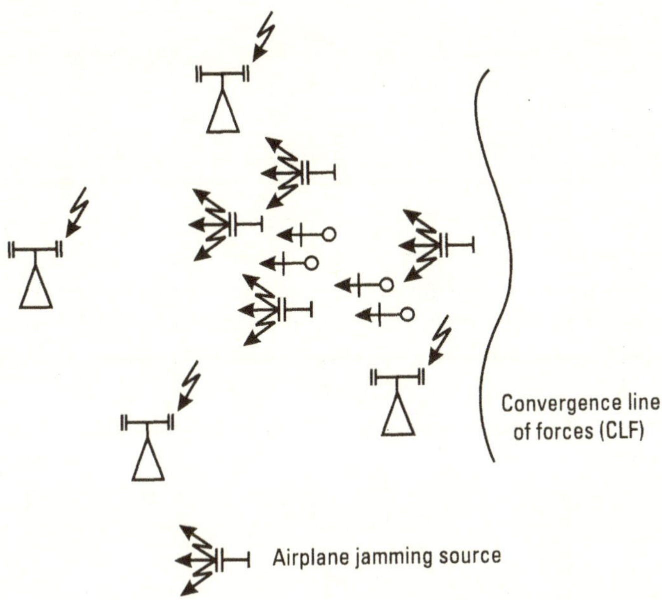

Figure 2.10 Methods of screening airplanes (helicopters) in battle formation using jamming.

features of electronic jamming techniques are essentially linked with specific jamming targets. We propose to analyze such models in the material that follows. Certain variants of employing automatic active jammers are exceptions. In particular, if the radar system of an automated system for the control of AAD forces is jammed by a group of automatic jammers of the same type, then, on the condition that their electromagnetic compatibility is provided for, the dynamics of changes in state of the group can be determined by representing it as a multichannel mass service system with rejects. Assuming the service process to be stationary, then, using (1.50), it is possible to determine the likelihood of jamming (servicing) the radar system of the automated control systems in the case that there is an automated control system for EW systems (EW ACS) in the group of jammers. If an EW ACS is absent from the group of jammers, then the likelihood of jamming can be determined with the help of (1.60). According to (1.56), in the case of a group of jammers with an ACS, the likelihood of jamming (servicing) the radar system P_{serv} is equal to

$$P_{serv} = \frac{\sum_{i=0}^{n-1} \frac{\alpha^i}{i!}}{\sum_{i=0}^{n} \frac{\alpha^i}{i!}} \tag{2.81}$$

where

$$\alpha = \frac{\lambda}{\mu} \tag{2.82}$$

λ is the average signal stream density for the ACS radar system to be jammed; $\mu = \frac{1}{\bar{t}_n}$ is the average signal stream density serviced by a single jammer; $\bar{t}_n$ is the average time to generate jamming by one of the jammers in the group; and n is the quantity of jammers in the group.

If there is no EW ACS and control of the jammers is decentralized, then, according to (1.65), the likelihood of jamming P'_{serv} of any of the radar stations in the ACS generating a stream of signals with an average density of λ, is equal to

$$P'_{serv} = \frac{1}{1 + \frac{\alpha}{n}} \tag{2.83}$$

where α is determined from (2.82).

One of the special features of attack using masking and deception jamming is its increased sensitivity when compared to other types of armed warfare and to countermeasures by the target being jammed. The latter is due to the fact that the jamming radiation being considered does not cause irreversible destructive effects in the jamming target. This permits the latter to take defensive countermeasures directly during the period of the jamming action, attenuating it or totally removing it. What has been said justifies the necessity of developing mathematical models of jamming techniques reflecting the conflict aspect of the problem and permitting us to find the optimum course of action, taking into consideration countermeasures by the jamming target.

At the present time, such problems are being analyzed in the theory of games and statistical solutions. Let us pause briefly to analyze the possibilities of solving the problem at hand using the approaches mentioned.

The theory of zero-sum matrix games is quite fully developed [18–20]. In a matrix game, two players participate, each with his own game plan, including a finite number of variants for implementing it. The latter are usually termed pure strategies or simply strategies. All of the player's possible courses of action in the given conditions are his strategies. Each pair of the players' strategies is assigned a number called the cost or the cost function. This is the number of conditional units to be paid by the losing side. This number can be either positive or negative. Let us designate the players using the letters A and B and consider the winning side to be A. Further, let us designate the strategies of the sides and the costs accordingly by $A_i, (i = \overline{1, \dots, m})$, $B_j, (j = \overline{1, \dots, n})$ and a_{ij}. The interests of the sides are opposite: whatever amount side B loses, the opposing side A gains exactly the same amount, so the total winnings of the sides are equal to zero. For this reason, games of this kind are termed antagonistic with a zero total sum. The mathematical model of the game is a matrix $m \times n$, where the rows correspond to the strategies of $A_i, (i = \overline{1, \dots, m})$ and the columns to the strategies of $B_j, (j = \overline{1, \dots, n})$:

$$\|a_{ij}\| = \left\| \begin{matrix} a_{11} & a_{12} & \dots & a_{1n} \\ a_{21} & a_{22} & \dots & a_{2n} \\ \dots & \dots & \dots & \dots \\ a_{m1} & a_{m2} & \dots & a_{mn} \end{matrix} \right\| \quad (k = \overline{1, n}) \tag{2.84}$$

$$a_{l1\,\max} \quad a_{l2\,\max} \quad \dots \quad a_{ln\,\max} \quad (l = \overline{1, m})$$

Each of the sides, having information about the strategies of the opponent and the corresponding costs, makes an analysis of the matrices (2.84). The winning side in each of the rows chooses a strategy A_{ik}, where winnings are minimal $a_{ik\,\min}$. From among all the minimal values $a_{ik\,\min}$, $(k = \overline{1, n})$, side A selects the maximum $a_{ik\,\max\min}$. In other words, side A chooses a strategy A_i, corresponding in the space of pure strategies to the greatest of all m minimal winnings. This strategy is commonly considered to be maximum–minimum $(A_{\max\min})$. Accordingly the cost is designated as $a_{\max\min}$.

Side B, having opposite interests, evaluates its strategies B_j on the basis of the columns. In each of the columns, the maximum loss value is recorded $a_{lj\,\max}$, $(j = 1, n)$, that occurs for one of the strategies of the opponent A_l, $(l = \overline{1, m})$. From all the maximum values $a_{lj\,\max}$, the least is chosen $a_{lj\,\min\max}$. The strategy and cost corresponding to it are called the minimum–maximum $(B_{\min\max})$, $a_{\min\max}$.

If $a_{ik\,\max\min} = a_{lj\,\min\max} = a$, then it is said that the game has a solution in the space of pure strategies and the cost a is called the price of the game. The strategies $A_{\max\min}$ and $B_{\min\max}$ for each of the players are optimum for the given game. Player A wins no more and B loses no less than a conditional units.

In the general case,

$$a_{ik\,\max\min} < a_{lj\,\min\max}$$

and the solution to the games is to be sought in the realm of combined strategies. As follows from the basic theorem of the theory of matrix games (the Von Neumann Theorem), such a solution exists for any matrix game. Each player has an optimum combined strategy corresponding to the price of the game a.

A combined strategy is understood to be a linear combination of pure strategies taken with definite weights $P_i \geqslant 0$.

The optimum combined strategy of player A assures a receipt of winnings equal to the game price a, if the opponent applies his optimal combined strategy. If side B applies some pure strategy B_j, then the winnings of A will be no less than a; that is, for the total set of pure strategies B_j, $(j = \overline{1, n})$, the following system of nonlinear inequalities can be expressed:

$$\sum_{i=1}^{m} P_i a_{ij} \geqslant a, (j = \overline{1, n}) \tag{2.85}$$

$$\sum_{i=1}^{m} P_i = 1 \tag{2.86}$$

The problem is, with the help of the system of inequalities (2.85) and equalities (2.86), to determine weights P_i so as to assure the maximum value of the game price a. Through a simple transformation, this problem is reduced to a linear programming task. If we divide the two parts of the inequality (2.85) into a and introduce the designations $\frac{P_i}{a} = x_i \geqslant 0$, then (2.85) and (2.86) are transformed to the format:

$$\sum_{i=1}^{m} a_{ij} x_i \geqslant 1 \quad j = (\overline{1, n}) \tag{2.87}$$

$$\sum_{i=1}^{m} x_i = \frac{1}{a} \tag{2.88}$$

In the given formulation, the problem is already expressed in the following way: find such $x_i \geqslant 0$ that satisfy the system of linear inequalities (2.87), where the linear form (2.88) assumes a minimum value. This is a classic problem in linear programming.

It is possible to obtain an analogous system of inequalities for side B:

$$\sum_{j=1}^{n} a_{ij} y_j \leqslant 1 \tag{2.89}$$

$$\sum_{j=1}^{n} y_j = \frac{1}{a} \tag{2.90}$$

and the linear programming task is formulated accordingly: determine such $y_j \geqslant 0$, satisfying the system of linear inequalities (2.89), where the linear form of (2.90) assumes the maximum value. Let us clarify what has been said using a most elementary classroom example: the domain of waging EW to support aircraft overcoming AAD.

Let us assume that it is a possibility, with the help of an active jammer, to create an area S_1 in space within the bounds of which not a single aircraft T_1, T_2 and T_3 is detected by even a single k of the land-based radar systems of the opposing side. Besides this, let us assume that a passive jamming band (area S_2) of length L_j has been set up in advance, within the limits of which the detection of targets is impossible (Figure 2.11).

Further, let us consider that, besides the active jammer, the offensive side has two strike craft (bombers) and AAD has only two intercept fighters.

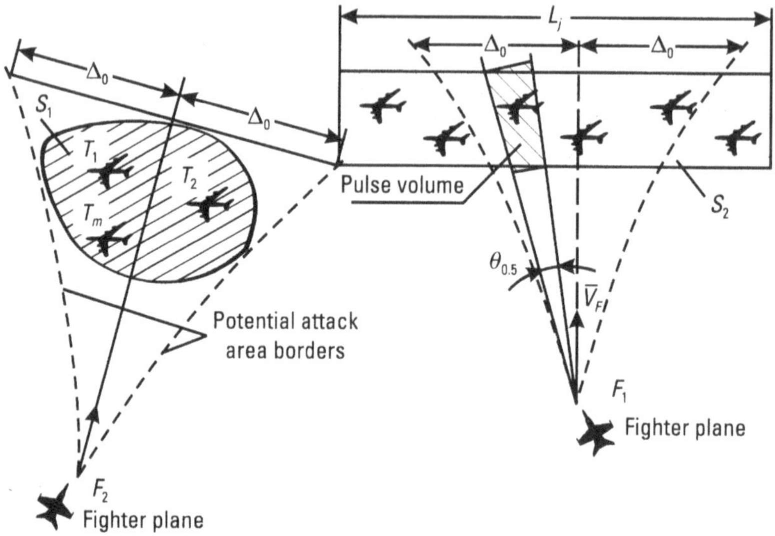

Figure 2.11 Using an active jammer to create an area S_1 in space where not a single aircraft T_1, T_2 or T_3 is detected.

Let us find a way of dislocating the strike craft in the areas S_1 and S_2 so that the average number of attacks on each bomber is minimum. In the conditions given, the following paths of action for both sides are possible.

Offensive side (Airforce)

- Action plan A_1: both bombers are in area S_1;
- Action plan A_2: both bombers are in area S_2;
- Action plan A_3: one bomber is in area S_1 and the other is on area S_2.

Defending side (AAD)

- Action plan B_1: both fighters (F_1 and F_2) are dispatched to area S_1;
- Action plan B_2: both fighters (F_1 and F_2) are dispatched to area S_2;
- Action plan B_3: one fighter (F_2) is dispatched to area S_2 and the other (F_1) to area S_1.

The average number of attacks $\overline{n}$ on the bombers serves as the quantitative measure of the effectiveness of the actions of both sides. In the instance being analyzed,

$$\overline{n} = \overline{n}_1 + \overline{n}_2 \tag{2.91}$$

where $\bar{n}_1$ is the average number of attacks on the bombers located in area S_1; and $\bar{n}_2$ is the average number of attacks on bombers located in area S_2.

To compile an initial effectiveness matrix for the action plans, it is necessary to establish a correspondence between each action plan comparison for the sides and the average number of attacks on the bomber aircraft.

Assuming there is an equal probability that fighter F_2, dispatched against a group being screened by active jamming, can attack any of the airplanes in the group, including the one generating the jamming; then, in the first approximation, it is possible to consider

$$\bar{n}_1 = \frac{n_b}{n_b + n_j} n_f \tag{2.92}$$

Here $\frac{n_b}{n_b+n_j}$ is the probability that a bomber will be selected from the overall number of aircraft in the jamming area S_1; n_b is the number of bombers in area S_1; and n_j is the number of jamming generators in the area S_1; and n_f is the number of fighters dispatched to area S_1.

The situation is analogous for one bomber aircraft screened by passive jamming (area S_2) and dislocated in a random fashion in a jamming area of length L_j:

$$\bar{n}_2 \approx \frac{2\Delta_0}{L_j} n_b n_f \tag{2.93}$$

where Δ_0 is the maximum miss for the fighter selected according to overload conditions during homing time; $\frac{2\Delta_0}{L_j}$ is the probability of a bomber reaching the area of possible attacks.

This formula is valid for small values of the ratio $\frac{2\Delta_0}{L_j}$ and a small value of n_f.

The miss magnitude Δ_0, selected during the homing period according to the overload conditions, is determined based on the homing range D_j, the relative velocity of the fighter (missile) V_{rel} and the maximum overload j_n. For movement in the horizontal plane,

$$\Delta_0 = \frac{1}{2} j_n \frac{D_j^2}{V_{rel}^2} \tag{2.94}$$

where D_j is determined by the range of operation of the onboard radar of the fighter with due consideration to the effect of passive jamming.

The range of operation of the radar on the aircraft located in a passive jamming cloud is determined by the effective diffusion area of the airplane and the quantity of chaff entering into the resolution volume of the radar station. Let $L_j = 100$ km, $D_j = 10$ km, $j_n = 5$g, and $V_{rel} = 500$ m/sec. Substituting this data into (2.94), we obtain $\Delta_0 = 10$ km.

Since we have all the required initial data, with the help of (2.91), (2.92), and (2.93), we can put together a results matrix for the actions of both sides (an action plan effectiveness matrix):

$$\begin{pmatrix} A & B & & \\ & B_1 & B_2 & B_3 \\ A_1 & 1.3 & 0 & 0.6 \\ A_2 & 0 & 0.8 & 0.4 \\ A_3 & 1 & 0.4 & 0.7 \end{pmatrix} \tag{2.95}$$

An analysis of the matrix does not indicate any obvious advantages for any of the action plans.

The most favorable appears to be plan A_2 (bombers in the passive jamming area), since, in this case, for any action plan of the opponent, the average number of attacks on the aircraft does not exceed 0.8. For all other action plan variants, the number of attacks can be greater. For example, for action plan A_3, it can be equal to 1.

The action plan A_3 would seem to be less advantageous in comparison to plan A_2, since, for the two action plans of the opponent B_1 and B_2, it gives a significantly greater number of attacks on the bombers.

However, if we select action plan A_2 as the only one, then we conscientiously give the opponent the possibility of using plan B_2, which gives him an average number of attacks on the bombers equal to 0.8. Therefore, it would seem to be convenient in the instance being analyzed to select not just one action plan, but several plans, whereas the choice of each of them should be made using a random law with a definite frequency (likelihood). In this case, the opponent is deprived of this known certainty in his choice of action plans and is forced to perform actions based on the assumption that we may implement any of the action plans A_1, A_2, or A_3.

The likelihood of choosing action plans must be determined so that the average number of attacks on the bombers is minimal. As was already noted before, in terms of the theory of games, this operation is called a transition from pure to mixed strategies.

We will conditionally designate the solution being sought by S. According to what has been said, the structure of the solution must be represented in the following way:

$$S_A = \begin{pmatrix} A_1, A_2, A_3 \\ P_1, P_2, P_3 \end{pmatrix}$$

Here A_1, A_2, A_3 are the action plans; and P_1, P_2, P_3 are the frequencies of use of action plans A_1, A_2, A_3.

Let us assume that we have found the optimum frequencies of use for P_1, P_2, P_3. Then given any of the opponent's action plans B_1, B_2, or B_3, the average number of attacks on the bombers will not exceed a certain number N, equal to the number of attacks when the action plan is optimum (in game theory, N is called the price of the game). As a result, if P_1, P_2, and P_3 are the optimum frequencies of use for action plans A_1, A_2, and A_3, then the following system of inequalities is valid:

$$\begin{aligned} P_1\overline{n_{11}} + P_2\overline{n_{21}} + P_3\overline{n_{31}} &\leqslant N \\ P_1\overline{n_{12}} + P_2\overline{n_{22}} + P_3\overline{n_{32}} &\leqslant N \\ P_1\overline{n_{13}} + P_2\overline{n_{23}} + P_3\overline{n_{33}} &\leqslant N \end{aligned} \tag{2.96}$$

The left-hand part of the first inequality determines the average number of attacks when the action opponent's plan is B_1, the second inequality correspondingly when the action plan is B_2, and the third when the action plan is B_3. It is necessary to write down the inequalities because, in the general case, not all action plans should be used (i.e., they are not always useful).

In those cases where it is known in advance that all action plans are useful, in the place of inequality (2.96) we can write down the corresponding equalities. However, at the present time, this can be said with complete certainty only in games where the number of strategies for at least one of the two sides does not exceed two (games 2×2 or $2 \times n$).

The sum of the frequencies P_1, P_2, and P_3 is equal to one; that is, there is always an equality:

$$P_1 + P_2 + P_3 = 1 \tag{2.97}$$

The joint solution of the inequalities (2.96) and the equalities (2.97) exhaust the problem posed.

For convenience, let us multiply all the numbers in matrix (2.95) by 10, thus increasing by 10 times the price of the game ($N' = 10N$). However, when doing this, the frequencies sought P_1, P_2, and P_3 do not change. The matrix (2.95) can now be expressed in the following way:

$$\begin{pmatrix} A & B & & \\ & B_1 & B_2 & B_3 \\ A_1 & 13 & 0 & 6 \\ A_2 & 0 & 8 & 4 \\ A_3 & 10 & 4 & 7 \end{pmatrix} \tag{2.98}$$

Thus, the system of inequalities (2.96) for specific values of the matrix (2.98) assumes the format:

$$\begin{aligned} &13P_1 + 10P_3 \leqslant N' \\ &8P_2 + 4P_3 \leqslant N' \\ &6P_1 + 4P_2 + 7P_3 \leqslant N' \end{aligned} \tag{2.99}$$

Next, we divide both parts of the inequalities (2.99) by N' and introduce the symbols:

$$\xi_1 = \frac{P_1}{N'} \quad \xi_2 = \frac{P_2}{N'} \quad \xi_3 = \frac{P_3}{N'}$$

Further, we add to the left-hand parts of the inequalities we have obtained certain nonnegative variables z_1, z_2, z_3 so as to obtain equalities. Then the inequalities (2.99) and the equality (2.97) can be transformed into a system of equations:

$$\begin{aligned} &13\xi_1 + 10\xi_2 + z_1 = 1 \\ &8\xi_2 + 4\xi_3 + z_2 = 1 \\ &6\xi_1 + 4\xi_2 + 7\xi_3 + z_3 = 1 \\ &\xi_1 + \xi_2 + \xi_3 = \frac{1}{N'} \end{aligned} \tag{2.100}$$

The task is now reduced to a problem of linear programming and, more specifically, to the determining of such values for ξ_1, ξ_2, ξ_3 where the value of N' will be minimum. Before finding ξ_1 $(i = 1, 2, 3)$, we determine z_1, z_2, z_3. These variables can be equal to zero, especially in those cases where all action plans are useful. The inequality to zero of any variable z_i $(i = 1, 2, 3)$ in the game being analyzed indicates that the use of one of the action plans is undesirable.

Let us express the values of ξ_1, ξ_2, ξ_3 by z_1, z_2, z_3:

$$\xi_1 = -z_1 - z_2 + 2z_3$$

$$\xi_2 = \frac{3}{40} - \frac{3}{5}z_1 - \frac{31}{40}z_2 - \frac{13}{10}z_3$$

$$\xi_3 = \frac{1}{10} + \frac{6}{5}z_1 + \frac{13}{10}z_2 - \frac{26}{10}z_3 \tag{2.101}$$

Substituting these expressions into the last equation of (2.100), we obtain

$$\frac{7}{40} - \frac{2}{5}z_1 - \frac{15}{40}z_2 + \frac{7}{10}z_3 = \frac{1}{N'} \tag{2.102}$$

It is reasonable to assume that $z_1 = z_2 = 0$ since, when any of them increases, the value of $1/N'$ decreases. When z_3 increases, the value of $1/N'$ grows, so therefore $z_3 \neq 0$. This circumstance justifies the supposition that one of the three action plans being analyzed is not useful.

In order to clarify which one it is, we must consecutively evaluate the nature of the effect of z_3 on ξ_i, $(i = 1, 2, 3)$ and $1/N'$. It is convenient to start the evaluation in the given case with ξ_3, which determines the frequency of use of plan A_3. At first glance, it is the least effective. As z_3 increases, the value of ξ_3 decreases and becomes zero when

$$z_3 = \frac{1}{26}$$

In order to be certain about our assumption that the equality $\xi_3 = 0$ is correct (action plan A_3 is unacceptable), it should be shown that the value of $\frac{1}{N_1}$ is at its maximum when $z_1 = z_2 = \xi_3 = 0$.

To this purpose, using (2.100), we should express $\frac{1}{N'}$ using z_1, z_2, ξ_3:

$$\frac{21}{104} - \frac{1}{13}z_1 - \frac{1}{8}z_2 - \frac{7}{26}\xi_3 = \frac{1}{N'}$$

The increase of any of the variable values z_1, z_2, ξ_3 decreases $\frac{1}{N'}$; so therefore,

$$\xi_3 = 0$$

From the first two equations (2.101) we obtain

$$\xi_1 = \frac{1}{13} \quad \xi_2 = \frac{1}{8}$$

From (2.102) we find

$$N' = 4.95$$

Further, we find the frequencies of use sought for the action plans A_1, A_2, and A_3, and the minimum attainable (assuming that the opponent acts in an optimal way) number of attacks on bombers:

$$P_1 = 0.38 \quad P_2 = 0.62$$
$$P_3 = 0 \quad N \approx 0.5$$

Thus, using plans A_1 and A_2 corresponding to frequencies $P_1 = 0.38$ and $P_2 = 0.62$, we obtain a reduction in the number of attacks on bombers to 0.5, independent of any possible countermeasures taken by the opponent in the given circumstances (given, of course, that he does not use any other action plans except B_1, B_2 and B_3). In an analogous manner, we can solve the problem of determining the optimum AAD action plans. The basic difference here is that, due to the opposing nature of the interests of the sides, the signs of the inequalities change and the value $1/N'$ should not be maximized, but minimized. The corresponding system of inequalities takes on the form:

$$q_1\bar{n}_{11} + q_2\bar{n}_{12} + q_3\bar{n}_{13} \geqslant N$$
$$q_1\bar{n}_{21} + q_2\bar{n}_{22} + q_3\bar{n}_{23} \geqslant N$$
$$q_1\bar{n}_{31} + q_2\bar{n}_{32} + q_3\bar{n}_{33} \geqslant N$$

Here, q_1, q_2, q_3 are the frequencies of use for action plans B_1, B_2, B_3.

For the example being analyzed: $q_1 = 0.38$, $q_2 = 0.62$, $q_3 = 0$, $N = 0.5$.

The decisions are implemented by each of the sides with the help of random selection, more specifically, using a random number table.

For example, if, as a result of solving the game matrix, we find the following frequencies of use for action plans A_1 and A_2:

$$P_1 = 0.38 \approx 0.4 \quad P_2 = 0.62 \approx 0.6$$

then implementation of a solution with the help of two-digit random number tables can be done in the following way.

The random number table is opened to any page. On this page we choose any number, located at the intersection of a row and column selected at random. If, in the two-digit number found in such a manner, the first digit is 0, 1, 2, or 3, then we select action plan A_1; if the number is 4, 5, 6, 7, 8, or 9, then we choose action plan A_2. Modern computer software permits such problems to be solved without any difficulty.

The general analysis conducted and the example given show that, for the use of matrix games to be practical, each of the sides (each player) must know a priori all the pure strategies A_i and B_j, and also each corresponding value of the cost a_{ij}. In conditions of armed conflict, measures for concealing operations, including hiding of intentions, disinformation, simulation etc., play a significant, and in many cases, decisive role. For this reason, more meaningful results can be obtained using positional (multistep) games with incomplete information. Practically speaking, when there is a large number of alternatives (strategies) and the level of information available to the sides is limited, the task is often reduced to making statistical decisions analogous to those taken when we are dealing with "a game with nature" [18].

The task of making a statistical decision in this case can be formulated using the same matrix $\|a_{ij}\|$ that was used in the matrix game. However, there is a difference in principle in that the side representing "nature" does not undertake deliberate countermeasures against the other side. If the side B represents "nature," then in matrix (2.84) its strategies B_j should be considered to be certain fixed states. Let us agree to denote them as S_j in the material that follows. The choice by side A of the most hypothetically reasonable strategy A_i in these conditions, to a great extent, is determined by the evolving uncertainties of the states of "nature." In order to account for possible losses, to the extent necessary, caused by uncertainty, we introduce the concept of risk r_{ij}. This is understood to be the difference between the maximum value of the winnings β_j, which occurs if the winnings are known, received from implementing the strategies of "nature," represented in column S_j, and the winnings a_{ij}, which correspond to the strategies of A_i:

$$r_{ij} = \beta_j - a_{ij}$$

According to what has been mentioned,

$$\beta_j = \max_j a_{ij}, r_{ij} \geqslant 0 \tag{2.103}$$

In the general case, the determining of risk in conditions of uncertainty is a fairly complex problem, the formulation of which is unclear. The task becomes much easier if we know the probabilities of the states of "nature":

$$P(S_j), j = (\overline{1, n}) \quad \sum_{j=1}^{n} P(S_j) = 1$$

In this case it is desirable not to orient ourselves toward the maximum values of particular winnings, but to the maximum of the mathematical expectation of winnings $\bar{a}_i$:

$$\bar{a}_i = \sum_{j=1}^{n} P[S_j]a_{ij}$$

Thus, it is possible to evaluate the average risk:

$$\bar{r}_i = \sum_{j=1}^{n} P[S_j]r_{ij} \tag{2.104}$$

As an optimum strategy, we should select $A^* = A_i$, where $\bar{a}_i$ attains the maximum. Naturally, when orienting ourselves towards an average risk of $\bar{r}_i$, we select such a strategy A_i so as to provide for a minimum $\bar{r}_i$. It can be shown [18] that a strategy that maximizes the average winnings $\bar{a}_i$ corresponds with a strategy that minimizes the average risk $\bar{r}_i$.

The distributions of the probabilities $P[S_j]$ are far from always being known. In such cases, a decision can be taken on the basis of other criteria. Let us consider three of the best known criteria: those of Wald, Savage and Gurwitz.

According to the Wald's minimum–maximum criterion, player A, based on an analysis of matrix (2.84) and, taking into consideration that S_j has been substituted in it for B_j, chooses a strategy where the minimum winnings (with respect to j) is maximum (with respect to i). In this case, under any conditions, player A receives winnings of no less than W:

$$W = \max_{j} \min_{i} a_{ij} \tag{2.105}$$

According to Savage's criterion of minimum risk in conditions of uncertainty, we should choose strategy A_i, which provides for a minimum risk in the most unfavorable circumstances (i.e., when we have a maximum risk with respect to j). Following this criterion, the side taking the decision avoids a big risk. In any case, the risk will be no larger than

$$\varepsilon = \min_{i} \max_{j} r_{ij} \tag{2.106}$$

Following Gurwitz's pessimism–optimism criterion, the player A can orient himself towards winnings of H, which is determined by the following expression:

$$H = \max_{i}\left(\chi \min_{j} a_{ij} + (1 - \chi) \max_{j} a_{ij}\right) \tag{2.107}$$

where χ is found by the researcher based on subjective considerations, $0 \leqslant \chi \leqslant 1$. If $\chi = 1$, then the Gurwitz criterion becomes the Wald pessimistic criterion. When $\chi = 0$, we have the maximum possible winnings (extreme optimism). Expert evaluations play an important role when determining χ.

In concluding this section, it should be noted that, in modern conditions, theoretical, game and statistical evaluations are obtained in the main using simulation modeling. An important place in these evaluations is played by methods of situation modeling, which is one of the trends in the theory of artificial intelligence [21]. Computer productivity and the level of specialists determine the intellectual potential of knowledge- and databases for EW systems.

References

[1] Panov, V. V., and A. P. Sarkis'yan, *Nekotorye aspekty problemy sozdaniya SVCH sredstv funktsional'nogo porazheniya* (*Certain Aspects of the Problem of Creating UHF Equipment for Functional Destruction*). Moscow: *Zarubezhnaya radioe'lektronika* (*Foreign Radio Electronics*), Vol. 10, 11, 12, 1993. pp. 3–10.

[2] Nebabin, V. G., and V. B. Kuznetsov, *Zashchita RLS ot protivoradiolokatsionnykh raket* (*Safeguarding Radar from Anti-Radar Missiles*). Moscow: *Zarubezhnaya radioe'lektronika* (*Foreign Radio Electronics*), Vol. 5, 1990, pp. 67–81.

[3] Fel'd, Ya. N., and L. S. Benenson, *Antenny santimetrovykh i detsimetrovykh voln* (*Centimeter and Decimeter Wave Antennas*). Moscow: N. E. Zhukovsky Airforce Academy, 1955, pp. 45–108, 167–202.

[4] Skolnik, M., editor, *Spravochnik po radiolokatsii* (*Radar Handbook*). Moscow: *Sov. radio* (*Soviet Radio*), Vol. 4, 1976, pp. 193–215.

[5] Helstrom, C. W., *Statisticheskaya teoriya obnaruzheniya signalov* (*Statistical Theory of Signal Detection*). Moscow: NIL, 1963, pp. 106–158.

[6] VanTrees, H. L., *Teoriya obnaruzheniya, ocenok i modulyatsii* (*Detection, Estimation and Modulation Theory*). Moscow: *Sov. radio* (*Soviet Radio*), 1972, pp. 203–249.

[7] Pugachev, V. S., *Teoriya sluchaynykh funktsiy i ee primenenie k zadacham avtomaticheskogo upravleniya* (*The Theory of Random Functions and its Application to Problems in Automatic Control*). Moscow: *Fizmatgiz* (*State Publishing House for Physics and Mathematics*), 1962, pp. 220–280.

[8] *Funktsii s dvoynoy ortogonal'nost'yu v radioe'lektronike i optike* (*Dually Orthogonal Functions in Radio Electronics and Optics*). Under the Editorship of M. K. Razmakhnin and V. P. Yakovlev. Moscow: *Sov. radio* (*Soviet Radio*), 1971, pp. 12–39, 232–237.

[9] Rice, S. O., *Teoriya fluktuatsionnykh shumov/Sbornik perevodov* (*Mathematical Analysis of Random Noise/Collection of Translations*). Under the Editorship of N. A. Zheleznov. Moscow: IIL, 1953, pp. 88–238, 118–132.

[10] Levin, B. R., *Teoreticheskie osnovy statisticheskoy radiotekhniki* (*Theoretical Bases of Statistical Radio Technology*). Moscow: *Sov. radio* (*Soviet Radio*), 1969, pp. 679–700.

[11] Davenport, W. B., and W. L. Root, *Vvedenie v teoriyu sluchaynyx signalov i shumov* (*Introduction to the Theory of Random Signals and Noise*). Moscow: IIL, 1960, pp. 110–146.

[12] Tikhonov, V. I., and Yu. N. Bakaev, *Statisticheskaya teoriya radiotekhnicheskikh ustroystv* (*The Statistical Theory of Radio Technical Devices*). Moscow: N.E. Zhukovsky Airforce Academy, 1976, pp. 5–119, 180–220, 263–416.

[13] Ryzhik, I., *Tablitsy integralov, summ, ryadov i proizvedeniy* (*Tables of Integrals, Sums, Series and Products*). Moscow: *Gostekhizdat* (*State Technical Publishing House*), 1948, 400 pp.

[14] Vakin, S. A., and L. N. Shustov, *Osnovy radioprotivodeystviya i radiotekhnicheskoy razvedki* (*The Basics of Radio Countermeasures and Radio Technical Intelligence*). Moscow: *Sov. radio* (*Soviet Radio*), 1968, pp. 9–122, 259–342.

[15] Vlasov, V. F., *Teoriya signalov i radiotsepey, chast' 1* (*The Theory of Signals and Radio Circuits, Part 1*). Moscow: N. E. Zhukovsky Airforce Academy, 1968, pp. 95–179.

[16] Van Brunt, L. B., *Applied ECM.*, EW Engineering Inc., Dunn Loring, VA, USA, 1978, 973 pp.

[17] Maher, Richard A., *An Introduction to Digital Avionics.* TRW Military Electronics Division. San Diego, California. John G. Weber, Veras. 1985, pp. 66–90.

[18] Venttsel', E. S., *Issledovanie operatsiy* (*Operations Research*). Moscow: *Sov. radio* (*Soviet Radio*), 1972, pp. 238–281, 291–357, 446–509.

[19] Owen, G., *Teoriya igr* (*Game Theory*). Moscow: *Mir* (*Peace*), 1971, pp. 35–117.

[20] Dresher, M., *Strategicheskie igry* (*Strategic Games*). Moscow: *Sov. radio* (*Soviet Radio*), 1964, pp. 32–43, 177–198.

[21] Pospelov, D. A., *Situatsionnoe upravlenie* (*Situation Management*). Moscow: *Nauka* (*Science*), 1986, pp. 9–130.

3

Electronic Warfare Effectiveness Criteria

3.1 General Characteristics of the Criteria

At present, four groups of criteria have been determined that describe the special characteristics of electronic warfare (EW) as an element of information warfare (IW) related to the military use of electromagnetic emissions. These criteria include: information; energy; operational and tactical; and military and economic indicators that permit the evaluation of the effects of the level of jamming signals, systems and techniques of EW. Besides this, electronic warfare systems must satisfy the general technical requirements placed on electronic equipment for military use. In the materials that follow, only those criteria are considered that describe the special characteristics of EW. The essence of the four indicators mentioned can be formulated as follows.

Information criteria describe the potential capabilities of jamming signals, systems and techniques to cause information damage in the electronic target being disabled. Information indicators of systems, considered as targets of jamming, also permit evaluation of their potential capabilities to protect against jamming.

Energy criteria of jamming signals, systems and techniques of jamming characterize the actual capabilities of causing information damage, taking into consideration the energy potential of the target being disabled. Energy indicators of target electronic systems permit evaluation of the degree to which they are, in practice, protected against specific jamming systems in the dynamics of EW.

Operational and tactical criteria permit the evaluation of techniques and systems of EW as elements of armed conflict, and the determination of operational and tactical norms for forces and systems of EW in the dynamics of armed warfare with regard to the electronic environment.

Military and economic indicators permit us to compare the effectiveness of systems and techniques of EW with their cost and, on this basis, to find optimum solutions.

3.2 Information Indicators of the Effectiveness of Jamming Signals, Systems and Techniques of Electronic Attack

3.2.1 Fundamental Concepts

As has already been noted (Section 3.1), information criteria must permit evaluation of the potential capabilities of jamming signals, systems and techniques for jamming to cause information damage in the dynamics of IW. The latter circumstance results in the necessity of having a criterion that permits evaluation not only of information damage, but also of information stability with respect to countermeasures taken in the dynamics of EW. It should be kept in mind that IW is more sensitive to countermeasures. In particular, this results from the fact that jamming signals (masking, deception) do not directly result in material destruction, since their effects cause only information damage. This permits the side being jammed, in a limited time, to adapt to the electronic environment and decrease the value of information damage caused, in the case where a priori uncertainty has not been assured. A priori uncertainty is achieved through a priori randomization of the parameters of jamming signals. This is valid not only for jamming signals, systems and techniques of electronic attack, but also for the targets of jamming.

The degree of uncertainty of random events can vary. Quantitatively, uncertainty about distributions of random events is usually evaluated using an entropy value [1–4]. We will briefly remind the reader of its definition and basic properties.

Average a priori uncertainty, as related to one of the states of the system and represented in the form of the model X with a finite number of states X_i and their related probabilities P_i,

$$X = \begin{pmatrix} X_i \\ P_i \end{pmatrix} \quad (i = \overline{1, m}) \tag{3.1}$$

$$\sum_{i=1}^{m} P_i = 1 \tag{3.2}$$

is defined by the expression:

$$H(X) = -\sum_{i=1}^{m} P_i \log P_i \tag{3.3}$$

The expression $H(X)$, represented in the form (3.3), is usually called the entropy. It quantitatively defines the degree of a priori uncertainty in the distribution of probabilities represented in the finite model (3.1). The adopted measure of uncertainty satisfies intuitive notions: to be more exact, it is equal to zero when one of the $P_i = 1$, $(i = \overline{1, m})$, and monotonically increases together with an increase in m. For a given m, the value of $H(X)$ reaches the maximum if all P_i are equal to one other:

$$P_i = 1/m \quad (i = \overline{1, m}) \quad H(X) = \log m \tag{3.4}$$

One of the virtues of adopting this measure of a priori uncertainty is its additivity. In particular, this means that, in the case that the product of two independent events X and Y is represented by finite probability models:

$$X = \begin{pmatrix} X_i \\ P_i \end{pmatrix} \quad (i = \overline{1, m}) \qquad Y = \begin{pmatrix} Y_j \\ P_j \end{pmatrix} \quad (j = \overline{1, n})$$

the entropy of the product $H(XY)$ is equal to the sum:

$$H(XY) = H(X) + H(Y) \tag{3.5}$$

The result given follows directly after a series of transformations from (3.3) and takes into account the normalization conditions (3.2). If events X and Y are independent, then,

$$H(XY) = H(X) + H_X(Y) \tag{3.6}$$

where $H_X(Y)$ is the conditional entropy (i.e., the distribution entropy of Y on the condition that event X occurs). In the materials that follow, X and Y will be understood to be random vectors:

$$X = X(x_i) \quad (i = \overline{1, m}) \qquad Y = Y(y_j) \quad (j = \overline{1, n})$$

For example, after removing the uncertainty of receiving a transmission in a channel without interference, the average quantity of information $I(X)$ transmitted by a single symbol is equal to $H(X)$:

$$I(X) = H(X) \tag{3.7}$$

The value of the entropy in a probability model with two equally probable states, defined when the logarithm radix is equal to two, is normally considered to be the unit of uncertainty. It is equal to 1 bit:

$$H = -\left(\frac{1}{2}\log_2\frac{1}{2} + \frac{1}{2}\log_2\frac{1}{2}\right) = 1 \text{ bit}$$

The quantity of information is measured in these same units.

In channels with interference, information loss occurs. Instead of the random vector X at the output of the channel, the vector Y is observed. The average quantity of information $I(X, Y)$ relative to one symbol is decreased by the value of the conditional entropy $H_{Y}(X)$:

$$I(X, Y) = H(X) - H_{Y}(X) \tag{3.8}$$

where $H(X)$ is defined in (3.3):

$$H_{Y}(X) = -\sum_{i=1}^{m}\sum_{j=1}^{n} P(x_i, y_j) \log P\left({}^{x_i}/_{y_j}\right) \tag{3.9}$$

The ratio (3.8), taking into consideration (3.3), (3.9) and the normalization conditions (3.2), is reduced to the form:

$$I(X, Y) = \sum_{i=1}^{m}\sum_{j=1}^{n} P(x_i, y_j) \log \frac{P(x_i, y_j)}{P(x_i)P(y_j)} \tag{3.10}$$

The entropy of the continuous random value X is found from (3.3) using a limit conversion:

$$H(X) = -\int p(x) \log p(x) dx - \lim_{\Delta x \to 0} \log \Delta x \tag{3.11}$$

where $p(x)$ is the probability density X.

For comparative evaluations involving a difference in entropies (and it is specifically these evaluations that are of interest), the entropy of the continuous random value X can be determined using the expression:

$$H(X) = -\int p(x) \log p(x) \mathrm{d}x \tag{3.12}$$

At the same time there is a normalization condition:

$$\int_{-\infty}^{\infty} p(x) \mathrm{d}x = 1 \tag{3.13}$$

By analogy with (3.10), the average quantity of reciprocal information $I(X, Y)$ contained in random vector Y about the random vector X is determined by the formula:

$$I(X, Y) = \int\int p(x,y) \log \frac{p(x,y)}{p(x)p(y)} \mathrm{d}x\mathrm{d}y \tag{3.14}$$

Here $p(x,y)$ is the multidimensional common probability density of vectors X and Y, and $p(x)$ and $p(y)$ are the multidimensional probability densities of vectors X and Y.

It is assumed that the vector components in (3.14) are statistically independent.

The representation of entropy in the form of a probability density expression (3.12) and the presence of limitations such as the normalization condition (3.13) permit us to formulate the problem of synthesizing an optimum probability density $p(x)$ that would provide the maximum entropy value. In the given case, we are dealing with the solution of an isoperimetric problem in variational calculus. In its general form, the problem can be formulated as follows [5].

We assume that we are given the integral expression:

$$\Psi = \int F(x, p) \mathrm{d}x \tag{3.15}$$

where p is a certain function x being sought.

Let us also assume that we are given m limitations, imposed on the variable x and the function p:

$$\int \varphi_i(x, p)\mathrm{d}x = C_i \ (i = \overline{1, m}) \tag{3.16}$$

where φ_i are certain given functions. It is necessary to define such a function $p(x)$ that would assure the maximum of the expression (3.15), taking into consideration the limitations (3.16). In the given case, the problem is reduced to finding a conditional extremum. Its solution can be found using indefinite Lagrange factors. The maximum of the expression occurs for those $p(x)$ that convert the following linear form (Lagrangian) to zero:

$$\frac{\partial F}{\partial p} + \lambda_1 \frac{\partial \varphi_1}{\partial p} + \lambda_2 \frac{\partial \varphi_2}{\partial p} + \cdots + \lambda_m \frac{\varphi_m}{\partial p} = 0 \tag{3.17}$$

Into the expression $p(x)$, obtained from (3.17), there enter indefinite factors $\lambda_i (i = \overline{1, m})$, determined from m equations (3.16).

As specific examples of the synthesis of optimum probability densities, let us consider three variants of sets of continuous random values:

1. The set of random values X, limited in maximum and minimum values;
2. The set of random values X with a given variance value σ^2;
3. The set of random values $X (0 \leqslant x \leqslant \infty)$ with a given mathematical expectation value a.

In the first instance, expression (3.15), limitation (3.16) and Lagrangian (3.17) are reduced to the form:

$$H = -\int_{-U_0}^{+U_0} p(x) \log p(x) \mathrm{d}x \tag{3.18}$$

$$\int_{-U_0}^{+U_0} p(x) \mathrm{d}x = 1 \tag{3.19}$$

$$\frac{\partial}{\partial p}(-p \log p) + \lambda_1 = 0 \tag{3.20}$$

In the second case:

$$H = -\int_{-\infty}^{+\infty} p(x) \log p(x) \mathrm{d}x \tag{3.21}$$

$$\int_{-\infty}^{+\infty} p(x) \mathrm{d}x = 1 \tag{3.22}$$

$$\int_{-\infty}^{+\infty} x^2 p(x) \mathrm{d}x = \sigma^2 \tag{3.23}$$

$$\frac{\partial}{\partial p}(-p \log p) + \lambda_1 + \lambda_2 x^2 = 0 \tag{3.24}$$

In the third case:

$$H = -\int_{0}^{\infty} p(x) \log p(x) \mathrm{d}x \tag{3.25}$$

$$\int_{0}^{\infty} p(x) \mathrm{d}x = 1 \tag{3.26}$$

$$\int_{0}^{\infty} x p(x) \mathrm{d}x = a \tag{3.27}$$

$$\frac{\partial}{\partial p}(-p \log p) + \lambda_1 + \lambda_2 x = 0 \tag{3.28}$$

Let us perform a total solution of the problem for the second case. Lagrangian (3.24) is converted to the following form:

$$- \log p - 1 + \lambda_1 + \lambda_2 x^2 = 0$$
$$p(x) = \mathrm{e}^{\lambda_2 x^2 + (\lambda_1 - 1)} \tag{3.29}$$

Using normalization condition (3.22), we obtain

$$1 = \exp(\lambda_1 - 1)2\int_0^\infty e^{\lambda_2 x^2}\,dx \qquad \lambda_2 < 0$$

Applying table integrals:

$$\int_0^\infty e^{-a^2x^2}\,dx = \frac{\sqrt{\pi}}{2a} \quad a > 0 \quad \text{and} \quad \int_0^\infty x^2 e^{-a^2x^2}\,dx = \frac{\sqrt{\pi}}{4a^3} \quad a > 0$$

and also limitation (3.23), we find

$$1 = \exp(\lambda_1 - 1)\sqrt{\frac{\pi}{-\lambda_2}} \quad \exp(\lambda_1 - 1) = \sqrt{\frac{-\lambda_2}{\pi}},$$

$$\sigma^2 = 2\sqrt{\frac{-\lambda_2}{\pi}}\int_0^\infty x^2 e^{\lambda_2 x^2}\,dx = \frac{1}{-2\lambda_2}$$

$$\lambda_2 = -\frac{1}{2\sigma^2} \quad \exp(\lambda_1 - 1) = \sqrt{\frac{1}{2\pi\sigma^2}}$$

Substituting the values of indefinite factors λ_1 and λ_2 into (3.29), we obtain the probability density sought:

$$p(x) = \frac{1}{\sqrt{2\pi\sigma^2}}\exp\left(-x^2/2\sigma^2\right) \tag{3.30}$$

Thus, among all the random values with a fixed variance, a random value distributed according to a normal (Gaussian) law has the maximum entropy. According to (3.21), the entropy of the random value X with the probability density represented in (3.30) is equal to

$$H(X) = \ln\sqrt{2\pi e\sigma^2} \tag{3.31}$$

It should be kept in mind that the variance of random electrical signals is equal to their average power.

In an analogous manner, it is possible to synthesize the probability densities in the first and third instances.

In the first case, the most entropic is a random value with a uniform probability density in the interval $(-U_0, U_0)$:

$$p(x) = {}^{1}\!/_{2U_0} \qquad (3.32)$$

accordingly,

$$H(X) = \log 2U_0 \qquad (3.33)$$

In the third case, maximum entropy occurs in the exponential distribution:

$$p(x) = \frac{1}{a}\exp\left(-{}^{x}\!/_{a}\right) \quad x \geqslant 0 \qquad (3.34)$$

$$H(X) = \ln(ea) \qquad (3.35)$$

The multidimensional entropy H_n of the sample function of a random process represented in the form of a series of uncorrelated random values is defined as the sum of the one-dimensional entropies H_i of the sample values of the sample function:

$$H_n = \sum_{i=1}^{n} H_i \qquad (3.36)$$

If the spectral density $S(\omega)$ of the sample function of a random process of duration T is constant, and the number of sample values is sufficiently large (i.e., $n = 2FT >> 1$), where F is the width of the spectrum, then it is possible to assume $H_i = H_1 = const$, $(i = \overline{1, n})$. Then, in the first approximation,

$$H_n = 2FTH_1 \qquad (3.37)$$

When the quantity of sample values is limited (not great), they are determined with the help of a canonical expansion (a Karhunen–Loeve series) (2.21).

The evaluation of the masking characteristics of a random sample noise function with a fixed average power P_j in comparison with the sample function of Gaussian noise with the same number of sample functions can be conducted by determining its entropic power $\overline{P}_j$. This is understood to be the power of Gaussian noise with a uniform spectrum, where the same entropy value H' is reached as occurs in the noise being considered. Assuming the number of sample values for the original and the standard

Gaussian noise to be identical, according to the definition and the formula (3.31), we obtain

$$H' = \ln\sqrt{2\pi e \overline{P}_j} \tag{3.38}$$

$$H' = \frac{1}{2}\ln(2\pi e \overline{P}_j)$$

The entropic power $\overline{P}_j$ is equal to

$$\overline{P}_j = \frac{1}{2\pi e}\exp(2H') \tag{3.39}$$

It is evident that

$$\overline{P}_j \leqslant P_j \tag{3.40}$$

At the suggestion of S. Kul'bak [6, 7], in the theory of statistical solutions, in a number of cases, instead of using the formula (3.14) defining the quantity of mutual information according to Shannon, we use formulas defining the average quantity of information $I(1{:}2)$ in favor of one of the competing parametric hypotheses Θ_0 or Θ contained in the sample function being considered:

$$I(1:2) = \int p(x,\Theta_0)\log\frac{p(x,\Theta_0)}{p(x,\Theta)}\,\mathrm{d}x \tag{3.41}$$

In the representation given, the amount of information in favor of hypothesis θ_0 is defined.

Here, x is a multidimensional random vector represented by the canonical expansion: $-p(x,\theta_0)$ and $p(x,\theta)$, the multidimensional probability densities corresponding to the values of parameters θ_0 and θ.

The quantity of information $I(1{:}2)$ defined in (3.41) can be represented as the difference of two entropies $H(\theta_0,\theta)$ and $H(\theta_0,\theta_0)$, depending on parameters θ_0 and θ:

$$I(1:2) = H(\theta_0,\theta) - H(\theta_0,\theta_0) \tag{3.42}$$

where

$$H(\theta_0,\theta) = -\int p(x,\theta_0)\log p(x,\theta)\mathrm{d}x \tag{3.43}$$

$H(\theta_0,\theta)$ reaches its extreme value (minimum) when $\theta = \theta_0$ [6].

$H(\theta_0, \theta_0)$ corresponds to the definition of entropy according to Shannon. When comparing two close hypotheses θ and θ_0, the relationship (3.42) can be simplified by expanding the function $H(\theta_0, \theta)$ into a Taylor series in the neighborhood of θ_0:

$$H(\theta_0, \theta_0 - \Delta\theta) = H(\theta_0, \theta_0) - \frac{\partial H(\theta_0, \theta_0)}{\partial \theta} \Delta\theta + \frac{1}{2} \frac{\partial^2 H(\theta_0, \theta_0)}{\partial \theta^2} \Delta\theta^2 + \cdots \tag{3.44}$$

The second term becomes zero at the point $\theta = \theta_0$, and the third term is positive, since the minimum occurs at $\theta = \theta_0$. Therefore, in the first approximation,

$$I(1:2) = H(\theta_0, \theta) - H(\theta_0, \theta_0) = \frac{1}{2} \frac{\partial^2 H(\theta_0, \theta_0)}{\partial \theta^2} \Delta\theta^2 + \cdots \tag{3.45}$$

In conclusion, it should be noted that maximizing entropy increases the information stability of jamming signals to countercountermeasures (ECCM) only to a limited extent. In this instance, we are speaking of excluding (or hindering) the enemy's implementing simple ECCM algorithms and technical techniques without significantly reducing the useful signal. Maximizing entropy forces the opposing side to seek unconventional solutions.

3.2.2 Information Quality Indicators of Masking Jamming Signals, Systems and Techniques of Generating Jamming

The masking properties of the jamming signal being analyzed are reflected in its ability to absorb the concealed (useful) signal that it is combined with. This can occur only in the case when the jamming signal is a sample function of a random process. The quality of masking, when all other conditions are equal, is determined by the degree of randomness in the jamming signal. The latter, in turn, is evaluated based on the entropy value. It would seem that, when all other conditions are equal, a jamming signal with greater entropy changes its parameters to a lesser degree when combining with a given useful signal. What has been said permits us to formulate the following statement.

With several fairly common limitations, the quality indicator of an additive masking jamming signal is the value of its entropy defined in the phase space of the concealed (useful) signal. A jamming signal is considered to be optimum when it provides maximum entropy in a given phase space. Maximizing the entropy in the information parameter of a masking jamming signal, the jamming side on the average minimizes the probability that the useful signal will be correctly detected in the victim receiver. Entropy in the given instance is considered to be an indicator of the potential capabilities of a disturbance to cause the maximum information damage to the target of jamming. With its help, we can only find the statistical characteristics of a masking jamming signal that, when other conditions are equal, corresponds to the lowest probability value that the useful signal will be detected in an additive combination with its disturbance. We will provide proofs of the validity of this statement formulated for a receiver that is optimum according to the Neyman–Pearson criterion.

Let us assume that we are verifying the hypotheses about parameters θ_0 and θ of two useful signals. Let $x = x(x_1, \ldots, x_n)$ be a random vector that defines the additive mixture of the masking jamming and useful signals in the phase space of the useful signal. We will assume the components of the random vector $x_1, \ldots, x_n$ to be statistically independent random values with identical distribution functions. This is assured for continuous noises if the spectrum is uniform and limited in the given frequency band, or when a canonical expansion (Karhunen–Loeve) exists. We will designate the multidimensional probability densities of the jamming signal, corresponding to the values of the parameters θ_0 and θ, by $p(x_1, \ldots, x_n, \theta_0)$ and $p(x_1, \ldots, x_n, \theta)$. The likelihood ratio for the competing hypotheses has the following form:

$$\Lambda(\theta_0, \theta) = \frac{p(x_1, \theta_0) \cdots p(x_n, \theta_0)}{p(x_1, \theta) \cdots p(x_n, \theta)} \tag{3.46}$$

Let us find the logarithm for the likelihood ratio:

$$\log \Lambda(\theta_0, \theta) = \sum_{i=1}^{n} \log \frac{p(x_i, \theta_0)}{p(x_i, \theta)}$$

The addends of this sum, $\log \frac{p(x_i, \theta_0)}{p(x_i, \theta)}$, $(i = \overline{1 \cdot n})$, form a totality of n independent random values with identical distribution functions. According to Khinchin's Theorem [8], the arithmetic average of these values converges in probability to their mathematical expectation (i.e., in the given case) to Kul'bak's information integral $I(1{:}2)$ (3.41). On the basis of Khinchin's Theorem we can write

$$\frac{1}{n}\left(\sum_{i=1}^{n}\log\frac{p(x_i,\theta_0)}{p(x_i,\theta)}\right)\approx I(1:2) \tag{3.47}$$

This equality occurs with a probability as near as one pleases to one, if n is sufficiently large.

The equality (3.47) can be converted to the form:

$$\log\frac{p(x_1,\theta_0)\cdots p(x_n,\theta_0)}{p(x_1,\theta)\cdots p(x_n,\theta)}\approx nI(1:2)$$

or

$$p(x_1,\theta_0),\ldots,p(x_n,\theta_0)\approx e^{nI(1:2)}p(x_1,\theta),\ldots,p(x_n,\theta) \tag{3.48}$$

Let us integrate both parts of equality (3.48) for $x_1,\ldots,x_n$ going from x_0 to ∞:

$$\int_{x_0}^{\infty}\cdots\int_{x_0}^{\infty}p(x_1,\theta_0)\ldots p(x_n,\theta_0)\mathrm{d}x_1\ldots\mathrm{d}x_n\approx e^{nI(1:2)}$$

$$\int_{x_0}^{\infty}\cdots\int_{x_0}^{\infty}p(x_1,\theta)\ldots p(x_n,\theta)\mathrm{d}x_1\ldots\mathrm{d}x_n$$

With respect to the problem of detection, the integrals in the left- and right-hand parts accordingly determine the probabilities of correct detection P_{det} and false alarm P_{fa}, as defined for the given threshold value x_0 (Figure 3.1). Therefore,

$$P_{\text{det}}\approx e^{nI(1:2)}P_{\text{fa}} \tag{3.49}$$

According to (3.45),

$$I(1:2)=\frac{1}{2}\frac{\partial^2 H(\theta_0,\theta_0)}{\partial\theta^2}\Delta\theta^2 \tag{3.50}$$

In order to prove the statement formulated earlier, it is necessary to show that the value $I(1{:}2)$ for more entropic distributions of probabilities is less than for less entropic ones. The latter, in particular, follows from Shannon's 15th Theorem [1], according to which the following inequality occurs:

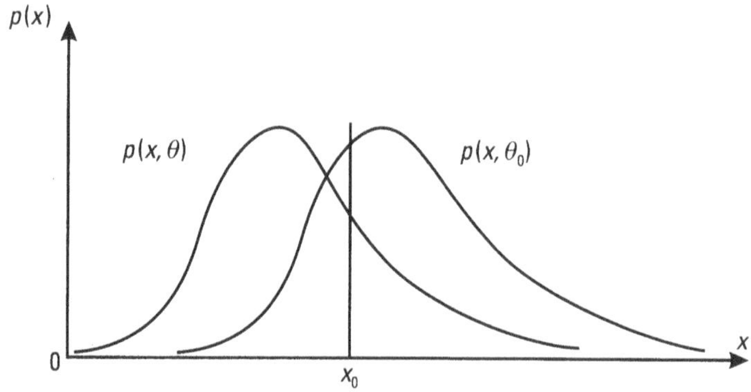

Figure 3.1 Probabilities of correct detection P_{det} and false alarm P_{fa} defined for a given threshold value x_0.

$$\overline{P}_1 + \overline{P}_2 \leqslant \overline{P}_3 \leqslant P_1 + P_2 \tag{3.51}$$

where P_1 and P_2 are the average powers of the two sample functions of the random processes; $\overline{P}_1$ and $\overline{P}_2$ are their entropic powers; and $\overline{P}_3$ is the entropic power of the sum of the two processes shown.

Let us consider two cases. In the first, both signals represent Gaussian noises with uniform spectrums and average powers of P_1 and P_2. In the second case, the first signal has an arbitrary probability density, an entropic power of $\overline{P}'_1$ and the same average power P_1; and the second signal is Gaussian noise with a uniform spectrum and power of P_2. Then the inequality (3.51) can be written down as follows. In the first case:

$$\overline{P}_1 + \overline{P}_2 = \overline{P}_3 = P_1 + P_2 \quad \overline{P}_3 - \overline{P}_1 = \overline{P}_2$$

In the second:

$$\overline{P}'_1 + \overline{P}_2 \leqslant \overline{P}'_3 \leqslant P_1 + P_2 \quad \overline{P}'_3 - \overline{P}'_1 \geqslant \overline{P}_2$$

Besides, by definition $\overline{P}'_1 \leqslant \overline{P}_1$ and $\overline{P}'_3 \leqslant \overline{P}_3$.

It follows from what has been said that

$$\overline{P}'_3 - \overline{P}'_1 \leqslant \overline{P}_3 - \overline{P}_1 \tag{3.52}$$

Based on the definition of entropic power (3.39) and using inequality (3.52), it is possible to show the validity of the following expression:

$$H_3' - H_1' \geqslant H_3 - H_1$$

Here, H_1, H_3, H_3' and H_1' are the respective entropies. The latter inequality, with consideration to (3.45), provides the basis to formulate the inequality sought:

$$\frac{1}{2}\frac{\partial^2 H'(\theta_0, \theta_0)}{\partial\theta^2}\Delta\theta^2 \geqslant \frac{1}{2}\frac{\partial^2 H(\theta_0, \theta_0)}{\partial\theta^2}\Delta\theta^2 \tag{3.53}$$

where $H'(\theta_0, \theta_0)$ is the entropic sample function of the random process with a lesser degree of uncertainty (less entropic); and $H(\theta_0, \theta_0)$ is the entropy of the sample set of the process with a greater degree of randomness.

The validity of the initial statement follows from (3.49) with due consideration to (3.50) and inequality (3.53). To be more precise, among all the masking jamming signals with a given average power, the most entropic jamming signal has the least probability of the useful signal being detected for a given false alarm probability.

The conclusions arrived at relate not only to jamming of optimum signal detection systems, but also to those directed at increasing the variance of parameter estimates, independent of the potential performance of the optimum estimator methods used. The validity of the statement made follows from the Cramer–Rao Inequality (1.20). Let us rewrite this inequality with consideration to the new designations, anticipating that the estimate shift will be equal to zero ($b = 0$). Then,

$$M[\Delta\theta^2] \geqslant \frac{1}{nm\left(\frac{\partial}{\partial\theta}\log p(x, \theta)\right)^2} \tag{3.54}$$

The mathematical expectation of the square of the logarithm of the probability density $p(x, \theta)$ by definition is equal to the Fisher information quantity [6]. The latter, in its turn, is determined by the Kul'bak information quantity (3.41) and (3.45), whence it follows that

$$M\left(\frac{\partial}{\partial\theta}\log p(x, \theta)\right)^2 = \frac{\partial^2}{\partial\theta^2}H(\theta_0, \theta)|_{\theta=\theta_0} \tag{3.55}$$

According to inequality (3.53), the Fisher information quantity will be less for more entropic distributions. Accordingly, for such distributions, the upper bound of the variance of estimates $M[\Delta\theta^2]$ turns out to be higher.

It is customary to directly evaluate the degree of entropicality of masking jamming signals and, consequently, their quality according to the information indicator, using the coefficient η_H, which is equal to the ratio of the entropic power of the jamming signal $\overline{P}_j$ to its average power P_j:

$$\eta_H = \frac{\overline{P}_j}{P_j} \tag{3.56}$$

The entropic power is calculated with the help of (3.39). In the general case, H' is understood to designate the average entropy falling at one degree of freedom of the sample function of the jamming signal:

$$H' = \frac{1}{n}\sum_{i=1}^{n} H_i \tag{3.57}$$

where H_i, $(i = \overline{1,n})$ is the entropy of the ith random value, defined by a canonical expansion of the sample function of the jamming signal in the phase space of the useful signal being concealed.

The information indicator for effectiveness η_H, when evaluating the quality of jamming signals, permits us at the same time also to evaluate the quality of masking jammers. The techniques for jamming generation can also be evaluated using the degree of proximity of a priori uncertainty (security) to that potentially attainable in the developing electronic environment.

3.2.3 Information Indicators of the Quality of Deception Jamming Signals

Just as in the case of masking jamming, the quality of deception jamming signals must be evaluated in the anticipation that the task of recognizing signals being emulated against a deception jamming background is performed using optimum methods of statistical solutions, as stated in Chapter 1.

Let us consider the case of deliberate emulation of one of the parameters of the distribution of probabilities of one of the attributes. Let us assume that the jamming side knows the multidimensional probability densities of the sample functions of attributes α, corresponding to the real $p(x, \alpha_0)$ and emulated $p(x, \alpha)$ signals. Further, let us assume that the side being jammed makes its decision regarding measurement (observation) results based on an analysis of the likelihood ratio Λ:

$$\Lambda = \frac{p(x, \alpha_0)}{p(x, \alpha)}$$

where $x = x(x_1, \ldots, x_i, \ldots, x_n)$, $(i = \overline{1, n})$, and x_i are independent random values with an identical probability distribution law.

Converting the likelihood ratio to logarithms, shifting our point of consideration to that of a sum of logarithms, and using the Khinchin Theorem, we obtain

$$\frac{1}{n}\left(\sum_{i=1}^{n} \log \frac{p(x_i, \alpha_0)}{p(x_i, \alpha}\right) \approx I(1:2) \tag{3.58}$$

where $I(1{:}2)$ is the Kul'bak information integral (3.41). The equality (3.58) occurs with a probability as close to one as you like, if n is great enough. It can be transformed in the following manner:

$$p(x_1, \alpha_0), \ldots, p(x_n, \alpha_0) = e^{nI(1:2)} p(x_1, \alpha), \ldots, p(x_n, \alpha)$$

If we designate the threshold value x, which corresponds to the allowable probability of errors of the first type P_{error}, by $x_0 = x_{i,0}$, $(i = \overline{1, n})$, then, integrating both parts of the latter equality within limits from x_0 to ∞, we obtain

$$P_{\text{rec}} = e^{nI(1:2)} P_{\text{error}} \tag{3.59}$$

where P_{rec} is the probability of recognition:

$$P_{\text{rec}} = \int_{x_0}^{\infty} \cdots \int_{x_0}^{\infty} p(x_1, \alpha_0) \ldots p(x_n, \alpha_0) \mathrm{d}x_1 \ldots \mathrm{d}x_n$$

$$P_{\text{error}} = \int_{x_0}^{\infty} \cdots \int_{x_0}^{\infty} p(x_1, \alpha) \ldots p(x_n, \alpha) \mathrm{d}x_1 \ldots \mathrm{d}x_n$$

The quantity of information $I(1{:}2)$ is equal to the difference of the entropies (3.42):

$$I(1:2) = H(\alpha_0, \alpha) - H(\alpha_0, \alpha_0) \tag{3.60}$$

The expressions $nH(\alpha_0, \alpha)$ and $nH(\alpha_0, \alpha_0)$ represent multidimensional entropies of the corresponding sample functions. Taking into consideration (3.60) from (3.59), it follows that, for a given probability of errors of the first type, the probability of recognition of the signal being emulated against the background of the deception signal will decrease together with a decrease in the differences of the corresponding multidimensional entropies defined based on the sample sets of the attribute being considered. In principle, such attributes can be quite large in number. In the general case, the condition of optimum deception based on the sample function of attribute α can be written down in the following way:

$$\min_{\alpha} |H_n(\alpha_0, \alpha) - H_n(\alpha_0, \alpha_0)| \tag{3.61}$$

where $H_n(\alpha_0, \alpha)$ is an n-dimensioned entropy defined based on the sample function of attribute α.

Having minimized the difference of the multidimensional entropies, the operator generating deception jamming can only state that the probability of recognition will be minimal. The numeric value is determined by solving the corresponding specific statistical problem.

The preceding can be considered to be initial considerations when evaluating the quality of deception jamming systems, false radar targets, thermal radar decoys, as well as the techniques of their use.

3.3 Energy Effectiveness Criteria of Jamming Signals and Techniques of Electronic Jamming

3.3.1 Fundamental Concepts

Information criteria permit us to determine the statistical characteristics of jamming signals that, for a given power, assure the realization of potential to cause maximum information damage to a jamming target. In the dynamics of IW, it is necessary to quantitatively evaluate the magnitude of the information damage caused and determine the energy potential required to provide the necessary degree of jamming of targets for given conditions of the electronic environment.

The problem comes down to establishing the dependency of the degree of jamming on the magnitude of the jamming-to-signal (J/S) ratio of the powers of the jamming and useful signals at the input to the receiver of the electronic device being jammed. This problem has been solved for specific targets and jamming signals. As a result of this solution, energy criteria for jamming have been established. Based on them, it is possible to determine the information damage caused by jamming.

Let us clarify the concept of information damage as provided in Chapter 1. Besides geometric measurements, information damage can be evaluated using the amount of information lost expressed in bits. For example, if radar jamming blocks out a volume of V_j, and the magnitude of the pulse volume (resolution element) of the radar is V_{pr}, and the detection (or nondetection) probability of a target in each resolution element is equal to 0.5, which gives a quantity of information of 1 bit, then for each scan cycle the radar loses N bits of information, where $N = \frac{V_j}{V_{pr}}$.

3.3.2 Energy Effectiveness Indicators of Masking and Deception Radar Jamming Signals

Following the recommendation defined in Chapter 1, we will select as the model jamming target a radar receiver that is optimal according to the Neyman–Pearson criterion. In the first stage, the task of detecting a totally unknown signal against a background of white Gaussian noise will be considered.

As was noted in Chapter 1, the probabilities of correct detection of a signal P_{det} and a false alarm P_{fa} are defined by the expressions (1.4), (1.5) and (1.6) using parameter q, which is equal to the signal/noise ratio (1.3), and the threshold h. For a given q, the threshold h is fully determined by the false alarm level assumed. In order to determine the dependency of the degree of jamming on the value of the J/S ratio at the input to the receiver being jammed, we must transform the expression (1.3) in the following way:

$$q = \sqrt{\frac{2E}{N_0}} = \sqrt{\frac{2(P_S)_{inp}\tau_S}{\frac{(P_j)_{inp}}{\Delta f_{jP}}}} \tag{3.62}$$

Here, $(P_S)_{inp}$ is the power of the useful signal at the input to the receiver; and τ_S is the duration of the useful signal; and $(P_j)_{inp}$ is the power of the jamming signal at the input to the receiver, defined in the band of its linear component; and Δf_{jP} is the equivalent width of the passband of the linear component of the receiver.

The spectral density of the jamming signal is assumed to be constant over a fairly broad band Δf_j.

From (3.62), let us determine the J/S ratio, as related to the passband of the linear component of the of the receiver being jammed:

$$K = \left(\frac{P_J}{P_S}\right)_{\text{inp}} = \frac{2\tau_S \Delta f_{jP}}{q^2} \tag{3.63}$$

If the receiver is optimum with respect to the Neyman–Pearson criterion and the task at hand is detecting a fully known signal against a background of white Gaussian noise, then the passband of the linear component of the receiver is matched to the duration of the pulse; that is,

$$\tau_S \Delta f_{jP} = 1 \tag{3.64}$$

and then

$$K = \frac{2}{q^2} \tag{3.65}$$

Using (1.4), (1.5) and (3.65), it is possible to determine the J/S ratio at the input to the receiver, such that the probability of correct detection for a given probability of a false alarm does not exceed required values. For example, in order to reduce the detection probability to the value $P_{\text{det}} = 0.1$ when $P_{\text{fa}} = 0.001$, the J/S ratio in the band of the linear component of the receiver must be 0.62 ($K = 0.62$). Based on (3.65), as well as (1.4) and (1.5), it is possible to determine the energy criteria that provide the necessary degree of jamming in the corresponding electronic systems. It is customary to refer to the energy criterion of jamming as the jamming coefficient K_j [9]. In view of the importance of this parameter K_j in the material that follows, we shall give its general and particular definitions.

In the general case, we understand the jamming coefficient of a electronic system to be the **minimum** required J/S ratio defined at the input to the receiver in the passband of its linear component, where the required degree is jamming is achieved. In the case of attacking an optimum device by means of masking jamming, this definition is formulated in the following way.

We understand the jamming coefficient of an optimum radar receiver using a masking jamming signal to be the **minimum** required J/S ratio defined at the input to the receiver in the passband of its linear component, for which the probability of correct detection for a given probability of false alarm does not exceed a given value.

The jamming coefficient in the latter case is unambiguously defined by the threshold value q_{thresh}. Accordingly, taking into consideration (3.65), we receive

$$K_j = \frac{2}{q^2_{\text{thresh}}} \tag{3.66}$$

The threshold value q_{thresh} is fully defined by (1.4) and (1.5) and does not depend on the form of the masking signal. At the same time, it is assumed that its parameters are fully known.

In those cases, where the number of pulses n being processed is small and the signal energy loss in the nonlinear components can be ignored, the ratio (3.62) can be transformed in the following way:

$$q = \sqrt{\frac{2n\Delta f_C \tau_C}{K}}$$

Accordingly, (3.66) converts to the form:

$$K_j = \frac{2}{q^2_{\text{thresh}}} n = K_{jo} n \tag{3.67}$$

where

$$K_{jo} = \frac{2}{q^2_{\text{thresh}}}$$

Practically speaking, in modern radar, the initial phase of the signal remains unknown. Besides, when reflected from a target of complex form, fluctuations occur in the envelope of the reflected signal. When defining energy criteria for parameters of masking jamming signals, these circumstances provide reason basically to give preference to equation (1.8), which presupposes a Rayleigh distribution of the envelope and a uniform phase distribution in the interval $(-\pi, \pi)$. Adhering to the earlier proposed transformation scheme for parameter q, we obtain

$$P_{\text{det}} = P_{\text{fa}}^{\frac{1}{1+\frac{\tau_s \Delta f_s}{\left(P_n / P_s\right)_{in}}}}$$

After converting to logarithms, we find

$$K = \left(\frac{P_n}{P_s}\right)_{in} = \tau_s \Delta f_s \frac{\lg P_{\text{det}}}{\lg P_{\text{fa}} - \lg P_{\text{det}}} \tag{3.68}$$

The ratio (3.68) is the starting point for defining energy criteria for assuring the required degree of jamming, using masking jamming, of various types of radar operating in the scan mode.

In noncoherent radar with a low duty cycle, the IF filter is matched to the spectrum of the single pulse (i.e., for a single pulse $\tau_s \Delta f_s = 1$). Let us designate by K_{jo} the radar jamming coefficient in the case where the decision is made by observing one pulse:

$$K_{jo} = \frac{\lg P_{\text{det}}}{\lg P_{\text{fa}} - \lg P_{\text{det}}} \tag{3.69}$$

If the decision is made according to the results of observing a train of n pulses, which is normally the case, then the corresponding K_n in the general case can not be obtained by multiplying K_{jo} by n. The determination of the threshold values K for large n must be made in noncoherent radar with consideration to losses in the useful signal in the nonlinear component [10]. This task is quite complex. In the first approximation for weakly fluctuating signals, the jamming coefficient can be determined using the formula:

$$K_j = K_{jo} \times \begin{cases} n, n \leqslant n_0 \\ n_0 + \sqrt{n - n_0} \end{cases} \tag{3.70}$$

where $n_0 = 25$ for weak fluctuating signals.

In radar with coherent processing (pulse-Doppler and quasi-CW radar), K_{jo} is determined by the duration of the train of coherent integrable pulses t_{qcr} matched with the passband of the linear receiver stages Δf_{qcr}:

$$\tau_s n_c = t_{\text{qcr}} \tag{3.71}$$

$$t_{\text{qcr}} \Delta f_{\text{qcr}} = 1 \tag{3.72}$$

$$\Delta f_{\text{qcr}} = \frac{1}{n_c \tau_s} \tag{3.73}$$

where n_c is the number of coherent integrable pulses in a train.

Just as in the preceding case, K_{jo} will be totally determined by expression (3.69), and the jamming coefficient is equal to

$$K_j = K_{jo} n_{cp} \tag{3.74}$$

Here, n_{cp} is the number of coherent integrable pulse trains:

$$n_{cp} = \frac{n}{n_c} \tag{3.75}$$

where n is the number of pulses in a train reflected from a target within the time of a single dwell; and

$$n_c = \frac{F_{cH}}{\Delta f_{\partial\Phi}} \tag{3.76}$$

where F_{cH} is the pulse repetition rate; and $\Delta f_{\partial\Phi}$ is the Doppler filter passband.

For broadband radar with a large base (product $\tau_s \Delta f_s$),

$$B = \tau_s \Delta f_s \tag{3.77}$$

$$K_{jo} = B \frac{\lg P_{\text{det}}}{\lg P_{\text{fa}} - \lg P_{\text{det}}} \tag{3.78}$$

In the cases analyzed previously, it was assumed that a masking jamming signal is white Gaussian nose, the entropy power of which is equal to its average power. True masking jamming signals, while possessing the same average power, have lower masking properties. To a certain extent, this deficiency is compensated for by an increase in the average power of the jamming signal, which increases the jamming coefficient. The quality of the masking jamming signal with respect to its energy indicator is estimated using the coefficient η_E, which we understand to be the ratio of the jamming coefficients $\overline{K}_{jo}$ and K_{jo}:

$$\eta_E = \frac{\overline{K}_{jo}}{K_{jo}} \tag{3.79}$$

where $\overline{K}_{jo}$ is the jamming coefficient using white Gaussian noise, K_{jo} is the jamming coefficient using the actual masking jamming signal, and $0 \leqslant \eta_E \leqslant 1$.

It is obvious that there exists a mutually unambiguous relationship between the information and energy indicators for masking jamming signal quality. The higher the coefficient η_H for a specific jamming signal, the higher its coefficient η_E will be, too.

The energy coefficient η_E permits the evaluation of the quality not only of jamming signals, but also of the corresponding masking jammers.

The energy quality indicator for deception jamming signals of radar working in the scan mode is, to a great extent, determined by the quality with which the jammer simulates the signal reflected from an airplane. On the average, the jamming coefficient in this case must be near to one. Accordingly, the level of the jamming signal must be selected with consideration to the change in the radar cross section (RCS) of the airplane (helicopter) depending on the relative bearing of the radar.

The required degree of radar jamming in the dynamics of EW is determined by the conditions of the operational and tactical environment and it's component part — the electronic environment, as well as the nature of the tasks being solved.

3.3.3 Energy Quality Indicators of Jamming Signals for Radar Operating in Automatic Target Tracking Mode

Electronic trackers, which provide the control system with information about the angular coordinates, range and velocity of the target in electronic systems for weapon (battle equipment) control are prime targets for jamming.

Just as before, the basic energy effectiveness indicator for jamming signals is the value of the J/S ratio, determined at the input to the receiver of the jamming target, at which the specified degree of jamming is achieved. The specified degree of jamming is defined, for specific conditions of the environment, as the deliberate change that is potentially achievable in the operating mode or output parameters of the jamming targets. The output parameters of trackers are their mathematical expectations and estimation variances. The energy indicator for masking jamming signals is defined, as a rule, by the degree of increase in the variance of tracker output estimates. Deception jamming signals basically lead to an increase in mathematical expectations for output estimate variances. As a result of their influence, both types of jamming signals can lead to tracking loss in trackers. The influence of jamming in specific conditions leads to a change in loop gain and, in particular, to a change in the loop gain of the of the angle measuring device for angular velocity in the line of sight K_{V}, which can potentially disrupt the stability of the homing loop of a missile at short ranges.

The qualitative definition of criteria for the energy indicator requires, as a rule, the specific analysis of the jamming target and the type of jamming signal, and the problem is not always solved analytically. In many cases, it is necessary to resort to simulation. All the same, certain general considerations as to the expected order of criteria can be made.

Trackers, providing a weapon control system with information about target range and velocity, can be disabled using jamming generated from a single point in space. For this purpose, for example, it is sufficient to implement a jamming signal that blocks out certain time or frequency intervals. In order to disable the angle measuring channel, it is necessary to implement angular noise blocking out a certain interval of angles. It is not possible to obtain such an implementation by generating jamming from a single point. In the simplest instance, for the generation of angle noise, it is necessary to have at least two jamming emissions from sources spaced in angle. The entropy power of the angle noise, generated from a single point, is equal to zero. The angle noise generated by two sources, modeled after a telegraph signal (Figure 2.5) with an average intensity of ν and an exponential distribution of switching times (3.34), is the most entropic among all distributions with a fixed average time $\bar{\tau} = \frac{1}{\nu}$. According to (3.35), the entropy of the exponential distribution, in the given case, is equal to

$$H' = \ln(e\bar{\tau}) \tag{3.80}$$

The criteria for angle noise spectral density are determined by two circumstances, specifically, by the required value of the output parameter estimate variance, for example, the variance of the rocket miss D_{bk} (1.131), and, in the given conditions, the value of the change $\Delta\Theta$ in the line of sight (Figure 3.2) implemented.

To determine energy criteria specifically, it is necessary to establish the dependency of both the angle noise spectral density and the angle $\Delta\Theta$ on the value of the J/S ratio. In its general form, the solution to this problem is quite complex and is optional for the determining of jamming criteria. Sufficiently close estimates can be obtained by limiting ourselves to the analysis of jamming effects on a discriminator with a linear characteristic. The discrimination characteristic can be considered linear on the condition that $\Delta\Theta_0 \leqslant 2\Theta_0$ (Figure 3.2).

As an example, let us consider an amplitude sensing angle tracker with sum-difference processing and astatism of the first order, which is influenced simultaneously by two noncoherent determinate signals S and J, spaced at an angle and with amplitudes U_s and U_j. The angular distance between the signal sources is equal to $\Delta\Theta_0$ (Figure 3.2). Furthermore, we shall suppose that the mode in the tracking direction finder has been established and that the voltage at the output to the phase detector can be assumed equal to zero. Accordingly, the equisignal direction of the antenna radiation pattern (ARP) is oriented towards a certain intermediate point between the sources

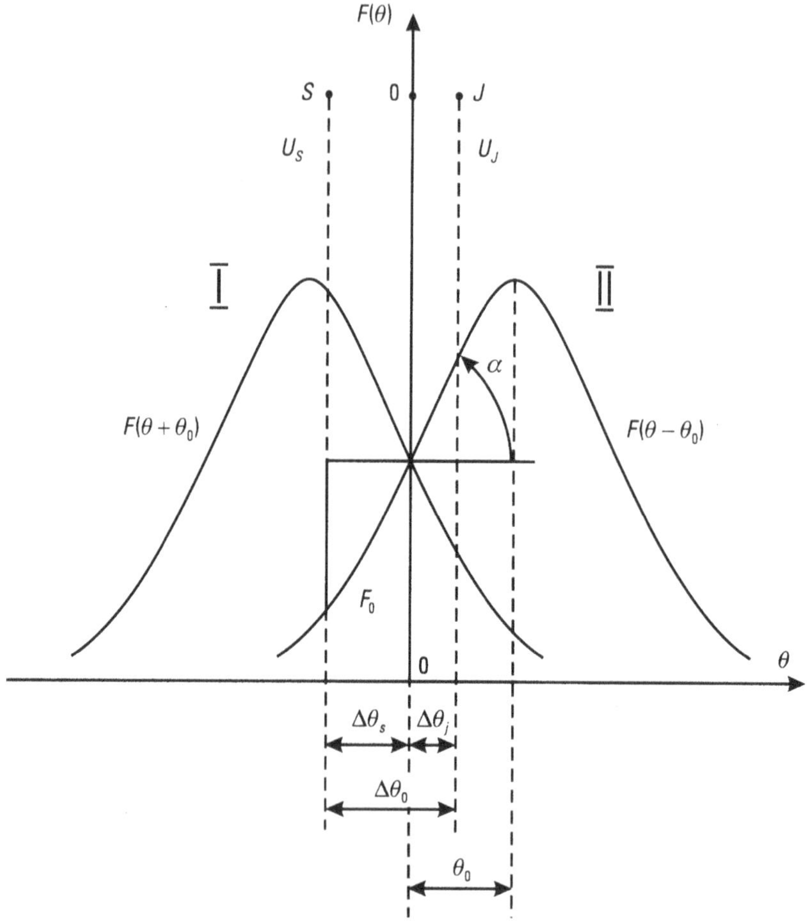

Figure 3.2 Value of the change $\Delta\Theta$ in the line of sight.

S and J. We set the origin of the coordinates at this 0 point. The main lobes of the antenna radiation pattern by field (I and II), relative to the adopted origin for the coordinates, are represented by the functions $F(\Theta + \Theta_0)$ and $F(\Theta - \Theta_0)$, where Θ_0 is the angle between the maximum ARP and the equisignal direction (squint angle). The areas $F(\Theta + \Theta_0)$ and $F(\Theta - \Theta_0)$ within the angle $\Delta\Theta_0$ of the ARP are assumed linear. The problem is to determine the angles $\Delta\Theta_s$ and $\Delta\Theta_j$ where the signal at the output to the phase detector of the direction finder difference channel is equal to zero. The voltage at the output to the phase detector of the difference channel is proportional to the difference in the voltages arriving from ARP II - $U_{\Sigma,II}$ and ARP I - $U_{\Sigma,I}$. In accordance with the designations given in Figure 3.2, we obtain

$$U_{\Sigma,II} - U_{\Sigma,I} = \left(U_j(F_0 + \Delta\Theta_j \mathrm{tg}\alpha) + U_s(F_0 - \Delta\Theta_s \mathrm{tg}\alpha)\right) - \left(U_j(F_0 - \Delta\Theta_j \mathrm{tg}\alpha) + U_s(F_0 + \Delta\Theta_s \mathrm{tg}\alpha)\right) = 0$$

After simple transformation, the following equilibrium condition is found for the tracking angle measuring device, which is simultaneously effected by both signals:

$$\Delta\Theta_s U_s = \Delta\Theta_j U_j \quad (3.81)$$

$$\Delta\Theta_s + \Delta\Theta_j = \Delta\Theta_0 \quad (3.82)$$

From (3.81) and (3.82), it follows that

$$\Delta\Theta_s = \Delta\Theta_0 \frac{b}{1+b} \quad (3.83)$$

where

$$b = \frac{U_j}{U_s} = \sqrt{\left(P_j/P_s\right)_{in}} = \sqrt{K} \quad (3.84)$$

The formula (3.83) defines the average displacement value of the dispersion center of the missile (projectiles) generated under the influence of a jamming signal with an amplitude of U_j in the case where the power ratio of the jamming and useful signals is equal to K. It also permits us to determine the dependency of the spectral density of the angle noise, generated by two sources and switched like a telegraph signal, on the J/S ratio.

Indeed, the formula (2.44), in the given case, is formulated in the following way:

$$S_{\Delta\Theta}(f) = \frac{8\Delta\Theta_C^2 \nu}{4\nu^2 + \omega^2} \quad (3.85)$$

The value of the spectral density $S_{\Delta\Theta}(\Delta f_{0.5})$ within the 3 dB bandwidth $(0, \Delta f_{0.5})$ is determined using (3.85):

$$S_{\Delta\Theta}(\Delta f_{0.5}) = {}^{\Delta\Theta^2}\!/_{\nu} \text{ rad}^2/\text{Hertz} \tag{3.86}$$

The angle $\Delta\Theta_s$ is determined from (3.83), where the parameter b is unambiguously related to the ratio mentioned (3.84). The frequency $\Delta f_{0.5}$ in the given case is equal to the equivalent of the angle noise spectrum width Δf_j:

$$\Delta f_j = \Delta f_{0.5} = {}^{\nu}\!/_{\pi} \text{ Hertz} \tag{3.87}$$

The maximum achievable value for the angle noise spectral density $S_{\Delta\Theta}(\Delta f_{0.5})$ is limited by the angular distance between the jamming sources $\Delta\Theta_0$ and the switching frequency ν. The latter determines the equivalent angle noise spectrum width Δf_j, which must not be less than the equivalent closed loop bandwidth of the tracking angle measuring device (1.94).

It is possible to assume that the threshold values of K, determining its minimum required value, occur when

$$\Delta\Theta_s = (0.7 \ldots 0.8)\Delta\Theta_0 \tag{3.88}$$

This corresponds to the jamming coefficient:

$$K_j = 10 \ldots 12 \, \text{dB} \tag{3.89}$$

It can also be less if a smaller value of the ratio $\frac{\Theta_s}{\Theta_0}$ is permissible in the given situation.

It was noted earlier that the value of the entropy power of the angle noise generated from a single point is equal to zero. Let us determine the entropy power for angle noise generated from two points on the model of a telegraph signal.

According to (3.39) and (3.80), the entry power sought $\overline{P}_j'$ is defined by the formula:

$$\overline{P}_j' = \frac{e}{2\pi}\overline{\tau}^2 \tag{3.90}$$

Let us compare it with the entropy power of a sample function of Gaussian noise with a constant spectral density and the same number of sampling values as the original angle noise.

The variance of the initial random value, distributed according to the exponential law, is determined by the equality:

$$D_{\overline{\tau}} = \overline{\tau}^2 \tag{3.91}$$

and accordingly, the entropy power $\overline{P}_j$ of the Gaussian noise in the given case is equal to $\overline{\tau}^2$:

$$\overline{P}_j = \tau^{-2} \tag{3.92}$$

The ratio of the entropy powers being compared is equal to

$$\frac{\overline{P}'_j}{\overline{P}_j} = \frac{e}{2\pi} = 0.43 \tag{3.93}$$

The entropy power in the given variant of two-point interference is less than half the Gaussian angle noise realized in the conditions analyzed, which indicates a potential possibility for its reduction using appropriate algorithms and circuit solutions.

The entropic nature of angle noise can be increased by raising the angle density of jammers. Tentatively, in the linear section of the location finder characteristic, there must be of the order of 5–6 jammers within this sector. Based on considerations stated earlier, the jamming coefficient using angle noise can be assumed equal 10–12 dB. It can be less if, in view of the situation, it is permissible to reduce the value of the ratio Θ_s/Θ_0.

The analysis of expressions (3.81) and (3.84) indicate a possibility of disrupting direction finder tracking by generating distraction jamming. A variant of a scheme where distraction jamming is implemented is given in Figure 3.3. At the moment in time t_0 the useful signal, reflected from a target with amplitude U_s, is placed in the center 0 of the aperture of the direction finder characteristic $\Delta\Theta_{adfc}$.

At the moment of time t_1 and under the simultaneous influence of useful and jamming signals with amplitudes U_s and $U_j = 3U_s$ ($b = 3$) in the mode established, the center 0 of the direction finder characteristic, according to formula (3.83), will shift from the useful signal in the direction of the jamming signal by $\Delta\Theta_s = \frac{3}{4}\Delta\Theta_0$.

At moment of time t_2, after additional shifting of the jamming source by angle $\Delta\Theta'_0$, the useful signal falls out of the aperture of the location finder characteristic $\Delta\Theta_{adfc}$. If, at this moment of time, the jamming influence ceases, then a disruption of tracking in the tracker will occur.

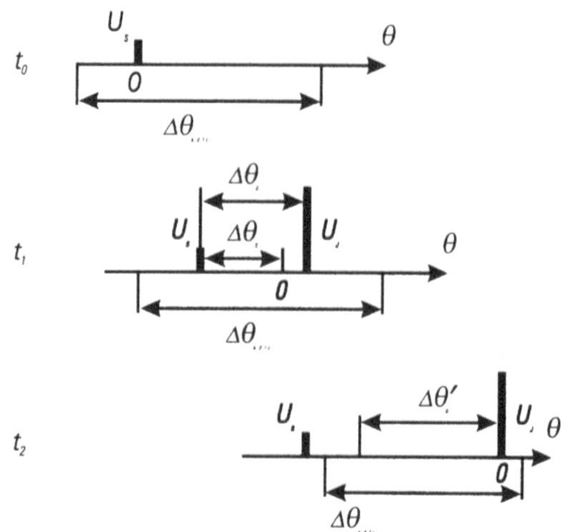

Figure 3.3 Variant of a scheme where distraction jamming is implemented.

The analyzed variant of tracking loss is achieved using a determinate jamming signal in a determined sequence. The absence of ambiguity makes it vulnerable to comparatively simple ECCM.

More stable with regard to ECCM is the angle noise considered previously, generated on the model of a telegraph signal. Implementation of this noise covers a range of angles, corresponding to the linear portion of the location finder, which is of the order of the 3-dB beamwidth ($\Theta_{0.5}$).

In order to ensure tracking loss, it is necessary to occupy the range of angles that correspond to the width of the direction finder characteristic $\Delta\Theta_{adfc}$ (Figure 3.3). This corresponds to an interval of angles on the order of three or four $\Theta_{0.5}$. The practical implementation of such a variant of angle noise requires three or four pairs of flickering sources with an angular distance between each pair of sources $\Delta\Theta_0$ that is approximately equal to $\Theta_{0.5}$. This is in view of the requirement in this case of using the whole angular interval between a pair of emission sources ($\Delta\Theta_0$) to the maximum extent. The jamming coefficient must be no less than 10–12 dB.

The determining of energy criteria for other sample function variants of angle noise where the angular density of jamming sources is larger, requires deeper research.

Some of the results received for an angle tracker, in principle, can be applied to range and velocity trackers. The values of corresponding jamming coefficients are obtained as a result of an indepth analysis of specific types of jamming and schematic solution variants. For tentative estimates of information damage caused to the control system when jamming tracking channels for range and velocity a jamming coefficient of 5–6 dB can be used.

Basically, the proposals stated above for criteria of jamming signals for electronic devices in weapons control systems assume the placement of jamming sources within the linear portion of the discriminative (direction finder) characteristic.

The jamming coefficient K_j, determined by the degree of jamming achieved when doing so, is somewhat too high. Due to the influence of additional jamming interference in nonlinear stages on the dynamics of EW, the degree of jamming increases, which permits, in principle, an insignificant decrease in the value of K_j. This circumstance gives cause to assume that criteria, determined in such a way, can be assumed to be robust (i.e., stable against change in environment conditions), although they are somewhat too high.

In conclusion, we will note once again that energy criteria of information damage are determined by the conditions of the operational, tactical and electronic environment. As concerns specific devices and techniques for generating jamming, the magnitude of information loss they evoke can be shown, provided space and time conditions are set strictly enough and specific electrical characteristics are present in targets and jammers.

3.4 Operational and Tactical Indicators of EW Effectiveness

3.4.1 Fundamental Concepts

The operational and tactical indicators of EW devices and techniques, as applied in the defense of an airforce, are determined in the main by the form and content of military actions and the character of the problems to be solved. Of the comparatively large number of variants of conducting military operations involving the air force, the conducting of military actions by airforce divisions (subdivisions) to penetrate AAD is taken as the basis for defining effectiveness indicators, above all, for jamming. Specifically, the content and particulars of jamming are reflected most fully when solving problems associated with the overcoming (penetrating) of AAD. Besides this, it should be borne in mind that the appearance of jamming as an element of armed warfare was a result of the development of means for attack and defense. It was first used massively by the English and American airforces in 1943 to overcome Germany's AAD.

Besides the variant of the military operations mentioned, we consider in addition the specifics of defining effectiveness indicators for the conducting of EW in the dynamics of combat involving homogeneous groupings of armed forces, as described by differential equations for the dynamics of averages. In particular, on this basis it is possible to define the required norms for EW forces and devices when conducting military operations using fighter aircraft.

3.4.2 Effectiveness Indicators of Jamming when Conducting Military Actions Using Airborne Forces (Subdivisions) to Penetrate AAD Systems

The jamming targets in the given case are electronic systems for control of AAD forces and weapons. As effectiveness indicators, we use the value of average losses in airforce troops (subdivisions) and the averaged probability of penetrating AAD, as related to a single aircraft in a battle formation and occurring as a result of using the jamming devices and the techniques considered [9]. In the first approximation, quantitative estimates for the indicators mentioned can be obtained with the help of an analytical model of AAD. When considering this problem, it is convenient to imagine a single AAD system in the form of a set of sufficiently independent subsystems with respect to operations, which we will subsequently term AAD areas. Examples are AAD areas of military operations, communications areas etc. There may be several such areas. In each area, there are various types of firepower against airborne targets, for example short- and medium-range antiaircraft missile installations, mobile antiaircraft installations and fighter aircraft.

The control of each type of AAD firepower device is performed by two types of automated control systems — for forces and for weapons.

The stream of airborne targets entering into the AAD system is assumed to be Poissonian fragmentally stationary. Its density λ is considered to be constant while flying through the kill area of the given AAD firepower system. It changes (decreases) in fixed amounts when the battle formation enters an adjacent area.

An operations model for each separate group of AAD fire channels (weapons system), together with its corresponding ACS, can be obtained, in the first approximation, using the methods of the mass service theory and the theory of radio control (sections (1.3) and (1.4)). In particular, the operations model for groups of n fire channels of antiaircraft missile installations with a medium or long range of operation and intercept fighters can be represented as an n-channel mass service system with rejects. By imposing appropriate limitations on a group, it is possible to represent the operations of other firepower channel types in the AAD area in the same manner.

Let us represent the unified AAD system in the form of a set of l areas. The arbitrary ith area in the AAD has M_i groups of various types of firepower systems. The group comprising each of j firepower devices ($j = \overline{1, M_i}$) can be represented as an n_{ij}-channel mass service system with rejects and with an exponential service time distribution in each channel. According to (1.43), the average stream density of serviced requests (in the given case, targets shot at) μ_{ij} and the average service time $\bar{t}_{ij}$ are linked by the equality:

$$\mu_{ij} = \frac{1}{\bar{t}_{ij}} \tag{3.94}$$

The average stream density of targets successfully shot at μ^*_{ij} and the averaged probability of hitting the target with one generalized shot from the jth device in the ith area P_{ij}, according to (1.66), are linked by the dependency:

$$\mu^*_{ij} = \mu_{ij} P_{ij} \tag{3.95}$$

Depending on the type of weapon, a generalized shot can be understood to be [11] a volley or several volleys from an antiaircraft artillery installation, a series of antiaircraft guided missiles, one attack by an intercept fighter and other firing variants. In the material that follows, the term "generalized successful shot" will also be used. Instead of the average stream density of targets successfully shot at, we will speak of the average stream density of generalized successful shots from the average intensity of generalized successful shots $\bar{\mu}^*_{ij}$ of one fire channel of the jth device in the ith AAD area. The average intensity of generalized successful shots of a group of j devices in this area $\bar{\mu}^*_{ij}$ is determined by the average intensity of generalized successful shots of one channel $\bar{\mu}^*_{ij}$ and the average number of channels $\bar{K}_{ij}$ in the group n_{ij} firing on the group of airborne targets in the given environment:

$$\bar{\mu}_{ij} = \bar{K}_{ij}\, \mu_{ij} \tag{3.96}$$

The value $\bar{K}_{ij}$ is determined using (1.59), which in the given case is formulated in the following way:

$$\overline{K}_{ij} = \alpha_{ij} P_{\text{serv}}(i, j) \tag{3.97}$$

Here,

$$\alpha_{ij} = \frac{\lambda_{ij}}{\mu^*_{ij}} \tag{3.98}$$

λ_{ij} is the stream density of airborne targets entering the ith AAD area,

$$\lambda_{ij} = \frac{m_{ij}}{T_{ij}} \qquad T_{ij} = \frac{L_{bf}(i, j)}{V_{bf}(i, j)} \tag{3.99}$$

m_{ij} is the average number of airplanes in a battle formation entering the kill area of the jth device; T_{ij} is the time for the battle formation of airplanes to pass outside the far bound of the kill area of the jth device; $L_{bf}(i, j)$ is the spread of the battle formation of m_{ij} airplanes; and $P_{\text{serv}}(i, j)$ is the service probability for (shooting at) the airborne target by an n_{ij}-channel mass service system with rejects.

In a system with centralized channel control, according to (1.56),

$$P_{\text{serv}}(i, j) = \frac{\sum_{k=0}^{n_{ij}-1} \frac{\alpha_{ij}^k}{k!}}{\sum_{k=0}^{n_{ij}} \frac{\alpha_{ij}^k}{k!}} \tag{3.100}$$

If centralized channel control is absent and the airborne targets are uniformly distributed among fire channels, then taking into consideration (1.65),

$$P'_{\text{serv}}(i, j) = \frac{1}{1 + \frac{\alpha_{ij}}{n_{ij}}} \tag{3.101}$$

The average losses in a battle formation of a group of m airplanes $M_j[m]$, after overcoming all l AAD areas, by definition is equal to [12, 13]

$$M_j[m] = \sum_{i=1}^{l} \sum_{j=1}^{M_i} m_{ij} \overline{P}_{ij} \tag{3.102}$$

Unlike P_{ij} in (3.95), which is the averaged probability that a target will be hit by a single generalized shot of the jth device in the ith area, $\overline{P}_{ij}$ is the averaged probability of firepower hitting a group of j devices in the ith area, as related to one airplane in the battle formation.

The probability of overcoming the kill area of the jth device in the ith area by one of the airplanes of the battle formation according to (1.38) is determined by the expression:

$$P_{aad}(i, j) = \exp\left(-\overline{\mu}_{ij}^{*'} \overline{t}_{ij}\right) \tag{3.103}$$

where $\overline{\mu}_{ij}^{*'} = \frac{\overline{\mu}_{ij}^{*}}{m_{ij}}$ is the average density of a Poissonian stream of successful shots by the group of j devices, as related to one airplane in the battle formation; and $\overline{t}_{ij}$ is the average time the battle formation spends in the kill area of the group of j devices:

$$\overline{t}_{ij} = \frac{\Delta D_{ij}}{V_{bf}(i, j)} \tag{3.104}$$

ΔD_{ij} is the route length of the battle formation in the kill area of a group of j devices; and $V_{bf}(i, j)$ is the averaged flight velocity of the battle formation in the kill area.

The sum of the averaged probability $\overline{P}_{ij}$ of hitting one airplane, randomly selected in a battle formation of m_{ij} airplanes, and the probability $P_{aad}(i, j)$ are equal to one; therefore,

$$\overline{P}_{ij} = 1 - P_{aad}(i, j) = 1 - \exp\left(-\overline{\mu}_{ij}^{*'} \overline{t}_{ij}\right) \tag{3.105}$$

Let us determine the average losses $M_j[m]$ in the case where $\overline{\mu}_{ij}^{*'} \overline{t}_{ij} << 1$. From (3.105), we obtain

$$\overline{P}_{ij} = \overline{\mu}_{ij}^{*'} \overline{t}_{ij} = \frac{\overline{\mu}^{*}}{m_{ij}} \overline{t}_{ij} \tag{3.106}$$

After substituting (3.106) into (3.102), we find the average losses we are looking for. For small $(\overline{\mu}_{ij}^{*'} \overline{t}_{ij})$, they are equal to the sum of the generalized successful shots in all areas from all the firepower devices present in them:

$$M_j[m] = \sum_{i=1}^{I} \sum_{j=1}^{M_i} K_{ij} \mu_{ij} P_{ij} t_{ij} \tag{3.107}$$

When performing practical calculations, it is always necessary to bear in mind the natural limitation $M_j[m] < m$.

Based on what was said previously, it is possible to sequentially determine the probabilities that an airplane from a battle formation will penetrate both an individual area, as well as the entire AAD system as a whole.

The probability of overcoming the ith AAD area $P_{aad}(i)$, as related to a single airplane in the battle formation, assuming the statistical independence of generalized shots and the absence of accumulated damage from previous firings at the airplane, can be determined as the product $P_{aad}(i, j)$

$$P_{aad}(i) = \prod_{j=1}^{M_i} P_{aad}(i, j) \tag{3.108}$$

Replacing $P_{aad}(i, j)$ by its value from (3.103), we obtain

$$P_{aad}(i) = \exp\left(-\sum_{j=1}^{M_i} \overline{\mu}_{ij}^{*'} \overline{t}_{ij} \right) \tag{3.109}$$

The probability of overcoming all l AAD areas (P_{aad}) is equal to

$$P_{aad} = \exp\left(-\sum_{i=1}^{l_i} \sum_{j=1}^{M_i} \overline{\mu}_{ij}^{*'} \overline{t}_{ij} \right) \tag{3.110}$$

The expressions (3.102), (3.103), (3.107), (3.109) and (3.110) permit us to obtain an estimate of the effectiveness of systems and techniques for jamming used in the interests of overcoming AAD using operational and tactical criteria at various levels of an individual subsystem for the control of military equipment (weapons) up to and including a system for the control of forces (troops) of the AAD as a whole. The jamming of electronic systems comprising components of the control system leads to increases in $P_{aad}(i)$ and P_{aad} in particular and, as a result of disruption in the system for centralized troop control, to the reduction of the time that a battle formation of airplanes spends in AAD kill areas $(\overline{t}_{ij})$ and the reduction of the number of firepower devices firing on airborne targets (reduction of M_i). The jamming of electronic systems for weapons control leads to a reduction in actively working fire channels, to a reduction of the probability an airplane will be hit by one generalized shot (P_{ij}), and to tracking loss in the automatic weapons control system. Below, we consider certain techniques for estimating the probability P_i when jamming influences trackers in electronic systems for weapons control.

The possibility of using analytical methods when evaluating the effectiveness of systems and techniques of jamming of electronic systems for weapons control is a result of the special character of the problem. For the jammer, for example, there are no limits to a miss by a antiaircraft missile from above or the probability of a hit by an antiaircraft missile (projectile) from below. It is only necessary that the miss by missile (projectile) h be greater that its kill radius and that the kill probability (P_{ij}) not exceed a certain value.

Let us assume that, as a result of deliberate actions at time t_j in the missile trajectory segment with length D_j, a change in the angular velocity in the line of sight $\dot{\varepsilon}_j$ occurs. Let us determine the missile miss h_j that has formed during the time segment indicated. From (1.74), it follows that a change in miss $h(t)$ over the small time interval dt, presuming $V_{rel} = const$, is equal to

$$\mathrm{d}h(t) = \frac{1}{V_{rel}}(D^2\ddot{\varepsilon} + 2D\dot{D}\dot{\varepsilon})\mathrm{d}t$$

or

$$\mathrm{d}h(t) = \frac{D}{V_{rel}}(D\ddot{\varepsilon} + 2\dot{D}\dot{\varepsilon})\mathrm{d}t \tag{3.111}$$

The sum in parentheses defines the acceleration component j_n perpendicular to the line of sight (normal acceleration) and it includes in itself the tangential and rotational (Coriolis) accelerations. Accordingly, assuming $V_{rel} = V_{appr}$ and $j_n = const$, we obtain

$$\mathrm{d}h(t) = \frac{D}{V_{rel}} j_n \mathrm{d}t$$

$$D = D_j - V_{rel}t \qquad \frac{D}{V_{rel}} = t_j \tag{3.112}$$

$$h_j = \int_0^{t_j} \frac{D_j - V_{rel}t}{V_{rel}} j_n \mathrm{d}t$$

Finally we have

$$h_j = \frac{1}{2} j_n t_j^2 \qquad (3.113)$$

If $h_j \geqslant R_j$, where R_j is the missile kill radius, then jamming may be considered to have been effective.

The formula (3.113) can be also used for the determination of the miss h_0 selected by an antiaircraft missile during homing t_{hom} with a constant normal overload j_{hom}:

$$h_0 = \frac{1}{2} j_{\text{hom}} t_{\text{hom}}^2 \qquad (3.114)$$

If jamming is not active at all phases of missile guidance, and in the last phase its homing occurs with a normal overload j_{hom}, then the effectiveness of the jamming influence should be evaluated using the value of the resulting miss a, equal to the difference of h_j and h_0:

$$a = h_j - h_0 \qquad a \geqslant 0 \qquad (3.115)$$

Jamming is effective if

$$a \geqslant R_j \qquad (3.116)$$

3.4.3 Examples of Determining Misses

Let us assume that the radar homing seeker of missile M (Figure 3.4) has been shut off as a result of the jamming action for the time t_j and the missile is being controlled by the autonomous inertial guidance system providing for its final approach to point 0. In order to exclude the possibility of a hit, the airplane being covered by jamming A performs a directional maneuver with overload j_a. Normal overload will be acting on the missile $j_n = j_a \cos q$ for relative movement during the time t_j. According to (3.113), the miss we are looking for h_j is equal to

$$h_j = \frac{1}{2} j_a (\cos q) t_j^2 \qquad (3.117)$$

In case it is possible subsequently to perform homing during the time

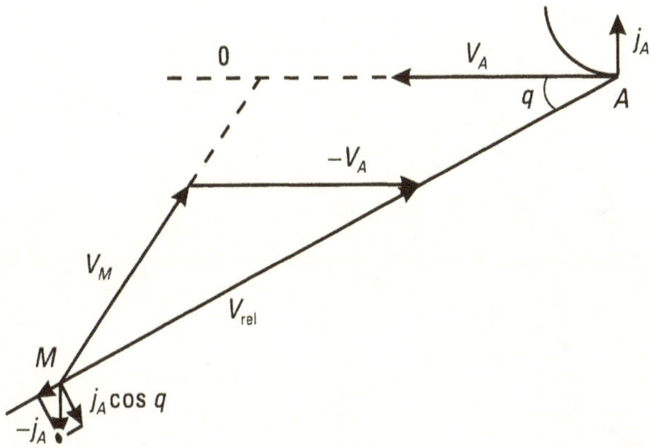

Figure 3.4 Miss when the radar homing seeker of missile M has been shut off due to jamming.

t_{hom}, the resulting miss a is determined by the formula (3.115). Let us note that a negative value a indicates the possibility of eliminating the miss h_j in a time less than t_{hom}.

In the second example we determine the miss of an artillery shell AS from an antiaircraft installation, which occurs as the result of an error in measuring the range to the target (Figure 3.5). The airplane A will be hit by shell AS, provided the following equality is satisfied [14]:

$$t_{as} = t_a \tag{3.118}$$

where t_{as} is the flight time of the shell to the predetermined intercept point 0:

$$t_{as} = \frac{D_{as}}{V_{as}}$$

t_a is the flight time of the airplane from the given point A, fixed at the moment of the shot, to that same predetermined intercept point 0:

$$t_a = \frac{D_a}{V_a}$$

The intercept angle ψ, corresponding to condition (3.118), is determined

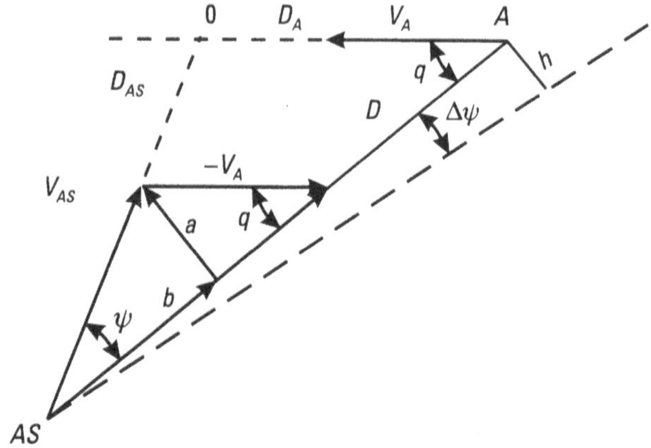

Figure 3.5 Miss of an artillery shell *AS* from an antiaircraft installation due to an error in measuring the range.

from the velocity triangle, obtained as a result of decomposing the average velocity vector of the shell into radial (b) and transversal (a) components (Figure 3.5):

$$\sin\psi = \frac{a}{\overline{V}_{as}} = \frac{\omega D}{\overline{V}_{FC}} \tag{3.119}$$

where ω is the averaged angle velocity in the line of sight AS, S.

Let us assume that the range to the target is measured with an error $\pm\Delta D$. Then an error appears in the determination of the intercept angle $\pm\Delta\psi$. The expression (3.119) is formulated in the following way:

$$\sin(\psi \pm \Delta\psi) = \frac{\omega(D \pm \Delta D)}{\overline{V}_{as}} \tag{3.120}$$

Assuming $\Delta\psi$ to be sufficiently small, we obtain

$$\sin\psi\cos\Delta\psi \pm \cos\psi\Delta\psi = \frac{\omega D}{\overline{V}_{as}} \mp \frac{\omega\Delta D}{V_{as}}$$

From here it follows that

$$\Delta\psi\cos\psi = \frac{\omega}{V_{as}}\Delta D$$

From the velocity triangle (Figure 3.5), we receive

$$a = V_a \sin q = \omega D$$

$$\omega = \frac{V_a \sin q}{D}$$

$$\Delta\psi = \frac{V_a \sin q}{V_{as} D \cos\psi} \Delta D \qquad (3.121)$$

For small $\Delta\psi$, it is possible to assume

$$h = D\Delta\psi \qquad (3.122)$$

According to the law of sines,

$$\frac{\sin\psi}{\sin q} = \frac{V_a}{V_{as}} = \xi$$

$$\cos\psi = \sqrt{1 - \xi^2 \sin^2 q}$$

The miss h being sought, resulting from the error in measuring the range ΔD, is determined using the equality:

$$h = \frac{\xi \sin q}{\sqrt{1 - \xi^2 \sin^2 q}} \Delta D \qquad (3.123)$$

In the general case, it is also necessary to determine the component of the miss resulting from the ballistic specifics of the shell trajectory in the vertical plane. It is possible that this will not occur if $h >> R_j$, which gives cause to assume a probability of hitting the airplane practically equal to zero.

In many cases jamming actions generate random changes in the miss. With an accuracy sufficient for practical calculations, it is possible to assume the probability densities for misses to be normal. In Figure 3.6, we show an instance where the center of dispersion of missile 0_{cd} in the picture plane is shifted relative to the center of gravity of the airplane 0_t along the axes 0_tX and 0_tY by values a and b. The probability densities $p(h_x)$ and $p(h_y)$ are assumed to be independent and Gaussian with mathematical expectations of a and b and rms deviations σ_x and σ_y. Bearing in mind the approximate

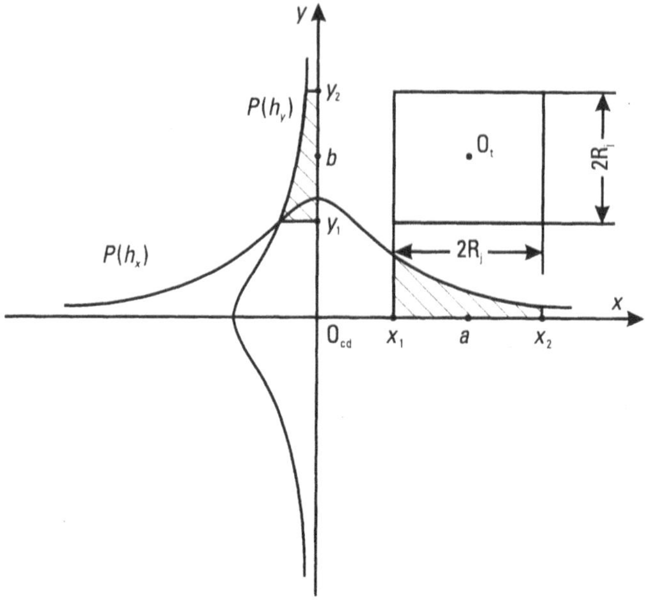

Figure 3.6 Missile 0_{cd} with center of dispersion in the picture plane shifted relative to the center of gravity.

nature of the calculations, it would seem possible to approximate the conditional hit $P(h_x, h_y)$ in (1.132) using a Rayleigh characteristic of the following form:

$$P(h_x, h_y) = P(h_x)P(h_y) \tag{3.124}$$

where

$$P(h_x) = \begin{cases} 1, & a - R_j \leqslant h_x \leqslant a + R_j, \\ 0, & h_x < a - R_j; \quad h_x > a + R_j \end{cases}$$

$$P(h_y) = \begin{cases} 1, & b - R_j \leqslant h_y \leqslant b + R_j, \\ 0, & h_y < b - R_j; \quad h_y > b + R_j \end{cases}$$

Here, R_j is the averaged kill radius. This permits, instead of the hit probability, one to determine the probability of hitting a square whose side length is 2 R_j. Then the expression (1.132), defining the probability of hitting an airplane (helicopter) with one missile, can be formulated in the following way:

$$P_{\text{hit}}(1) = \int_{a-R_j}^{a+R_j} p(h_x)\mathrm{d}h_x \int_{b-R_j}^{b+R_j} p(h_y)\mathrm{d}h_y \tag{3.125}$$

where

$$\int_{a-R_j}^{a+R_j} p(h_x)\mathrm{d}h_x = \frac{1}{\sqrt{2\pi}\sigma_x}\int_{-\infty}^{a+R_j} e^{-\frac{h_x^2}{2\sigma_x^2}}\mathrm{d}h_x - \frac{1}{\sqrt{2\pi}\sigma_x}\int_{-\infty}^{a-R_j} e^{-\frac{h_x^2}{2\sigma_x^2}}\mathrm{d}h_x$$

or

$$\int_{a-R_j}^{a+R_j} p(h_x)\mathrm{d}h_x = \Phi_0\left(\frac{a+R_j}{\sigma_x}\right) - \Phi_0\left(\frac{a-R_j}{\sigma_x}\right) \tag{3.126}$$

Accordingly,

$$P_{\text{hit}}(1) = \left(\Phi_0\left(\frac{a+R_j}{\sigma_x}\right) - \Phi_0\left(\frac{a-R_j}{\sigma_x}\right)\right) \left(\Phi_0\left(\frac{b+R_j}{\sigma_y}\right) - \Phi_0\left(\frac{b-R_j}{\sigma_y}\right)\right) \tag{3.127}$$

where

$$\Phi_0(x) = \frac{1}{\sqrt{2\pi}}\int_0^x e^{-\frac{t^2}{2}}\mathrm{d}t \tag{3.128}$$

The expression (3.127) permits us, given a fixed variance, to determine the values of a and b for the shift in the center of dispersion in the missiles (projectiles) to assure the required reduction in the probability of hitting the aircraft with one missile (one volley from an antiaircraft artillery system). Once again, attention should be drawn to the necessity of fixing the dispersion variance. If, when shifting the center, an increase in the dispersion variance occurs, then the hit probability $P_{\text{hit}}(1)$ in individual cases can also increase. Changes in the hit probability, generated solely by an increase of variances σ_x^2 and σ_y^2, are determined using (3.127), and a and

b in it are assumed to be equal to zero:

$$P_{\text{hit}}(1) = 4\Phi_0\left(\frac{R_j}{\sigma_x}\right)\Phi_0\left(\frac{R_j}{\sigma_y}\right) \tag{3.129}$$

If the dispersion law in the picture plane is symmetrical $(\sigma_x = \sigma_y = \sigma)$, then

$$P_{\text{hit}}(1) = 4\Phi_0^2\left(\frac{R_j}{\sigma}\right) \tag{3.130}$$

Earlier, for a coordinate law for hits that is symmetrical with respect to x and y and defined by a Gaussian curve (1.133), we obtained a relatively simple formula (1.135) defining the hit probability using a single missile, which, in the new notation $(D_x = \sigma^2, R_{\ni\Phi} = R_j)$ can be formulated in the following manner:

$$P_{\text{hit}}(1) = \frac{1}{1 + \dfrac{\sigma^2}{R_j^2}} \tag{3.131}$$

A comparison of the results of probability calculations for $P_{\text{hit}}(1)$ using formulas (3.129) and (3.130) shows that they are quite close and, moreover, a greater value for $P_{\text{hit}}(1)$ is obtained when calculating using formula (3.131). This permits us to use it when making calculations, bearing in mind that the results obtained limit the probability of $P_{\text{hit}}(1)$ from above. If the dispersion is asymmetric, $(\sigma_x \neq \sigma_y)$, then $P_{\text{hit}}(1)$ is determined using formula (3.129).

3.4.4 Effectiveness Indicators for Conducting EW in the Dynamics of a Battle between Homogeneous Groups of Armed Forces

In the model variants of military operations considered before, the operational and tactical effectiveness indicators for EW were determined, basically, without consideration to the dynamics of active countermeasures. This makes it difficult to consider in full volume all elements of EW, both offensive (jamming) and defensive (ECCM) in nature. As an example permitting us to consider countermeasures, below we analyze two model variants of the dynamics of military operations by two conflicting sides with homogeneous kill systems (firepower systems). Each of the sides

A and B is armed with weapons of the same type. Each of the sides can have its own firepower systems, different from analogous systems of the enemy. The firepower productivity for firepower systems is characterized by the Poissonian stream density for successful shots:

$$\Lambda = \lambda P$$

where λ is the stream density of generalized shots (the number of shots per unit of time); P is the probability of hitting the target with one generalized shot.

Accordingly, for each of the sides we determine Λ_A and Λ_B:

$$\Lambda_A = \lambda_A P_A \qquad \Lambda_B = \lambda_B P_B \tag{3.132}$$

In the model variants considered for the dynamics of military operations it is assumed that the following conditions are fulfilled [11].

The conflicting sides have a finite number of firepower systems (channels) N_A and N_B. Each firepower channel, if not hit, can fire using a Poissonian stream of successful shots at any firepower system of the enemy.

The average flight time for a missile (projectile) is much less than the average duration of a battle.

Total military might is determined by the average number of fire channels not hit, and not their random number at a given moment in time.

In the first of the models considered, it is assumed that the sides are waging a highly organized battle. Each shot is aimed. With one shot it is possible to hit only one target and, after it has been hit, firing is, practically speaking, instantly shifted to a target that has not been hit.

In the second model, the battle is not organized in a proper fashion. The Poissonian stream of successful shots for each of the groups is distributed uniformly among all firepower systems of the enemy. Information is not received about the hitting of a target. Both hit and unhit targets may be fired at. One successful shot may hit only one target.

EW measures performed by each of the sides can lead to a reduction in both the average density of the Poissonian stream of generalized shots (λ_A, λ_B) and the probability of hitting a target with one shot (P_A, P_B). Accordingly, the stream densities of successful shots Λ_A and Λ_B prove to be quite small. For this reason, the average time necessary to hit a single target proves to be much greater than the flight time of a missile (projectile) to a target. Everything mentioned permits us to state that the basic assumption that the determining influence on total military might for each of the groups is the average number of firepower systems that have not been hit when waging EW, is borne out even in the case where they are relatively few in number.

Let us determine the change over time in the average values of numbers (strengths) for the groups m_A and m_B in the case of a highly organized battle (Model I). At the moment in time $t = 0$, let the strengths of the groups comprise

$$m_A(0) = N_A \qquad m_B(0) = N_B \tag{3.133}$$

At the end of time t, the average strengths will comprise $m_A(t)$ and $m_B(t)$. Let us determine the change (reduction) $m_A(t)$ over the time Δt that has occurred as a result of firing by side B. According to the assumptions made,

$$\Delta m_A = -\Lambda_B m_B(t) \Delta t \tag{3.134}$$

After dividing both parts of equality (3.134) by Δt and limit conversion at $\Delta t \to 0$, we obtain the first differential equation for the combat dynamics:

$$\frac{\mathrm{d}m_A}{\mathrm{d}t} = -\Lambda_B m_B \tag{3.135}$$

Reasoning in an analogous manner, we obtain a second differential equation for combat dynamics:

$$\frac{\mathrm{d}m_B}{\mathrm{d}t} = -\Lambda_A m_A \tag{3.136}$$

Differential equations (3.135) and (3.136) are solved with regard to initial conditions (3.133) for m_A and m_B, as well as the initial conditions for their derivatives:

$$\left.\frac{\mathrm{d}m_A}{\mathrm{d}t}\right|_{t=0} -\Lambda_B N_B \qquad \left.\frac{\mathrm{d}m_B}{\mathrm{d}t}\right|_{t=0} -\Lambda_A N_A \tag{3.137}$$

In order to obtain an analytical solution to the equations for combat dynamics, let us differentiate (3.135) by time. Then, with consideration to (3.136), we obtain

$$\frac{d^2 m_A}{dt^2} = -\Lambda_A \Lambda_B m_A \tag{3.138}$$

If Λ_A and Λ_B do not depend on time, then the solution of linear differential equation (3.138) can be formulated in the following way [15]:

$$m_A = C_1 e^{s_1 t} + C_2 e^{s_2 t}$$

Here, S_1 and S_2 are the roots of the characteristic equation. In the given case $S_{1,2} = \pm\sqrt{\Lambda_A \Lambda_B}$. The arbitrary constants C_1 and C_2 are found from the initial conditions (3.133) and (3.137):

$$C_1 = \frac{1}{2}\left(N_A - N_B\sqrt{\frac{\Lambda_B}{\Lambda_A}}\right)$$

$$C_2 = \frac{1}{2}\left(N_A - N_B\sqrt{\frac{\Lambda_B}{\Lambda_A}}\right)$$

Accordingly,

$$m_A = \frac{1}{2}\left(N_A - N_B\sqrt{\frac{\Lambda_B}{\Lambda_A}}\right)e^{\tilde{t}} + \frac{1}{2}\left(N_A + N_B\sqrt{\frac{\Lambda_B}{\Lambda_A}}\right)e^{-\tilde{t}}$$

where

$$\tilde{t} = \sqrt{\Lambda_A \Lambda_B t} \tag{3.139}$$

Reasoning in an analogous manner, we obtain the second level solution:

$$m_B = \frac{1}{2}\left(N_B - N_A\sqrt{\frac{\Lambda_A}{\Lambda_B}}\right)e^{\tilde{t}} + \frac{1}{2}\left(N_B + N_A\sqrt{\frac{\Lambda_A}{\Lambda_B}}\right)e^{-\tilde{t}}$$

Using the well-known definitions of hyperbolic functions ch$t$ and sh$t$:

$$\text{ch}\tilde{t} = \frac{e^{\tilde{t}} + e^{-\tilde{t}}}{2} \qquad \text{sh}\tilde{t} = \frac{e^{\tilde{t}} - e^{-\tilde{t}}}{2}$$

we formulate the final solution to (3.135) and (3.136):

$$\frac{m_A}{N_A} = \text{ch}\tilde{t} - \frac{N_B}{N_A}\sqrt{\frac{\Lambda_B}{\Lambda_A}}\,\text{sh}\tilde{t} \tag{3.140}$$

$$\frac{m_B}{N_B} = \text{ch}\tilde{t} - \frac{N_A}{N_B}\sqrt{\frac{\Lambda_A}{\Lambda_B}}\,\text{sh}\tilde{t} \tag{3.141}$$

The solution received permits us to obtain the change over time of the average strength of firepower systems of the conflicting groups depending on the ratio of their combat capabilities. The latter, as is known [16], are determined by the quantitative and qualitative indicators of each of the sides' firepower systems. In the case under consideration, the quantitative indicators are the strengths of groups' firepower systems (N_A and N_B). The qualitative indicators are their firepower productivities Λ_A and Λ_B, determined in their turn by the quantitative and qualitative characteristics of the EW systems and techniques. It is convenient to perform a quantitative evaluation of the relationship of combat capabilities with the help of the corresponding coefficient K_{cc}, determined using (3.141):

$$K_{cc} = \frac{N_A}{N_B}\sqrt{\frac{\Lambda_A}{\Lambda_B}} \tag{3.142}$$

It would seem to be convenient to call K_{cc} the coefficient of initial combat capabilities of the conflicting sides A and B. If $K_{cc} > 1$, then it is possible to speak of the initial defense superiority of side A over side B. In the opposite case ($K_{cc} < 1$), the initial superiority is on side B.

In the model without transfer of firepower (Model II), the change in the average strength of group A is determined not only by the average intensity of successful shots on the part of group B, but also by the probability that, due to the fact that firepower is not transfered, the shot considered is fired namely at a target that has not yet been hit. This probability, given the assumptions made, is equal to the ratio $\frac{m_A}{N_A}$, where m_A is the average number of targets not hit at the moment of time t. Then,

$$\Delta m_A = -m_B \Lambda_B \frac{m_A}{N_A} \Delta t \tag{3.143}$$

After limit conversion at $\Delta t \rightarrow 0$, we obtain the first differential equation for Model II:

$$\frac{\mathrm{d}m_A}{\mathrm{d}t} = -\Lambda_B m_B \frac{m_A}{N_A} \tag{3.144}$$

In an analogous manner we obtain the second differential equation:

$$\frac{\mathrm{d}m_B}{\mathrm{d}t} = -\Lambda_A m_A \frac{m_B}{N_B} \tag{3.145}$$

The solution of (3.144) and (3.145) for initial conditions at $t = 0$: $m_A = N_A$, $m_B = N_B$, permits us to obtain the following expressions for the relative numbers of the groups A and B, occurring at the moment of time t [11]:

$$\frac{m_A}{N_A} = \frac{u_2 - u_1}{u_2 \mathrm{e}^{(u_2 - u_1)t} - u_1} \tag{3.146}$$

$$\frac{m_B}{N_B} = \frac{u_1 - u_2}{u_1 \mathrm{e}^{(u_1 - u_2)t} - u_2} \tag{3.147}$$

where

$$u_1 = \frac{\Lambda_A N_A}{N_B} \qquad u_2 = \frac{\Lambda_B N_B}{N_A} \tag{3.148}$$

The coefficient of the initial combat capabilities of sides K_{cc} in the given case, as before, can be determined in the following manner:

$$K_{cc} = \sqrt{\frac{u_1}{u_2}} = \frac{N_A}{N_B}\sqrt{\frac{\Lambda_A}{\Lambda_B}} \tag{3.149}$$

Before we start comparing Models I and II, let us consider one of the variants for combat dynamics that has a direct relationship with EW effectiveness indicators. We are speaking of the instance where one of the sides has an opportunity to deliver and delivers a preemptive strike against enemy targets. Let us assume that the strike is delivered by side A. This means that, for a certain time Δt_A, only side A fires at targets of side B with a density $\Lambda_A N_A$ of successful shots per unit of time, leading to a decrease in the average number of group B by a value of $\Lambda_A N_A \Delta t_A$ before the moment two-sided battle begins.

In the case of Model I, the coefficient of initial combat capabilities of the sides, defined by formula (3.142) for the moment of time Δt_A, is formulated in the following way:

$$K_{cc}(\Delta t_A) = \frac{N_A}{N_B^*}\sqrt{\frac{\Lambda_A}{\Lambda_B}} \tag{3.150}$$

where

$$N_B^* = N_B - N_A \Lambda_A \Delta t_A \tag{3.151}$$

The average relative number of group B at the moment of time t according to (3.141) is defined by the following expression:

$$\frac{m_B}{N_B^*} = \text{ch}\tilde{\tau} - K_{cc}(\Delta t_A)\text{sh}\tilde{\tau} \tag{3.152}$$

where

$$\tilde{\tau} = \tilde{t} - \Delta\tilde{t}_A \tag{3.153}$$

Accordingly,

$$\frac{m_A}{N_A} = \text{ch}\tilde{\tau} - \text{sh}\tilde{\tau}/K_{cc}(\Delta t_A) \tag{3.154}$$

If Model II occurs (combat without transfer of firepower), then during the time $0 < t \leqslant \Delta t_A$, in expressions (3.146) and (3.147), $u_2 = 0$. Therefore,

$$m_A(\Delta t_A) = N_A$$

$$N_B^* = m_B(\Delta t_A) = N_B \exp\left(-\Lambda_A \frac{N_A}{N_B} \Delta t_A\right) \tag{3.155}$$

Accordingly, the average relative numbers of groups A and B at the moment of time $t > \Delta t_A$ are determined using the following expressions:

$$\frac{m_A}{N_A} = \frac{u_2(\Delta t_A) - u_1(\Delta t_A)}{u_2(\Delta t_A)e^{(u_2(\Delta t_A) - u_1(\Delta t_A))(t - \Delta t_A)} - u_1(\Delta t_A)} \tag{3.156}$$

$$\frac{m_B}{N_B^*} = \frac{u_1(\Delta t_A) - u_2(\Delta t_A)}{u_1(\Delta t_A)e^{(u_1(\Delta t_A)-u_2(\Delta t_A))(t-\Delta t_A)} - u_2(\Delta t_A)} \tag{3.157}$$

The coefficient of initial combat capabilities $K_{cc}(\Delta t_A)$, according to (3.149) and with consideration to (3.155), is defined by the expression:

$$K_{cc}(\Delta t_A) = \sqrt{\frac{u_1(\Delta t_A)}{u_2(\Delta t_A)}} = \frac{N_A}{\sqrt{N_B m_B(\Delta t_A)}}\sqrt{\frac{\Lambda_A}{\Lambda_B}} \tag{3.158}$$

Let us explain the results received using a specific training example in the dynamics of long-range missile combat between two groups of fighters A and B of equal strength ($N_A = N_B = 30$) with onboard radar of the same type (Figure 3.7). On board each fighter, active jammers are installed destined for the attack of radar homing seekers of air-to-air missiles. Masking jamming for the onboard radar of the fighters is not generated. The fighters of side A have higher EW effectiveness indicators than the fighters of side B. The RCS of fighter A is two times less than the RCS of fighter B. The RHH of fighter A has a higher level of protection from deception jamming than the RHH of fighter B. The deception jamming

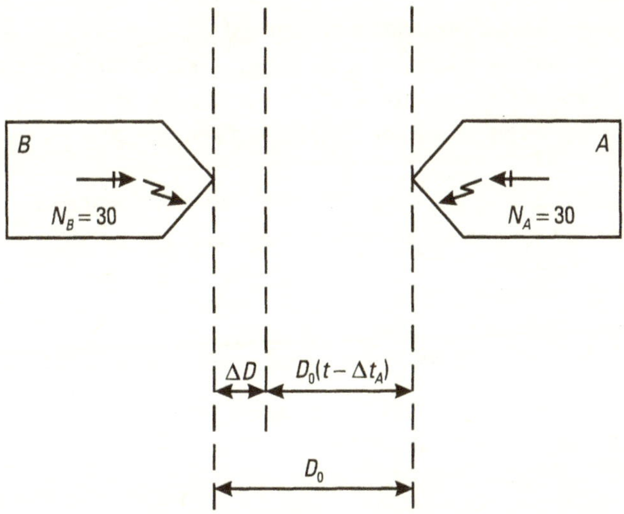

Figure 3.7 Long-range missile combat between two groups of fighters *A* and *B* of equal strength with onboard radar of the same type.

signals of the jammer in fighter A surmount the protection circuits of the seeker of fighter B.

Quantitatively, this is expressed in that the probability of fighter B being hit by one generalized shot (by one missile) from fighter A is $P_{\text{hit}}(B) = P(B/A) = 0.2$, whereas the probability of fighter A being hit by a single missile from fighter B is $P_{\text{hit}}(A) = P(B/A) = 0.1$. If we assume that the average stream densities for missiles launched by sides A and B are identical and launches are made every 40 sec (i.e., $\lambda_A = \lambda_B = \frac{1}{40}\, 1/s$), then the average stream densities of successful shots (launches) for each side is, accordingly,

$$\Lambda_A = \lambda_A P\left({}^{B}\!/_{A}\right) \qquad \Lambda_B = \lambda_B P\left({}^{B}\!/_{A}\right)$$

$$\Lambda_A = \lambda \frac{0.2}{40} 1/s \qquad \Lambda_B = \lambda \frac{0.1}{40} 1/s$$

As a result of the greater radar visibility, the fighters of side B are detected at a range that exceeds the detection range of their onboard radar by 1.2 times. This permits side A to make preemptive air-to-air missile launches from a range of $D_0 = 120$ km, while side B can make aimed counter launches only from a range of $D(t - \Delta t_A) = 100$ km (Figure 3.7).

It is assumed that the missile battle ends when the range between the groups reaches 20 km. If the flight velocity of fighters A and B is 250 m/sec (the closing velocity is 500 m/sec), the length of a preemptive strike would be $\Delta t_A = 40$ sec. The length of a two-sided combat would be equal to 160 sec. The overall length of the battle between the two groups would be 200 sec.

The coefficient of combat capabilities of the sides and the strength of the groups when implementing Model I according to (3.150) and (3.154) is equal to

$$K_{\text{cc}}(\Delta t_A) = \frac{5}{4}\sqrt{2} = 1.77 \qquad N_B^* = 24 \qquad \tilde{\tau} = 0.566s$$

$$\frac{m_B}{N_B^*} = 0.11 \qquad m_B = 2.64 \qquad \frac{m_A}{N_A} = 0.83 \qquad m_A = 24.9$$

In the case Model II is implemented, according to (3.156) and (3.157), we obtain

$$\frac{m_A}{N_A} = 0.797 \qquad m_A = 23.9 \qquad \frac{m_B}{N_B^*} = 0.49 \qquad m_B = 12.1$$

The training example given shows that the coefficient of initial combat capabilities K_{cc} enables us to quantitatively evaluate the effectiveness of various EW measures. In the example analyzed, it was shown that, if we reduce by a factor of two the RCS of an airplane and the probability of hitting it with a single shot during air combat in comparison with an enemy airplane ($K_{cc} = 1.41$), side A can, when waging a highly organized battle (Model I), practically speaking, totally crush the enemy in the long-range missile combat stage ($m_B = 2.64$), at the same time suffering relatively small average losses ($m_A = 24.9$). When waging a less organized battle (Model II), the effectiveness of military actions for side A is noticeably less $m_B = 12.1$, $m_A = 23.9$.

Thus, in the example analyzed, side A by means of increasing the quality of control (implementing Model I instead of Model II) increases the effectiveness of his firepower by more than four times. It is worthwhile noting the heightened sensitivity of the coefficient of initial combat capabilities to comparatively small differences in the quality of EW measures both as a part of jamming and ECCM. For example, if in the previous example, when all other conditions are equal, the small quantitative difference in the effectiveness of EW measures consists merely in the magnitude of the probability of a hit with one shot ($P(B/A) = 0.2$, $P(A/B) = 0.16$, $K_{cc} = 1.12$), then, when waging a highly organized battle, the average strength of the groups after 200 sec would be: $m_B = 8.4$, $m_A = 15.3$. The battle ends practically speaking with a two-fold superiority of side A.

Equations for the dynamics of averages can be used in a quite large number of variants of conducting military operations. Model I was first researched in the works of Staff-Captain Osipov of the Russian Army (1915) and the English scientist Lanchester (1916).

Model II was proposed and researched by E. S. Venttsel'. She also established the link between the models analyzed and Markov processes [11, 17].

It is worth bearing in mind that the reliability of the method considered can prove to be insufficient in the final stage of a battle when the average strengths of the groups are small and the process of their changing becomes continuous. In this case, it is convenient to supplement analytical investigations with simulation modeling of combat dynamics. As was already noted earlier, under the influence of jamming Λ_A and Λ_B are small and the results of estimates are acceptable for small values of m_A and m_B.

3.5 Military and Economic Indicators of EW Effectiveness

The criteria considered earlier do not allow us in full measure to perform a comprehensive evaluation of the effectiveness of EW systems and methods. In addition, it is necessary to know the required costs for the production of equipment and its operation, as well as during combat use. In particular, for EW systems, the issue of ensuring their stability with respect enemy ECCM is fundamental. Indicators of such stability can be the average lifetime of a system or technique in standard conditions of the environment. Effectiveness must be ensured for the duration of their lifetime, which increases the cost of the system (technique).

An important place when analyzing the economic side of EW is the evaluation of prevented damage occurring in the dynamics of EW.

In principle, there exists an optimal relationship between effectiveness and cost. As a rule, the more effective a system, the higher its cost. Cheap EW systems are more frequently than not either ineffective or have a short lifetime.

The existence of an optimum relationship between the effectiveness and the cost is illustrated by Figure 3.8. Here N is the quantity of homogeneous elements assuring the EW effectiveness (for example, this may be the number of microcircuits), and e is the effectiveness expressed in appropriate units (for example, the averaged probability of penetrating AAD, the average value of

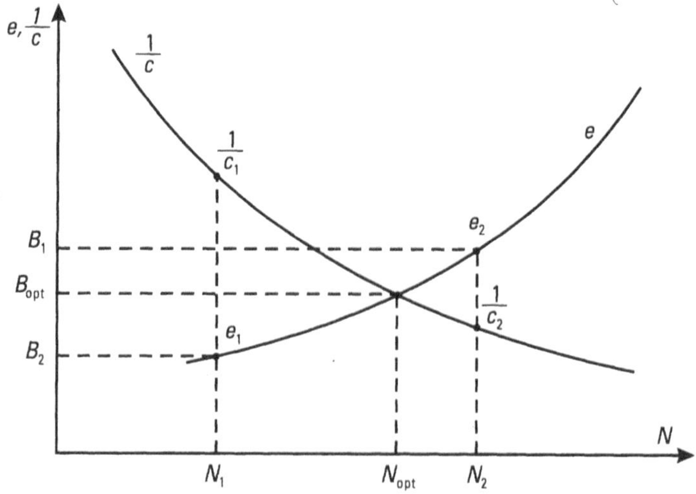

Figure 3.8 Optimum relationship between effectiveness and cost.

prevented damage). c is the cost of the corresponding quantity of selected elements. It is apparent that the product $e\frac{1}{c}$, as a function of N, has an extremum. The corresponding variant of the EW system or complex V_{opt} and the value N_{opt} are optimum. In practice, the optimum value of N is not always selected. Depending on the specific task, the conditions of the environment, and bearing in mind available experience and real possibilities, one or another variant (B_1 or B_2) may be selected, together with a number of elements N_1 or N_2, with the effectivenesses e_1, e_2 and costs (c_1 and c_2) corresponding to them. In the general case, military and economic estimates require the solution of quite complex problems, which are of independent interest. Practically speaking, these estimates are performed at the stage of equipment design. Military and economic analysis is also required when determining the strength of forces and means required to implement the specific technique of waging EW.

The military and economic indicator is, in many cases, the fundamental one for the decision maker.

References

[1] Shannon, C. E., *Raboty po teorii informatsii i kibernetike* (*Works on the Theory of Information and Cybernetics*). Moscow: IIL, 1963, pp. 243–488.

[2] Khinchin, A. Ya., *Ponyatie e'ntropii v teorii informatsii i kibernetike. Uspekhi matematicheskikh nauk* (*The Concept of Entropy in the Theory of Information and Cybernetics. Successes in Mathematical Sciences*), Vol. V.VIII, 1953, Issue 3(55).

[3] Kolmogorov, A. N., *Tri podkhoda k opredeleniyu ponyatiya "Kolichestvo informatsii"/Problemy peredachi informatsii* (*Three Approaches to Defining the Concept of "Information Quantity"/Problems of Data Transmission*). Vol. V.I, 1965, Issue. I, pp. 25–38.

[4] Brillyue'n, L., *Nauka i teoriya informatsii* (*Science and the Theory of Information*). Moscow: *Fizmatgiz* (*State Publishing House for Science and Mathematics*), 1960, pp. 19–260.

[5] Goldman, S., *Teoriya informatsii* (*The Theory of Information*). Moscow: IIL, 1957, pp. 149–210.

[6] Wilks, S., *Matematicheskaya statistika* (*Mathematical Statistics*). Moscow: *Nauka* (*Science*), 1967, pp. 353–390, 402–426.

[7] Kul'bak, S., *Teoriya informatsii i statistika* (*The Theory of Information and Statistics*). Moscow: *Nauka* (*Science*), 1967, pp. 12–46.

[8] Cramer, H. P., *Matematicheskaya statistika* (*Mathematical Methods of Statistics*). Moscow: IIL, 1948, pp. 277–287, 513–540.

[9] Vakin, S. A., and L. N. Shustov, *Osnovy radioprotivodeystviya i radiotekhnicheskoy razvedki* (*The Basics of Radio Countermeasures and Radio Technical Intelligence*). Moscow: *Sov. radio* (*Soviet Radio*), 1968, pp. 9–122, 259–342.

[10] Helstrom, C. W., *Statisticheskaya teoriya obnaruzheniya signalov* (*Statistical Theory of Signal Detection*). Moscow: NIL, 1963, pp. 106–158.

[11] Venttsel', E. S., *Issledovanie operatsiy* (*Operations Research*). Moscow: *Sov. radio* (*Soviet Radio*), 1972, pp. 238–281, 291–357, 446–509.

[12] Tikhonov, V. I., and Yu. N. Bakaev, *Statisticheskaya teoriya radiotekhnicheskikh ustroystv (The Statistical Theory of Radio Technical Devices)*. Moscow: N.E. Zhukovsky Airforce Academy, 1976, pp. 5–119, 180–220, 263–416.

[13] Venttsel', E. S., *Teoriya veroyatnostey (The Theory of Probability)*. Moscow: *Fiz.-mat.literat. (Physics and Mathematics Literature)*, 1962, pp. 504–547.

[14] *Aviatsionnoe vooruzhenie (Aircraft Arms)*. Under the Editorship of D. I. Gladkov. Moscow: *Voenizdat (Military Publishing House)*, 1987, pp. 170–198.

[15] Bronshteyn, I., and K. Semendyaev, *Spravochnik po matematike (Mathematics Handbook)*. Moscow: *Nauka (Science)*, 1986, pp. 72–73.

[16] *Voennyy e'ntsiklopedicheskiy slovar' (Military Encyclopedic Dictionary)*. Moscow: *Voenizdat (Military Publishing House)*, 1983, p. 89.

[17] Venttsel', E. S., *Vvedenie v issledovanie operatsiy (Introduction to Operations Research)*. Moscow: *Sov. radio (Soviet Radio)*, 1964, pp. 132–272.

4

Active Jamming of Radar – The Jamming Equation

4.1 Fundamental Concepts

The generation of jamming causes information damage to radar if the ratio of interference power in the radar receiver to the power of the useful (reflected) signal, determined in the passband of its linear part, is equal to or exceeds the jamming coefficient. The latter is determined by the type and parameters of the jamming and useful signals, as well as by the radar circuit implemented to process them. In screening jamming, we introduce an energy quality coefficient η_E given by (3.79). The magnitude of the information damage inflicted depends on the energy indicators of both the target and the jamming system. Quantitatively, the information damage inflicted on a radar is determined by the size of the screened area, which in many cases is a certain segment within the bounds of the detection zone of the victim radar. Along its borders, the ratio of the power of the jamming to that of the useful signal is equal to the jamming coefficient. Within the area, the ratio exceeds it. The jamming equation for radar permits us to analytically determine the form and dimensions of the radar area jammed.

We understand the jamming equation to be the dependence of the ratio K of the power of the jamming to that of the useful signal within the passband of the linear part of the receiver, on a set of factors determined by the electric, frequency and space/time characteristics of the radiation generated by the targets and jamming systems; the propagation conditions; and the radar detectability of the target being covered. Below, we derive and analyze the equation for active jamming of one monostatic radar. We consider the specifics of active jamming of various levels on radar for various purposes. We determine parameter optimization conditions in screening jamming. We do not separately consider deception jamming of the radar. We merely draw attention to the specifics of the jamming equation encountered in this case ($\Delta f_j = \Delta f_s$, $K_j = 1$, etc.).

4.2 The Jamming Equation for Monostatic Radar Using Active Jamming (the General Case)

We assume that jamming is generated by jammer JS (a point source) to screen an airplane battle formation BF (Figure 4.1). We must find the ratio K of the power of the jamming $(P_j)_{in}$ to that of the useful signal $(P_s)_{in}$ at the input to the receiver; that is,

$$K = \left(\frac{P_j}{P_s}\right)_{in} \tag{4.1}$$

The jamming conditions determining the screened area and its borders can be written down in the following way:

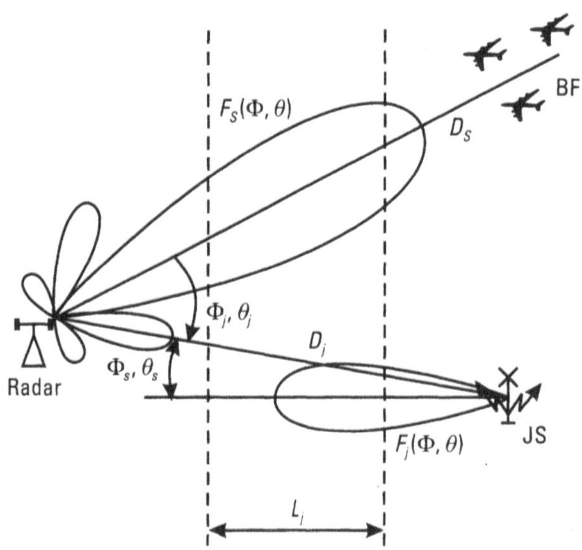

Figure 4.1 Point source jammer JS screening an airplane battle formation BF.

$$K \geqslant K_j \tag{4.2}$$

where K_j is the jamming coefficient. In order to write the jamming equation, we must determine $(P_j)_{in}$ and $(P_s)_{in}$ for the corresponding target and jammer parameters.

Let us introduce symbols to designate the parameters that characterize the jammer and the victim radar (Figure 4.1).

Jammer parameters: P_j is the jamming power at the input to the antenna (average or peak, depending on the type of jamming); G_j is the maximum gain of the jammer antenna; Δf_j is the effective spectrum width of the jamming; γ_j is a coefficient accounting for the polarization difference in the antennas of the jammer and the radar being jammed; σ_{BF} is the radar cross section of the battle formation (BF) screened by the jamming; $F_j(\Phi, \Theta)$ is the normalized pattern of the jammer antenna; D_j, Φ_j, and Θ_j are the polar coordinates of the jammer. The origin of the coordinates is at the radar site. The angles Φ_j and Θ_j are measured in the corresponding planes relative to the beam axis of the victim radar; D_s, Φ_s, and Θ_s are the polar coordinates of the radar. The origin of the coordinates is at point JS; D_s is the distance to the battle formation being screened.

Parameters of the victim radar: P_s is the power of the victim radar (pulse or average, depending on the type of radar), considering transmission line loss; G_s is the maximum gain of the radar antenna; $F_s(\Phi, \Theta)$ is the normalized antenna pattern of the victim radar; Δf_s is the equivalent noise bandwidth of the receiver; A_s is the effective receiving antenna area for the radar:

$$A_s = \frac{G_s \lambda^2}{4\pi}$$

Let us determine the power $(P_j)_{in}$ of the jamming at the input to the receiver within its passband, using the well-known formula for an ideal radio transmission as a guide, according to which the flux density p of the power emitted per unit solid angle (the modulus of the Umov–Poynting vector) is

$$p = \frac{P_\Sigma}{4\pi D^2} = \frac{1}{2} EH \tag{4.3}$$

where P_Σ is the power of the isotropic emitter; E and H are the amplitudes of the electric and magnetic fields; and $\frac{E}{H} = Z_0 = 120\pi$ ohm is the impedance in free space.

In the general case the flux density for the jamming power at the radar antenna is found using the formula:

$$p_j = \frac{P_j G_j F_j^2(\Phi_s, \Theta_s)}{4\pi D_j^2} \Gamma^2_{\text{JS,radar}} 10^{-0,1\alpha L_j} \quad (4.4)$$

where α is the attenuation coefficient (dB/km) over the attenuating portion of the path of length L_j (km); $\Gamma_{\text{JS,radar}}$ is the propagation factor that takes into consideration the influence of the earth (sea) beneath the path between the jammer and the radar [1, 2].

The power at the input to the radar receiver is

$$(P_j)_{in} = p_j A_s F_s^2(\Phi_j \Theta_j)\gamma_j + P_n \quad (4.5)$$

Here, P_n is the power of the intrinsic receiver noise in the passband Δf_{rec}:

$$P_n = kTN_n \Delta f_{\text{rec}} \quad (4.6)$$

where $k = 1.38 \cdot 10^{-23}$ Joules/Kelvin is Boltzmann's constant; T is the temperature in Kelvin; and N_n is the noise factor of the receiver.

Normally, the jamming power significantly exceeds the power of the intrinsic noise. This permits us to ignore the second term in (4.5) and assume

$$(P_j)_{in} = p_j A_s F_s^2(\Phi_j \Theta_j)\gamma_j \quad (4.7)$$

Only a part of the jamming power enters the receiver, as determined by the ratio of the spectrum width of the jamming to the passband of the receiver. Assuming a rectangular approximation for the spectrum of the masking jamming and the passband of the receiver of the victim radar, the jamming power at the input to the receiver is given by

$$(P_j)_{in} = p_j A_s F_s^2(\Phi_j \Theta_j)\gamma_j \frac{\Delta f_{\text{rec}}}{\Delta f_j}$$

Finally, using (4.4), we obtain

$$(P_j)_{in} = \frac{P_j G_j}{4\pi D_j^2} A_s F_s^2(\Phi_j \Theta_j) F_j^2(\Phi_s, \Theta_s)\gamma_j \frac{\Delta f_{\text{rec}}}{\Delta f_j} \Gamma^2_{\text{JS,radar}} 10^{-0,1\alpha L_j} \quad (4.8)$$

It is assumed that $\Delta f_j \geqslant \Delta f_{rec}$. In the case of a deception jamming, $\Delta f_j = \Delta f_{rec}$.

In an analogous fashion, for the power of a signal at the input to the receiver of the victim radar, we can write

$$(P_s)_{in} = \frac{P_s G_s}{4\pi D_s^2} \frac{\sigma_{BF}}{4\pi D_s^2} A_s \Gamma^4_{\text{radar,BF}} 10^{-0{,}2\alpha L_j} \tag{4.9}$$

Here, $\Gamma^4_{\text{radar,BF}}$ is the propagation factor, taking into consideration the influence of the earth (underlying terrain) in the area between the radar and the battle formation, and L_j is the length of the attenuating portion of the path from radar to target.

Substituting (4.8) and (4.9) into (4.1), we obtain the jamming formula for active radar jamming we are looking for

$$K = \left(\frac{P_j}{P_s}\right)_{in} = \frac{P_j G_j}{P_s G_s} \frac{\Delta f_{rec}}{\Delta f_j} \gamma_j F_s^2(\Phi_j \Theta_j) F_s^2(\Phi_j, \Theta_j) \frac{4\pi}{\sigma_{BF}} \frac{D_s^4}{D_\Pi^2} \frac{\Gamma^2_{\text{JS,radar}}}{\Gamma^4_{\text{radar,BF}}} 10^{0{,}1\alpha L_j} \tag{4.10}$$

In Figure 4.2, we show the dependency of the coefficient K on D_s for various fixed values of D_j, according to (4.10). As we approach the radar, the

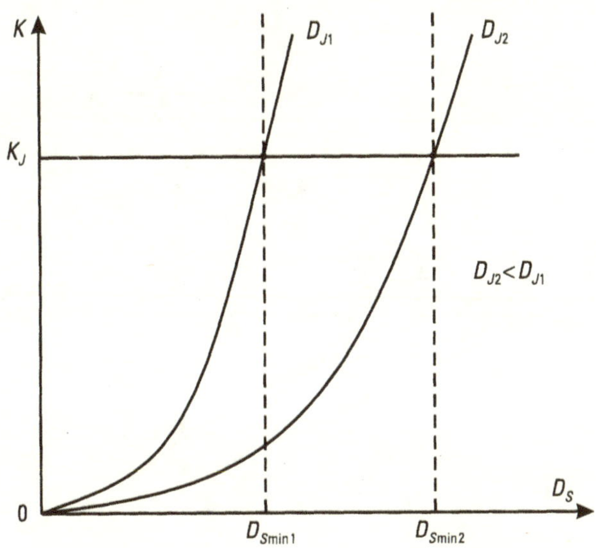

Figure 4.2 Dependency of the coefficient K on D_s for various fixed values of D_j.

value of the ratio K of the jamming power to the signal power at the input to the victim receiver decreases. When K reaches a value less than K_j, the required degree of jamming is not achieved. The value of D_s, corresponding to the equation $K = K_j$, determines the border of the jamming area. In Figure 4.2 there are two such borders — $D_{s\min 1}$ and $D_{s\min 2}$. The coefficient K_j (defined below) is called the jamming coefficient.

In Figure 4.3, we show the jamming area for various bearings q of the targets being covered, relative to the victim radar and for fixed coordinates of the jammer. When the jammer, the battle formation being covered, and the victim radar are all located at the bearing $(q = 0)$, $F_s^2(0, 0) = 1$ (0 dB), the coefficient K is at its maximum value and the corresponding detection range is minimum. It is customary to call it the minimum jamming range $D_{s\min}$. For bearings q_1 and q_2, $F_s^2(\Phi_j, \Theta_j) < 1$ and, therefore, the jamming effect decreases and the target detection range increases. The bearing q_1 corresponds to detection range D_{s1}, which occurs, for example, when jamming is on the level of $F_s^2(\Phi_j, \Theta_j)$ in the order of -10 dB. The bearing q_2 corresponds to the range D_{s2}, which occurs at value $F_s^2(\Phi_j, \Theta_j)$ approximately equal to -20 dB. The borders of the jamming area are shown symbolically on Figure 4.3 using striped lines.

It follows that the width of the jamming area $\Delta\Phi_j$ is determined by the jamming range provided when attacking the side lobes of the radar antenna. If, for example, the jamming area along the main lobe $(\Phi_j = \Phi_{0.5})$ corresponds to a range of $D_{s\min} = 10$ km, then the jamming area obtained

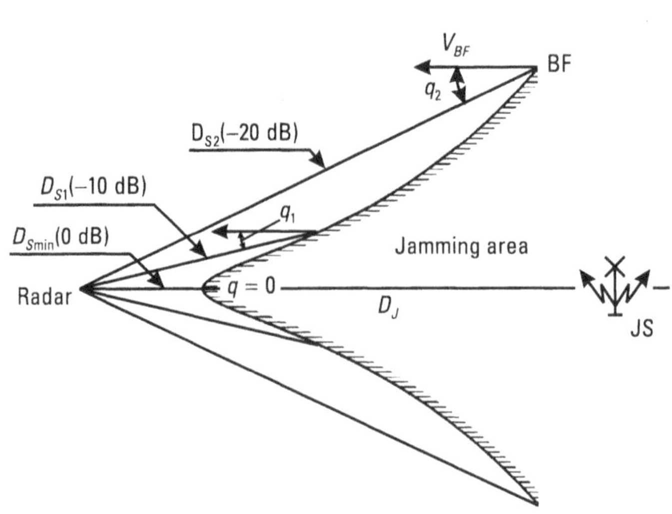

Figure 4.3 Generalized jamming area.

when attacking with a jamming level of $F^2(\Phi_j, \Theta_j) = -20$ dB represents a range of $D_s(-20 \text{ dB}) = 32$ km. This issue is stated more clearly in the material that follows.

In the variants of the jamming equation analyzed so far, the jammer was placed outside the airplane battle formation. If the jamming is generated from within the airplane battle formation and the jamming systems are installed directly onboard the aircraft being covered (self-screening), then the ranges D_s and D_j coincide ($D_s = D_j$) and (4.10) looks as follows:

$$K = \frac{P_j G_j}{P_s G_s} \frac{\Delta f_{\text{rec}}}{\Delta f_j} \gamma_j F_s^2(\Phi_j \Theta_j) F_j^2(\Phi_s, \Theta_s) \frac{4\pi}{\sigma_{BF}} D_s^2 \frac{1}{\Gamma^2_{radarBF}} 10^{0.1\alpha L_j} \quad (4.11)$$

Qualitatively, the dependency of K on D_s, as determined by (4.11), is shown in Figure 4.4. As the aircraft with the active jammer on board approaches the victim radar, the ratio of jamming to signal falls and, as a result, jamming effectiveness is reduced. This is because the increase in the power of the signal on approach is inversely proportional to the 4th power of the distance, whereas the increase in the power of the jamming is inversely proportional only to the square of the distance.

The minimum jamming range is equal to $D_{s\min}$ and, in the case of self-screening, it is found at $K = K_j$. Screening may also occur due to the limited dynamic range of the receiver. Starting at the range D_{j2}, the absolute power

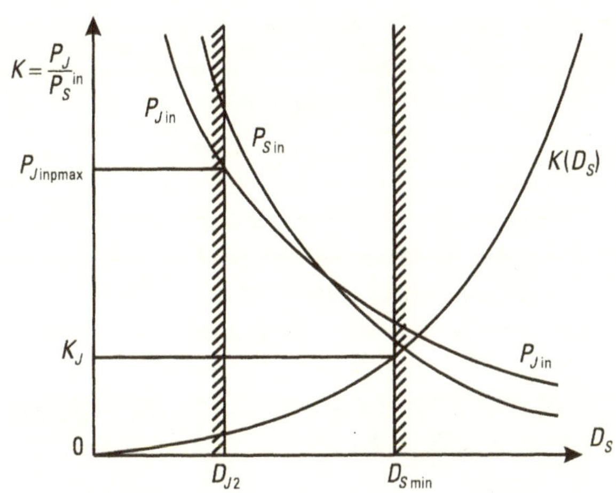

Figure 4.4 Qualitative dependency of K on D_s.

value of the jamming at the input to the receiver $(P_j)_{in}$ exceeds the dynamic range of the AGC or the logarithmic amplifier. In this case, the receiver loses its ability to detect the signal, even when its power is greater than the interference level. Let us briefly look at the characteristics and methods of determining certain parameters entering into the jamming equation (4.10). These include: the polarization pattern γ_j, the RCS of the battle formation σ_{BF}, and propagation factors.

4.2.1 The Polarization Coefficient

In the general case, the coefficient γ_j accounts for the difference in the polarization of the receiving and transmitting antennas. It is defined as the square of the absolute value of the inner product of two complex unit vectors $\bar{n}_E$ and $\bar{n}_A$, which represent the polarization patterns of the electric field $\overline{E}$ incident on the receiving antenna and the field generated by the receiving antenna when emitting, respectively [3]:

$$\gamma_j = |\bar{n}_E \bar{n}_A|^2 \tag{4.12}$$

Here,

$$\bar{n}_E = \frac{\overline{E}}{\sqrt{|E_\Phi|^2 + |E_\Theta|^2}} \qquad \bar{n}_A = \frac{\overline{F}}{\sqrt{|F_\Phi|^2 + |F_\Theta|^2}} \tag{4.13}$$

E_Φ and E_Θ, F_Φ and F_Θ are the projections of vectors E and F in spherical coordinates; and $\overline{F}$ is the radiation vector:

$$\overline{F} = \int_V \bar{j} e^{ik\rho \cos \vartheta} \mathrm{d}V \tag{4.14}$$

where $\bar{j}$ is the density vector of the electric current in the antenna when emitting; ρ is the absolute value of the position vector for the current element in the integration domain V; ϑ is the angle between the position vectors of the current element in the integration domain $\bar{\rho}$ and the observation point $\overline{R}$; and $k = \frac{2\pi}{\lambda}$ is the wave number.

In the Fraunhofer diffraction region $R >> \rho$, the vector $\overline{F}$ is defined by the Hertzian electric vector $\overline{\Pi}$ [3, 4]:

$$\overline{\Pi} = \frac{1}{4\pi i \omega \varepsilon} \frac{e^{-ikR}}{R} \overline{F}$$

where ε is the dielectric permittivity.

The electric vector of field $\overline{E}$ is accordingly defined by the expressions:

$$\overline{E} = \frac{\omega\mu}{4\pi i}\frac{e^{-ikR}}{R}\overline{F}_{\perp} \qquad \overline{F}_{\perp} = F\bar{i}_{\Theta} + F_{\Phi}\bar{i}_{\Phi}$$

where $\bar{i}_{\Theta}$ and $\bar{i}_{\Phi}$ are unit vectors in a spherical system of coordinates; and μ is the magnetic conductivity.

$$\overline{F} = -I_{\partial}\bar{l}$$

where I_{∂} is the dipole current and $\bar{l}$ is a vector pointing along the axis of the dipole, with an absolute value equal to its length.

The power P_{rec}, arriving from the receiving antenna at the matched load, is equal to [3]

$$P_{\text{rec}} = \frac{\lambda^2 G(\Phi,\Theta)}{4\pi}\frac{E^2}{120\pi}|\bar{n}_E\bar{n}_A|^2 \tag{4.15}$$

Here,

$$\frac{E^2}{120\pi}$$

is the power flux density of the plane wave incident on the receiving antenna; $G(\Phi,\Theta)$ is the gain of the receiving antenna; and γ is the wavelength.

In the case where the polarization of both antennas is linear, the vectors $\bar{n}_E$ and $\bar{n}_A$ are real, and their inner product is equal to $\cos\psi$, where ψ is the solid angle between the vectors. In these conditions, the polarization coefficient is equal to

$$\gamma_n = \cos^2\psi \tag{4.16}$$

If the polarizations are elliptic, then it is necessary to take into account the complex character of unit vectors $\bar{n}_E$ and $\bar{n}_A$ representing vectors $\overline{E}$ and $\overline{F}$. By definition (4.13), the unit vectors satisfy the following equalities:

$$\begin{aligned} \bar{n}_E\bar{n}_E^* &= 1 \qquad |\bar{n}_E|^2 = 1 \\ \bar{n}_A\bar{n}_A^* &= 1 \qquad |\bar{n}_A|^2 = 1 \end{aligned} \tag{4.17}$$

$\bar{n}_E^*$ and $\bar{n}_A^*$ are the matching complex conjugate vectors.

As follows from (4.13), they differ from $\overline{E}$ and $\overline{F}$ only in their real coefficients and, therefore, they have identical polarization patterns. This permits us, in the material that follows, to limit ourselves solely to the analysis of the unit vectors mentioned when discussing polarization patterns. In the plane perpendicular to the Umov–Poynting vector, each of them can be represented as a linear combination of the unit position vectors $\bar{i}_\Phi$ and $\bar{i}_\Theta$:

$$\bar{n}_E = (n_E)_\Phi \bar{i}_\Phi + (n_E)_\Theta \bar{i}_\Theta \tag{4.18}$$

$$\bar{n}_A = (n_A)_\Phi \bar{i}_\Phi + (n_A)_\Theta \bar{i}_\Theta$$

where $(n_E)_\Phi$, $(n_E)_\Theta$ and $(n_A)_\Phi$, $(n_A)_\Theta$ are complex quantities that are, in exponential form, written in the following way:

$$(n_E)_\Phi = \frac{|E_\Phi|}{\sqrt{|E_\Phi|^2+|E_\Theta|^2}} e^{i \arg E_\Phi}$$

$$(n_E)_\Theta = \frac{|E_\Theta|}{\sqrt{|E_\Phi|^2+|E_\Theta|^2}} e^{i \arg E_\Theta}$$

$$(n_A)_\Phi = \frac{|F_\Phi|}{\sqrt{|F_\Phi|^2+|F_\Theta|^2}} e^{i \arg F_\Phi}$$

$$(n_A)_\Theta = \frac{|F_\Theta|}{\sqrt{|F_\Phi|^2+|F_\Theta|^2}} e^{i \arg E_\Theta} \tag{4.19}$$

Let us introduce a new notation for the absolute values of the complex quantities $(n_E)_\Phi$, $(n_E)_\Theta$ $(n_A)_\Phi$ and $(n_A)_\Theta$ into (4.19). These quantities correspond to

$$\cos\alpha = \frac{|E_\Phi|}{\sqrt{|E_\Phi|^2+|E_\Theta|^2}} \qquad \cos\beta = \frac{|F_\Phi|}{\sqrt{|F_\Phi|^2+|F_\Theta|^2}} \tag{4.20}$$

$$0 \leqslant \alpha,\ \beta \leqslant \frac{\pi}{2}$$

Taking into consideration (4.19) and (4.20), let us write down (4.18) in a form more convenient for further transformations:

$$\bar{n}_E = \left(\cos\alpha\,\bar{i}_\Phi + \sin\alpha e^{i\gamma}\,\bar{i}_\Theta\right)e^{i\arg E_\Phi}$$

$$\gamma = \arg E_\Theta - \arg E_\Phi$$

$$\bar{n}_A = \left(\cos\beta\,\bar{i}_\Phi + \sin\beta e^{i\delta}\,\bar{i}_\Theta\right)e^{i\arg F_\Phi}$$

$$\delta = \arg F_\Theta - \arg F_\Phi \tag{4.21}$$

Assuming $\arg E_\Phi = \arg F_\Phi = 0$, which can be done by selecting the time origin, but retaining, of course, the phase differences γ and δ, we convert (4.21) to the following form:

$$\bar{n}_E = \left(\cos\alpha\bar{i}_\Phi + \sin\alpha e^{i\gamma}\,\bar{i}_\Theta\right) = \cos\alpha\bar{i}_\Phi + \sin\alpha(\cos\gamma + i\sin\gamma)\bar{i}_\Theta$$

$$\bar{n}_A = \cos\beta\bar{i}_\Phi + \sin\beta(\cos\delta + i\sin\delta)\bar{i}_\Theta \tag{4.22}$$

Bearing in mind that $\bar{i}_\Phi\bar{i}_\Phi = 1$, $\bar{i}_\Theta\bar{i}_\Theta = 1$, and $\bar{i}_\Phi\bar{i}_\Theta = 0$, we define the inner product of unit vectors $\bar{n}_E$ and $\bar{n}_A$:

$$\begin{aligned}(\bar{n}_E\bar{n}_A) &= \cos\alpha\cos\beta + \sin\alpha\sin\beta(\cos\gamma + i\sin\gamma)(\cos\delta + i\sin\delta)\\ &= \cos\alpha\cos\beta + \sin\alpha\sin\beta\cos(\gamma+\delta) + i\sin\alpha\sin\beta\sin(\gamma+\delta)\end{aligned}$$

We find the polarization coefficient sought according to (4.12), using the following expression:

$$\gamma_j = |\bar{n}_E\bar{n}_A|^2 = \cos^2\alpha\cos^2\beta + \sin^2\alpha\sin^2\beta + \frac{1}{2}\sin 2\alpha\sin 2\beta\cos(\gamma+\delta)$$

$$0 \leqslant \alpha \quad \beta \leqslant \frac{\pi}{2} \tag{4.23}$$

In order to determine the sequence of calculations for γ_j using the formula obtained, let us clarify the physical meaning and parameters of vectors $\bar{n}_E$ and $\bar{n}_A$. Toward this end, using (4.22), we determine the real part of the vector $\bar{n}_E$ at the point of time t:

$$\begin{aligned}\mathrm{Re}(\bar{n}_E e^{i\omega t}) &= \mathrm{Re}(\cos\alpha e^{i\omega t}\,\bar{i}_\Phi + \sin\alpha e^{i(\omega t+\gamma)})\\ &= \cos\alpha\cos\omega t\bar{i}_\Phi + \sin\alpha\cos(\omega t + \gamma)\bar{i}_\Theta\end{aligned} \tag{4.24}$$

From (4.24), which represents the instantaneous value of $\bar{n}_E$, it follows that the vector $\bar{n}_E$ consists of two orthogonal sinusoidal waves, unequal in amplitude, but with identical frequency ω and phase shift γ. Over the period $T = \frac{2\pi}{\omega}$, the end of the vector $\mathrm{Re}(\bar{n}_E e^{i\omega t})$ describes a polarization ellipse analogous to that represented by (2.5). In the given case, $\cos\alpha$ corresponds to $E_{\Phi 0}$, and $\sin\alpha$ to $E_{\Theta 0}$. The ratio of amplitudes $E_{\Phi 0}$ and $E_{\Theta 0}$ is equal to

$$\frac{E_{\Phi 0}}{E_{\Theta 0}} = \frac{\cos\alpha}{\sin\alpha} = \mathrm{ctg}\alpha$$

Analyzing (2.5) for a polarization ellipse, it is possible to show that the following dependencies exist between the parameters of the polarization ellipse:

$$\left(\frac{E_{\Phi 0}}{E_{\Theta 0}}\right)^2 = \frac{m^2\,\mathrm{tg}^2\,\psi - 1}{m^2 + \mathrm{tg}^2\,\psi} \tag{4.25}$$

$$\cos\delta = \frac{\mathrm{tg}2\psi}{2\dfrac{E_{\Phi 0}}{E_{\Theta 0}}}\left(\left(\frac{E_{\Phi 0}}{E_{\Theta 0}}\right)^2 - 1\right) \tag{4.26}$$

In the case of the polarization ellipses for vector $\mathrm{Re}(\bar{n}_E e^{i\omega t})$, the formulas (4.25) and (4.26) correspond to the expressions:

$$\mathrm{ctg}^2\alpha = \frac{m^2\,\mathrm{tg}^2\,\psi - 1}{m^2 + \mathrm{tg}^2\,\psi} \tag{4.27}$$

$$\cos\gamma = \frac{\mathrm{tg}^2\,\psi}{2\mathrm{ctg}\alpha}(\mathrm{ctg}^2\alpha - 1) \tag{4.28}$$

Here, m is the ellipticity of the polarization ellipse:

$$m = \frac{b}{a} \tag{4.29}$$

ψ is the angle between the semimajor axis of the polarization ellipse and a certain direction (Figure 4.5).

The parameters m and ψ can be determined from the polarization figure. The latter is formed by rotating a linearly polarized indicator antenna, pointed in the direction of the arriving signal, around the longitudinal axis. This curve, plotted in a polar system of coordinates, does not constitute a

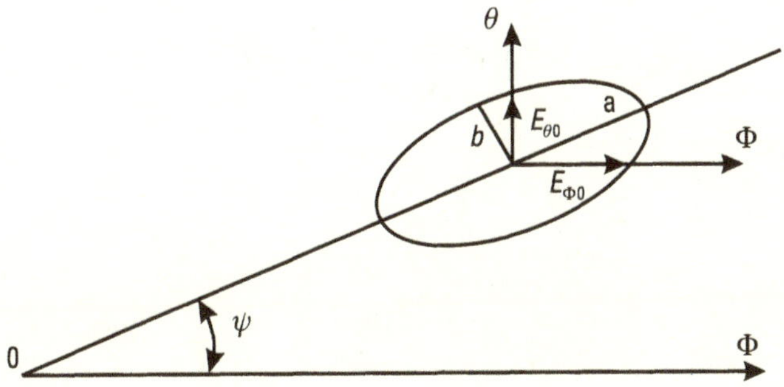

Figure 4.5 The angle ψ.

polarization ellipse, but rather the locus of maximum projections of the rotating field vector on the revolving line that corresponds to the polarization of the indicator antenna being rotated (Figure 4.6).

If α and γ are known, then the ratios (4.27) and (4.28) can be solved for ψ and *m*:

$$\mathrm{tg}2\psi = \frac{2\mathrm{ctg}\alpha\cos\gamma}{\mathrm{ctg}^2\alpha - 1} \tag{4.30}$$

$$m^2 = \frac{1 - \mathrm{ctg}^2\alpha\,\mathrm{tg}^2\psi}{\mathrm{ctg}^2\alpha - \mathrm{tg}^2\psi} \tag{4.31}$$

Analogous formulas can be written down for the unit vector $\overline{n}_A$, as well:

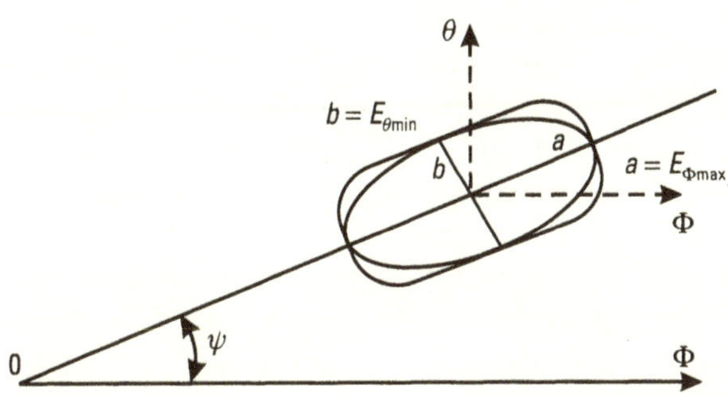

Figure 4.6 Determination of the parameters *m* and ψ from a polarization diagram.

$$\text{ctg}^2\beta = \frac{(m')^2\text{tg}^2\psi' + 1}{(m')^2 + \text{tg}^2\psi'} \tag{4.32}$$

$$\cos\delta = \frac{\text{tg}2\psi'}{2\text{ctg}\beta}(\text{ctg}^2\beta - 1) \tag{4.33}$$

$$\text{tg}2\psi' = \frac{2\text{ctg}\beta\cos\delta}{\text{ctg}^2\beta - 1} \tag{4.34}$$

$$(m')^2 = \frac{1 - \text{ctg}^2\beta\text{tg}^2\psi'}{\text{ctg}^2\beta - \text{tg}^2\psi'} \tag{4.35}$$

Let us now turn to the analysis the analytic expression (4.23). We first determine the conditions under which γ_j reaches it's maximum and minimum values. In the first case, $\gamma_j = 1$ and, in the second, $\gamma_j = 0$. Taking into account the limitations imposed on α and β, it follows from (4.23) that the value of γ_j is at its maximum when $\cos(\gamma + \delta) = 1$ and at its minimum when $\cos(\gamma + \delta) = -1$. After substituting the values mentioned into (4.23) and making several angular transformations, we obtain the expressions for the maximum and minimum values of γ_j:

$$\max\gamma_j = \cos^2(\alpha - \beta) \tag{4.36}$$

$$\min\gamma_j = \cos^2(\alpha + \beta) \tag{4.37}$$

The absolute maximum $\gamma_j = 1$ corresponds to the condition $\alpha - \beta = 0$, or $\alpha = \beta$. Phase shifts $\gamma + \delta = 0$, $\gamma = -\delta$. For the case at hand, let us set the mutual orientations of the polarization ellipses for vectors $\bar{n}_E$ and $\bar{n}_A$ using (4.30), (4.31), (4.34) and (4.35). For example, let us assume that the ellipse corresponding to vector $\bar{n}_E$ and defined by (4.24) has an ellipticity m and that its semimajor axis forms the angle ψ with a certain fixed direction. Using (4.34) and (4.35), we find the corresponding values m' and ψ' for the ellipse defined by vector $\bar{n}_A$ for the receiving antenna. In follows from (4.34) and (4.30) that $\text{tg}2\psi' = \text{tg}2\psi$. Accordingly, it follows from (4.35) and (4.31) that the two ellipses are identical in form and have an identical orientation (their semimajor axes are parallel). The direction of rotation of the electric vectors is determined by the ratio of the phase shifts γ and δ. In the given instance, $(\gamma_j = 1)\,\gamma = -\delta$. In accordance with the rule agreed upon for the determining of the direction of rotation of the electric vector, it follows from this that there is either a right-hand or a left-hand rotation in both cases. Furthermore, for one and the same observer, the directions of rotation of the electric vector in the incoming and outgoing wave will be opposite. Following the rules established, the observer which the Umov–Poynting vector is pointing at determines the direction of rotation. In the given instance, we are speaking of the polarization ellipses of two waves propagating in opposite directions. Therefore, an observer located at a single point where waves arrive from opposite directions will record opposite signs of rotation of the electric vectors in the plane perpendicular to the Umov–Poynting vector.

Reasoning in an analogous fashion, it is possible to show [3], that $\gamma_j = 0$, if:

- The polarization ellipse of the field incident on the receiving antenna is analogous in form to the polarization ellipse of the receiving antenna; but
- Its semimajor axis is rotated relative to the semimajor axis of the receiving antenna by 90 degrees;
- The direction of rotation of the electric vector for the incident field is determined based on the condition $\delta = \pi - \gamma$.

Physically speaking, this means that the orthogonal components of a signal with elliptic polarization arriving at the input to the receiver have opposite phases and identical amplitudes, and their sum is equal to zero.

Of practical interest is the determination of the polarization coefficient in the case where the transmitting antenna has elliptic polarization and the receiving antenna (the jamming target) has linear polarization. Then the unit vector $\bar{n}_A$ is real, not complex, allowing us to assume that $\beta = 0$. According to (4.23),

$$\gamma_j \cos^2 \alpha \tag{4.38}$$

where $\cos^2 \alpha$ is expressed as an identity by $\text{ctg}^2\alpha$:

$$\cos^2 \alpha = \frac{\text{ctg}^2\alpha}{1 + \text{ctg}^2\alpha} \tag{4.39}$$

Here, $\text{ctg}^2\alpha$ is calculated using (4.27) and a polarization figure.

We mention in conclusion that polarization coefficients γ_j vary within the limits of the antenna directional pattern depending on the angles Φ and Θ. These differences can be significant.

In principle, the use of a deterministic polarization pattern for the transmitting antenna of the jammer permits the victim to reduce jamming interference by using receiving antennas with adaptive polarization patterns.

It is possible to achieve a greater resistance to countermeasures by randomizing the polarization of jamming radiation. For example, if the angle ψ in the plane of rotation of the jamming radiation electric vector is a random value with a uniform probability density equal to $p(\psi) = \frac{1}{2\pi}$ in the range of angles 0.2π, then the mathematical expectation $\gamma_j = \cos^2\psi$ is equal to

$$M(\gamma_j) = \frac{1}{2\pi}\int_0^{2\pi} \cos^2\psi d\psi = \frac{1}{2} \tag{4.40}$$

4.2.2 The Propagation Factor

The requirement to take into consideration the effect of the surface of the earth or water on the effectiveness of active jamming becomes evident when jamming land- and ship-based radar to screen aircraft flying at low altitudes. The effect and the methods of its quantitative evaluation can be demonstrated with the example of a radar detecting airborne targets at long range (in the Fraunhofer diffraction region) and at quite a low altitude. The electric center of the radar antenna is assumed to be located at point 0 at altitude h. The land surface is assumed to be locally flat. If observation point A is sufficiently remote, then the grazing angle ψ can be assumed equal to the angle Θ (Figure 4.7). The intensity of the electric field $E_\Sigma(A)$ at the observation point A is defined as the sum of the fields generated by the antenna 0 and its mirror image 0*:

$$E_\Sigma(A) = E(D_1) + RE(D_2) \tag{4.41}$$

Here, R is the complex reflection coefficient. If the underlying surface is an ideal conductor and the antenna polarization is horizontal, then $R_h = -1$. When the polarization is horizontal, it has this same value for all types of soil. This is true for vertical polarization as well, if soils are dielectric or semiconductor and the grazing angle ψ is sufficiently small [2]. Thus, in the first approximation, it is possible to write $R_h = R_v = -1$, where R_v is the reflectivity of a vertically polarized field.

Assuming the wave to be plane at observation point A, the intensity of

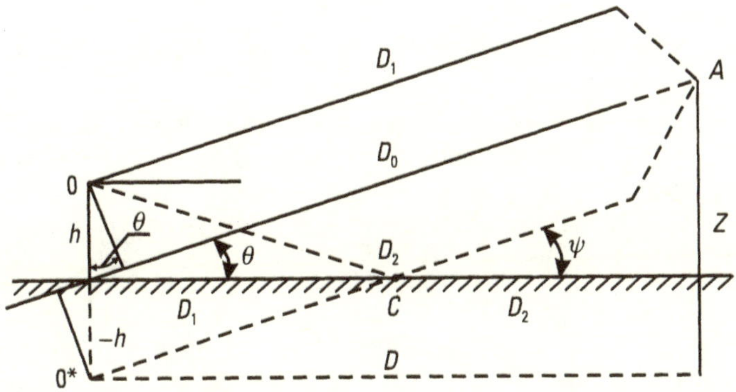

Figure 4.7 Radar detecting airborne targets at long range and at a quite low altitude.

a monochromatic electric field with a carrier frequency ω can be determined according to (4.3) using the following expression:

$$E = \frac{E_0}{D} e^{i(\omega t - kD)} \tag{4.42}$$

where $k = \frac{2\pi}{\lambda}$ is the wave number, $E_0 = \sqrt{60 P_\Sigma}$, and P_Σ is the radiating power.

Taking into account the dependencies of D_1 and D_2 on D_0 and Θ: $D_1 = D_0 - h \sin \Theta$, $D_2 = D_0 + h \sin \Theta$ (Figure 4.7); assuming $D_0 >> h \sin \Theta$; and considering (4.42), it is possible to write expression (4.41) in the following fashion:

$$E_\Sigma(A) = \frac{E_0}{D_0} e^{i(\omega t - kD_0)} 2i \sin(kh \sin \Theta) \tag{4.43}$$

The factor:

$$\Gamma_{OA} = 2|\sin(kh \sin \Theta)| \tag{4.44}$$

is termed the propagation factor for the path OA. It varies within limits from 0 to 2. In the general case, the propagation factor depends on the roughness of the surface, the angle of slide ψ, the complex dielectric permittivity of the material, and the polarization. Specific data is provided in applicable reference books [1]. The requirement to take propagation factors into account arises, to the greatest extent, when grazing angles are small and we are performing active jamming of radar in the meter wave band.

When grazing angles ψ are small and, accordingly, the values of Θ are small as well, the formula (4.44) can be transformed in the following way.

If $\Theta = \frac{Z}{D_0} << 1$ and $kh \sin\Theta \leqslant \frac{\pi}{6}$, then the value of Γ_{OA} will be equal to

$$\Gamma_{OA} = \frac{4\pi}{\lambda} h \frac{Z}{D_0} \tag{4.45}$$

The values of h, λ, Z and D_0 must satisfy the condition:

$$\frac{2\pi}{\lambda} h \frac{Z}{D_0} \leqslant \frac{\pi}{6} \tag{4.46}$$

Substituting (4.45) into (4.43), we can obtain B. A. Vvedensky's famous quadratic formula, valid when angles of slide are sufficiently small, which determines the amplitude of the electric field at point A:

$$E_\Sigma(A) = \frac{4\pi h Z}{\lambda} \frac{E_0}{D_0^2} \tag{4.47}$$

The substitution of the values for the propagation factors, determined according to (4.45), into the jamming equations (4.10) and (4.11) permits us to transform them to the following format:

$$K = A \frac{D_s^8}{D_j^4} \left(\frac{Z_j^2}{Z_s^4} \right) \frac{1}{\left(\frac{4\pi}{\lambda} h \right)^2} \tag{4.48}$$

provided that the jamming is generated from a remote point with coordinates D_j and Z_j. Then, in the case of self-screening, we obtain

$$K = B D_s^4 \frac{1}{\left(\frac{4\pi}{\lambda} h Z_s \right)^2} \tag{4.49}$$

Here, A and B represent all remaining parameters entering into formulas (4.10) and (4.11).

An analysis of formula (4.44), which defines the propagation factor Γ_{OA}, shows its lobe nature. This results from the factor $\sin(kh \sin\Theta)$, which can significantly influence the power ratios K of the jamming to the signal when there are changes in the angular coordinates in the vertical plane of either the jammer or the screened airplane (helicopter). In Figure 4.8, we (qualitatively) show the dependency of the propagation factor on the angle Θ in the case when we assume the surface to be flat. It is important that the jammer know the space angle Θ_1 for the radar antenna pattern resulting from the effect of the propagation factor.

The angle Θ_1 corresponds to the maximum of the first lobe of the propagation factor. In this instance, it follows from (4.44) that

$$\frac{2\pi}{\lambda} h\Theta_1 = \frac{\pi}{2}$$

Therefore,

$$\Theta_1 = \frac{\lambda}{4h} \tag{4.50}$$

In the first approximation, when it is necessary to take the spherical nature of the Earth's surface into consideration in the formula for the propagation factor (4.45), we substitute the adjusted altitudes for h' and Z' in the place of h and Z. It is normally possible to assume (Figure 4.9)

$$h' = h \qquad Z' = Z - \Delta Z \tag{4.51}$$

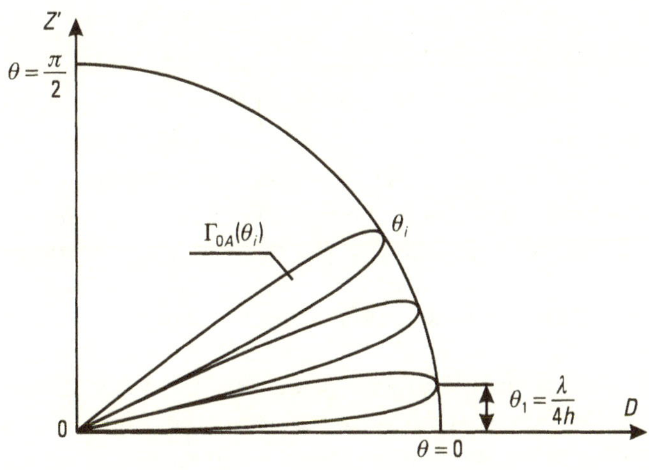

Figure 4.8 The dependency of the propagation factor on the angle Θ.

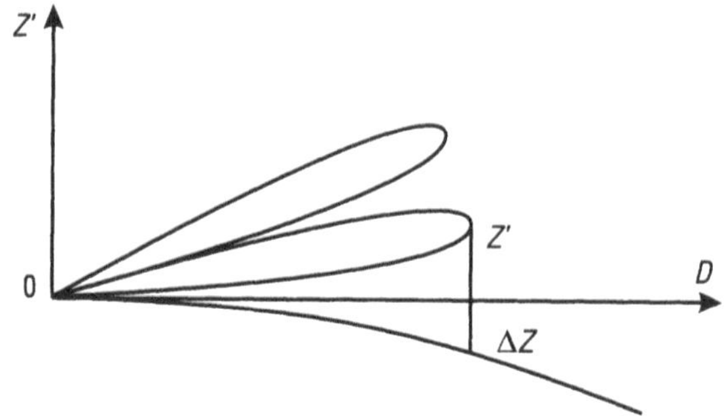

Figure 4.9 Taking the spherical nature of the Earth's surface into consideration.

where

$$\Delta Z = \frac{D_s^2}{2a}$$

D_s is the distance to the observation point; and $a = 6370$ km is the radius of the Earth.

4.2.3 The Radar Cross Section (RCS) of an Aircraft Battle Formation

The radar cross section σ_s of a target being detected is understood to be the ratio of the power of the signal, reradiated from the illuminated target in the direction of the receiving point, to the density of the power flux p of a locally illuminating plane wave. In accordance with what has been said,

$$\sigma_s = \frac{P_2}{p} G(\Phi, \Theta) \tag{4.52}$$

where P_2 is the power scattered by the target; and $G(\Phi, \Theta)$ is the gain of an antenna which we could view as the equivalent of the given scattering target. $G(\Phi, \Theta)$ is defined by the angles Φ, and Θ, which correspond to the receiving point of the signal reflected.

Let us clarify this definition using the example of a flat, ideally conducting plate, located in the Fraunhofer diffraction region, illuminated by a monostatic radar. Let us analyze the case where the Umov–Poynting vector of the incident wave p is perpendicular to the illuminated plate, and its characteristic overall measurements a and b are much larger than the wavelength λ. The electric field of the incident wave generates electric currents in the plate that, in the case being analyzed, can be considered to be in-phase and uniformly distributed over the surface of the plate. Under these conditions, the latter can be considered to be a flat cophasal antenna with a uniform current density over its surface. The equivalent antenna area in the conditions analyzed can be assumed equal to its geometric area S. The gain $G(\Phi, \Theta)$ should be determined for the angles $\Phi = \Theta = 0$ (Figure 4.10). By definition, under the conditions being considered,

$$P_2 = pS \qquad G(\Phi, \Theta) = G(0, 0) = 4\pi \frac{S}{\lambda^2} \tag{4.53}$$

Accordingly,

$$\sigma_s = 4\pi \left(\frac{S}{\lambda} \right)^2 \tag{4.54}$$

Expression (4.54) defines the maximum RCS value for a flat, ideally conducting plate. When it is turned in space by angles Φ and Θ (position S' in Figure 4.10), the RCS in the initial receiving point decreases and becomes equal to

$$\sigma_s(\Phi, \Theta) = \sigma_s F^4(\Phi, \Theta) \tag{4.55}$$

where $F(\Phi, \Theta)$ is the normalized directional pattern of the plate with respect to field.

In the first approximation, an aircraft, as a reflecting object, can be represented as a certain set of reflecting plates, positioned at various angles to the incident plane wave. Depending on the constructional specifics of the aircraft and its relative bearing and orientation relative to the radar, the quantity of reflecting plates, their sizes and orientations, and consequently

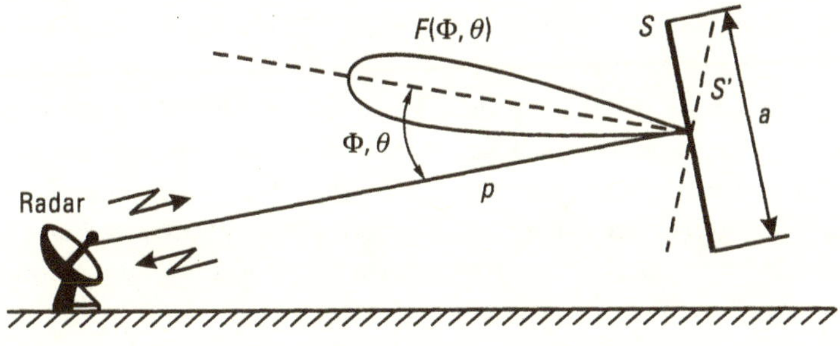

Figure 4.10 Determining the gain $G(\Phi, \Theta)$ for the angles $\Phi = \Theta = 0$.

their RCS, will vary. Due to the sharp change in the RCS of the plate, resulting from the 4th power of the dependency of the RCS on the angles Φ and Θ, the scattering sources, which are the greatest contributors to the total RCS, are often called "specular points." Experience shows that, for a modern airplane, the number of "specular points" is small (about 10).

A signal, reflected from an aircraft, is formed as the result of the combining of signals reflected by the totality of the randomly situated specular points. The amplitudes and phases of the signals reflected from them are also random. The circumstances mentioned also dictate the random nature of the changes (fluctuations) in the envelope and phase of the resulting reflected signal, which have to be taken into consideration when generating deception jamming. Accordingly, the RCS of an aircraft is also a random value with an exponential probability density [5, 6]:

$$p(\sigma) = \frac{1}{\sigma_{ave}} \exp\left(-\frac{\sigma}{\sigma_{ave}}\right) \tag{4.56}$$

Here, σ_{ave} is the average RCS of the airplane.

In the material that follows, if there is no statement to the contrary, we will take the RCS of an aircraft to be $\sigma_s = \sigma_{ave}$. The RCS of a battle formation (BF) σ_{BF} is taken to be equal to the sum of the RCS of the targets located in the radar resolution element:

$$\sigma_{BF} = \sum_{i=1}^{M_{pv}} \sigma_i \tag{4.57}$$

where σ_i is the RCS of an individual target; and M_{pv} is the number of targets in the resolution element (pulse volume).

The RCS value of an aircraft depends essentially on the relative bearing with respect to the illuminating radar. It has a lesser value when the aircraft is illuminated from the forward hemisphere. The greatest value is for radars located broadside.

Up to this point, we have been speaking of reflection from flat, ideally conducting elements. If the surface of the reflecting element is curved, then a depolarization of the scattered signal occurs. The degree of this depolarization can be determined using of a polarization dispersion matrix [6, 7]. Depolarization can occur for other reasons, too, leading to a change in the direction of current in the scattering element.

Providing the number of specular points is sufficiently large and their contribution to the resulting reflected signal is about the same, then the mathematical model of the random process that corresponds to the fluctuations of the signal reflected from the aircraft can be represented by the Rayleigh probability density of the envelope, (1.7), and its correlation time, or the corresponding spectrum width. In aircraft that do not use "STEALTH" technology, the latter does not exceed 10 Hz [1]. To a greater extent, in the signal reflected from a battle formation of planes (helicopters), elements of angle noise are apparent. Its spectrum width may increase.

4.3 Reduction of the Jamming Equation to Canonical Form — Methods of Determining Information Damage

The equation for radar jamming (4.10) defines the power ratio of the jamming to the useful signal when delivered to the input and the passband of the linear part of a receiver. Written down in its general form, this equation is rather unwieldy. It is difficult to use it directly to evaluate the electronic environment and determine the information damage caused (to calculate the radar area disabled by active jamming). The solution of these problems can be simplified by reducing the jamming equation to its canonical form. With that end, let us introduce two indicators that reflect the quantitative and qualitative aspects of jamming.

The first indicator K_{REJ} is equal to the ratio of the effective radiated power density (ERPD) of the target to that of the jammer and characterizes the quantitative aspect:

$$K_{\text{REJ}} = \frac{P_s G_s}{\Delta f_s} \Bigg/ \frac{P_j G_j}{\Delta f_j} \tag{4.58}$$

where $\Delta f_s \approx \Delta f_{\text{rec}}$.

The second indicator η_{REJ} reflects the qualitative aspect of the process and is defined by the following expression:

$$\eta_{\text{REJ}} = 4\pi \frac{\gamma_j}{K_j} \frac{F_j^2(\Phi_s \Theta_s)}{\sigma_{\text{BF}}} \qquad \eta_{\text{REJ}} > 0 \tag{4.59}$$

Here, K_j is the radar jamming coefficient for a given type of jamming.

The jamming side strives to assure the required quality of jamming measures by increasing γ_j and F_j^2 and reducing K_j and σ_{BF} (i.e., by increasing jamming quality, improving the subsystem for intelligence and control of the jammer complex, and decreasing the radar detectability of aircraft and other targets).

Using the indicators introduced, we transform (4.10) to the following format, more convenient for calculating the radar jamming area, after replacing K in the equation with K_j and solving it for D_s:

$$D_s(\Phi_j, \Theta_j)\sqrt{F_s(\Phi_j, \Theta_j)} = \sqrt{D_j}\,\sqrt[4]{\overline{K}_{\text{REJ}}}\,\frac{\Gamma_{\text{radar,BF}}}{\sqrt{\Gamma_{\text{BF,radar}}}}10^{-\frac{1}{4}(0,1\alpha L_j)} \tag{4.60}$$

$$\overline{K}_{\text{REJ}} = \frac{K_{\text{REJ}}}{\eta_{\text{REJ}}} = \frac{P_sG_s\,\Delta f_j}{P_jG_j\,\Delta f_s}\frac{K_j\sigma_{\text{BF}}}{4\pi\gamma_jF_j^2(\Phi_s, \Theta_s)} \tag{4.61}$$

In the materials that follow, $\overline{K}_{\text{REJ}}$ is called the reduced ERPD. Precisely this indicator, and not the value of the ERPD of the radar to the jammer in its pure form, is the indicator of the potential capacity of a jammer complex to deliver information damage to a jamming target.

When jamming originates from within battle formations (self-screening), the jamming equation, in its canonical form, is written down in the following format:

$$D_s(\Phi_j, \Theta_j)F_s(\Phi_j, \Theta_j) = \sqrt{\overline{K}_{\text{REJ}}}\,\sqrt{\Gamma_{\text{radar,BF}}}\,10^{-\frac{1}{2}(0,1\alpha L_j)} \tag{4.62}$$

In both cases, the minimum jamming range occurs when $F_s(\Phi_j, \Theta_j) = F(0, 0) = 1$ (i.e., when jamming acts on the main lobe of the radar). Thus, when jamming originates from within the area,

$$D_{s\min} = \sqrt{D_j}\,\sqrt[4]{\overline{K}_{\text{REJ}}}\,\frac{\Gamma_{\text{radar,JS}}}{\sqrt{\Gamma_{\text{BF,radar}}}}10^{-\frac{1}{4}(0,1\alpha L_j)} \tag{4.63}$$

In the case of self-screening,

$$D_{s\min} = \sqrt{\overline{K}_{\text{REJ}}}\,\sqrt{\Gamma_{\text{radar,BF}}}\,10^{-\frac{1}{2}(0,1\alpha L_j)} \tag{4.64}$$

According to (4.60) and (4.63), as well as to (4.62) and (4.64), we calculate the jamming area using the formulas:

$$D_s(\Phi_j, \Theta_j) = \frac{D_{s\min}}{\sqrt{F_s(\Phi_j, \Theta_j)}} \tag{4.65}$$

when jamming originates from within the area; and

$$D'_s(\Phi_j, \Theta_j) = \frac{D_{s\min}}{\sqrt{F_s(\Phi_j, \Theta_j)}} \tag{4.66}$$

in the case when jamming originates from within battle formations.

In the first approximation, it is most convenient to analyze the electronic environment and calculate jamming areas, in the dynamics of EW, using logarithmic units of measure, decibels (dB). Let us write down (4.63) and (4.64) in decibels. To this end, we convert both sides of the equation to logarithms and multiply by 10. To reduce the size of the equation, we assume that the propagation factors are equal to one.

In the case when jamming originates from within the area,

$$10\lg D_{s\min} = \frac{1}{2} 10\lg D_j + \frac{1}{4} 10\lg \overline{K}_{\text{REJ}} - \frac{1}{4}\alpha L_j$$

When jamming originates from within battle formations,

$$10\lg D_{s\min} = \frac{1}{2} 10\lg \overline{K}_{\text{REJ}} - \frac{1}{2}\alpha L_j$$

Accordingly, (4.63) and (4.64) are written down in decibels in the following manner.

When jamming originates from within the area,

$$D_{s\min} = \frac{1}{2} D_j + \frac{1}{4}\overline{K}_{\text{REJ}} - \frac{1}{4}\alpha L_j \tag{4.67}$$

When jamming originates from within battle formations,

$$D'_{s\min} = \frac{1}{2}\overline{K}_{\text{REJ}} - \frac{1}{2}\alpha L_j \tag{4.68}$$

In (4.67) and (4.68), ranges D_s and D_j are expressed in *decibels relative to meters* (dBm), and the reduced specific ERPD ratio $\overline{K}_{\text{REJ}}$ is expressed in *decibels relative to m^2* (dBm^2). The attenuation coefficient α is expressed in decibels per *km* or *meter* and, correspondingly, the length of the sector over which radiation attenuation L_j occurs is expressed in *km* or *meters*. The product αL_j defines the attenuation that occurs in decibels. $\overline{K}_{\text{REJ}}$ is recorded in decibels according to (4.61), where the powers P_s and P_j are defined in dBW, the antenna gains G_s and G_j in dB, the ratio $\Delta f_j/\Delta f_s$ in dB, the jamming coefficient K_j in dB, the RCS σ_{BF} in dB(m^2), the polarization coefficient γ_j in dB, and the value of F_j^2 in dB. Formulas (4.65) and (4.66), where jamming areas are calculated directly, are recorded in dB in the following manner.

When jamming originates from within the area,

$$D(\Phi_j, \Theta_j) = D_{s\min} - \frac{1}{4} F_s^2(\Phi_j, \Theta_j) \tag{4.69}$$

Here, $F_s^2(\Phi_j, \Theta_j)$ is the level in decibels of the normalized radar beam pattern with respect to power in the direction of the source of jamming radiation.

When jamming originates from within battle formations,

$$D(\Phi_j, \Theta_j) = D'_{s\min} - \frac{1}{2} F_s^2(\Phi_j, \Theta_j) \tag{4.70}$$

Based on two examples, let us clarify the method of calculating jamming areas using (4.69) and (4.70).

First, let us assume that the radar is being attacked with active jamming by jammer JS from within an area located at a distance $D_j = 64$ km (48 dB(m)) from the radar. We determine the reduced ERPD ratio $\overline{K}_{\text{REJ}}$ and the attenuation coefficient α using the expressions $\overline{K}_{\text{REJ}} = 10^8$ m^2 (80 dBm^2), $\alpha = 0$. We represent the normalized radar antenna pattern in the azimuthal plane with regard to field using the following conditional values. The 3-dB beam width $\Phi_{0.5} = 10$ degrees, and 10-dB beam width $\Phi_{0.1} = 2\,\Phi_{0.5}$. The width values of the first, second and other sidelobes are identical and equal to $\Phi_{0.5}$. The averaged levels of the first, second, third and fourth sidelobes are accordingly equal to -20, -24, -28, -32 dB. An analog of the pattern $F(\Phi)$ is shown in Figure 2.4. In Table 4.1, we show the results of calculating the detection range $D_s(\Phi_j)$ of an aircraft battle formation (BF), depending on the width of the sector covered $2\Phi_j$. The value $D_s(\Phi_j)$ is assumed to be equal to $D_{s\min}$ within the limits of the first lobe of the beam.

The corresponding jamming area is shown in Figure 4.11. In many respects, it shows in greater detail the generalized jamming area presented in Figure 4.3. In the example being analyzed, the level of the radar antenna sidelobes is quite large. In principle, this permits us to deliver sizeable information damage to the radar using one jammer (from a single point).

In the second case, we assume that jamming originates from battle formations. The reduced ERPD ratio and the victim radar pattern

Table 4.1
Aircraft Battle Formation Detection Range: Jamming from Outside Formation

$2\Phi_j$ (degrees)	10	20	40	60	80	100
$F_s^2(\Phi_j)$ (dB)	−3	−10	−20	−24	−28	−32
$D_s(\Phi_j)$ (dBm)	44	46.5	49	50	51	52
$D_s(\Phi_j)$ (km)	25	45	80	100	130	160

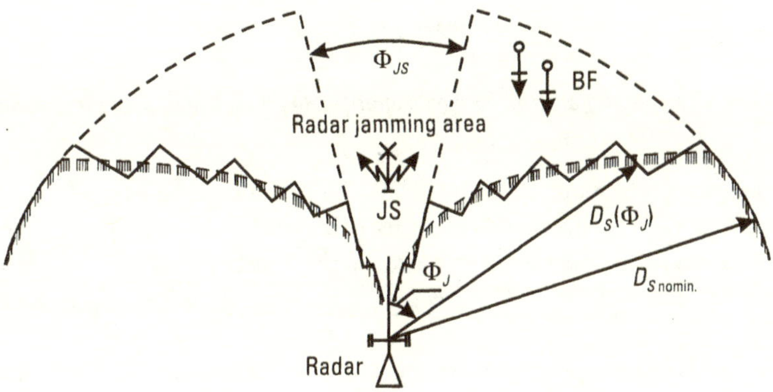

Figure 4.11 The jamming area when jamming originates from outside of battle formations.

parameters are the same. The corresponding calculation results are presented in Table 4.2.

The jamming area is shown in Figure 4.12. It follows from the results given that, in the second case, a great depth of radar jamming is achieved, but less along the front. The detection ranges $D_s(\Phi_j)$ obtained using (4.70) are presented in Table 4.2. They are equal to 250 and 400 km and exceed the

Table 4.2
Aircraft Battle Formation Detection Range: Jamming from Inside Formation

$2\Phi_j$ (degrees)	10	20	40	40	80	100
$F_s^2(\Phi_j)$ (dB)	−3	−10	−20	−24	−28	−32
$D_s(\Phi_j)$ (dBm)	40	45	50	52	54	56
$D_s(\Phi_j)$ (km)	10	30	100	160	250	400

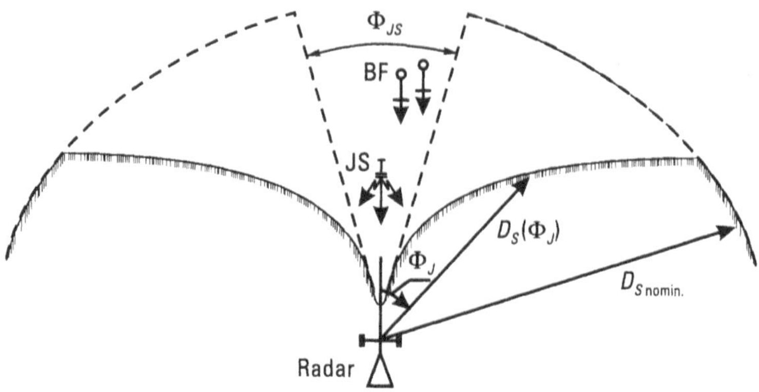

Figure 4.12 The jamming area when jamming originates from within battle formations.

nominal range D_s, which is equal to 200 km. In the given case, this indicates a requirement to perform calculations using more exact formulas that take into consideration the intrinsic noise of the receiver, which is determined using (4.6). Besides this, it is necessary to keep in mind the line-of-sight range.

In the examples analyzed, the level of the radar sidelobes is determined assuming a uniform distribution of the field amplitudes for the antenna aperture and that it is cophasal in nature (Figure 2.4). Such antennas were used in first and second generation radar. At present, to assure information stability in jamming targets, measures have been taken to significantly reduce the signal levels received along the sidelobes. Basically, this is achieved in two ways: through the tapered distribution of field amplitudes in the antenna aperture, and the cancellation of emissions received by the sidelobes.

In order to decrease the level of the sidelobes, the field amplitude along the aperture is tapered from the center to the edges. As sidelobes are reduced the main lobe becomes wider. If it is necessary to hold the value of the 3-dB beam width, then the external dimensions of the antenna have to be increased accordingly. In Tables 4.3 and 4.4, we show the values of the 3-dB antenna beam width ($\Phi_{0.5}$), the position of the first zero of the pattern (Φ_0), the maximum of the first sidelobe (MSL) in dB with respect to the main maximum. Two types of cophasal antennas are analyzed: with rectangular and circular apertures. The field amplitude distribution along the $0X$ coordinate of a rectangular aperture is provided by the expression [4]:

$$f(\xi) = \cos^n \frac{\pi\xi}{2} \quad \xi = \frac{2x}{a} \tag{4.71}$$

Table 4.3
Rectangular Aperture

N	$\Phi_{0.5}$ (rad)	Φ_0 (rad)	MSL (dB)
0	$0.88\frac{\lambda}{a}$	$\frac{\lambda}{a}$	−13.2
1	$1.2\frac{\lambda}{a}$	$1.5\frac{\lambda}{a}$	−23
2	$1.45\frac{\lambda}{a}$	$2\frac{\lambda}{a}$	−32
3	$1.66\frac{\lambda}{a}$	$2.5\frac{\lambda}{a}$	−40

Table 4.4
Circular Aperture

N	$\Phi_{0.5}$ (rad)	Φ_0 (rad)	MSL (dB)
0	$1.02\frac{\lambda}{D}$	$1.22\frac{\lambda}{D}$	−17.6
1	$1.26\frac{\lambda}{D}$	$1.63\frac{\lambda}{D}$	−24.6
2	$1.47\frac{\lambda}{D}$	$2.03\frac{\lambda}{D}$	−30.6

where a is the antenna aperture length along the $0X$ axis.

The field amplitude distribution along the radius of a circular aperture is given by the function:

$$f(r) = (1 - r^2)^p \quad r = \frac{2\rho}{D}$$

D is the diameter of the antenna aperture.

The tables provided demonstrate that, for an even distribution, as the sidelobes attenuate, the ratio of the width at the base of the directional pattern (at a level of the order −16 dB) to the 3-dB beamwidth increases from 2.3 to 2.8. At the same time, the maximum of the first sidelobe attenuates to a level of −30 dB. When setting up jamming areas, this should be considered to be the maximum attainable result.

It is possible to cancel jamming acting on sidelobes by introducing additional receiving channels. In the particular case when one coherent signal with the same carrier frequency ω as the useful signal, but with an unknown initial phase ψ_0 and amplitude U_0, is present in the sidelobe of the main radar antenna, it is necessary to have one additional receiving channel to cancel it. The complex transmission coefficient $\dot{K}$ for this additional channel (Figure 4.13) is determined based on two conditions, assuring that the signal at the output to the adder unit is equal to zero $(U_\Sigma = 0)$. The absolute value of the transmission coefficient is

$$K = \frac{U_s}{U_c} \tag{4.72}$$

and its phase characteristic is

$$\psi = \pm\pi + (\psi_0 + \psi_c) \tag{4.73}$$

Here U_c and ψ_c are the amplitude and initial phase of the signal in the cancellation channel.

In the general case, when dealing with jamming with signals of a random structure, we should speak not of a voltage U_Σ equal to zero, but of the minimum variance of canceled jamming at the output to the adder unit. In Figure 4.14, we show the envelopes $U_0(t)$ and $U_c(t)$ of canceled jamming along the main and cancellation channels arriving at the input to a single-channel canceler. The variance of the jamming σ_Σ^2 at the output to the adder unit, in the given case, is defined as the result of time averaging the square of sum of the signals $U_0(t)$ and $KU_K(t)$ [8, 9]:

$$\sigma_\Sigma^2 = \overline{(U_0(t) + KU(t))^2} \tag{4.74}$$

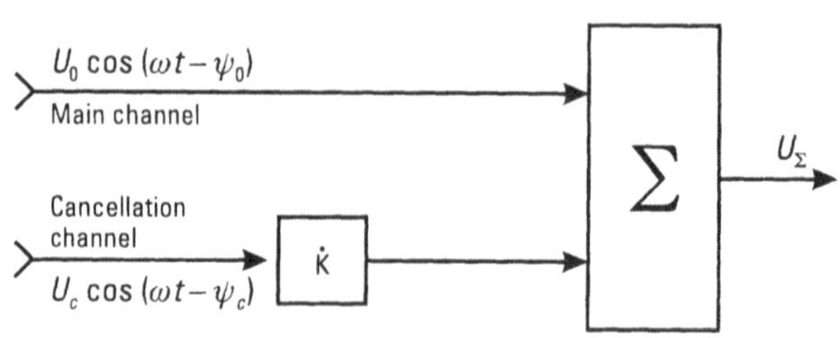

Figure 4.13 Determining the complex transmission coefficient $\dot{K}$ for an additional channel.

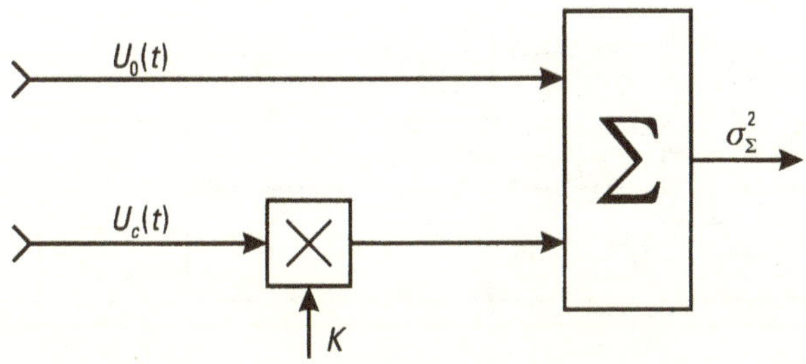

Figure 4.14 The envelopes $U_0(t)$ and $U_c(t)$ of cancelled jamming.

Let us determine the transfer coefficient K that minimizes σ_Σ^2. In this case,

$$\frac{d\sigma_\Sigma^2}{dK} = 2\overline{U_0(t)U_c(t)} + 2K\overline{U_c^2} = 0$$

The transfer coefficient sought:

$$K = -\frac{\overline{U_0(t)U_c(t)}}{\overline{U_c^2}} \tag{4.75}$$

is defined as the normalized cross-correlation function r_{ec} of the envelopes of canceled jamming in the main and cancellation channels over the time interval observed; that is,

$$k = r_{ec} \tag{4.76}$$

In order to determine the value that interests us of the minimum jamming variance at the output of the adder unit, let us replace the value k in (4.74) by its value from (4.76). Taking into consideration (4.75) we obtain

$$\sigma_\Sigma^2 = P_j(1 - r_{ec}^2) \tag{4.77}$$

where $P_j = \overline{U_0^2(t)} = \overline{U_c^2(t)}$ is the power of the jamming in the main and cancellation channels (which is assumed to be identical). In order to assure a coherent mode of operations in the canceler, it is provided with a quadrature channel. In Figure 4.15, we show a single-channel correlation quadrature (coherent) canceler, operating in automatic mode [8, 9]. The

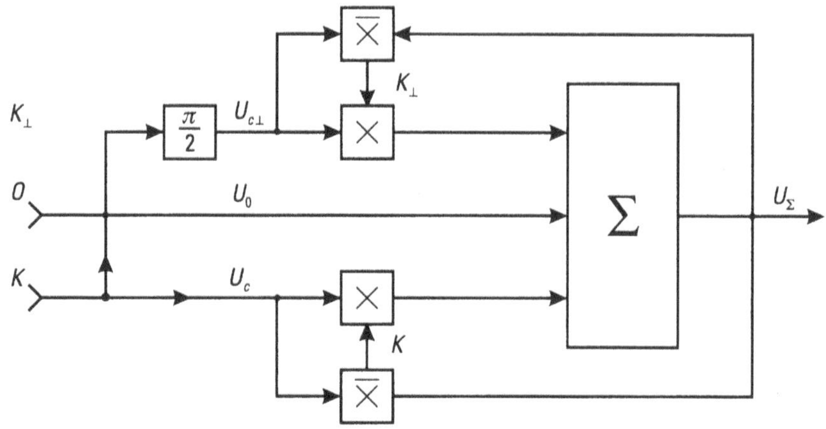

Figure 4.15 Single-channel correlation quadrature (coherent) canceller, operating in automatic mode.

quadrature channel $(K_\perp)$ is included parallel to the channel K through a phase shifter $\left(\frac{\pi}{2}\right)$.

In the given case, in order to evaluate the jamming attenuation that we are interested in, it is necessary to take into account the normalized cross-correlation function of the quadrature channel r_{eqc}. Assuming K and $K_\perp$ to be independent, it is possible to show that, in analogy to (4.77), the value of σ_Σ^2 being sought is determined by the equality:

$$\sigma_\Sigma^2 = P_j\left(1 - \left(r_{ec}^2 + r_{eqc}^2\right)\right) \tag{4.78}$$

The sum $\left(r_{ec}^2 + r_{eqc}^2\right)$ is equal to the absolute value of the normalized cross-correlation function $\dot{r}_{ec}$, the complex amplitudes of the waves being canceled, as determined by the formula:

$$\dot{r}_{ec} = \frac{M\left|\dfrac{U_0 U_c^*}{\sigma_\Sigma^2}\right|^2}{P_j} \tag{4.79}$$

$$\dot{r}_{ec}^2 + \dot{r}_{eqc}^2 = \left|\dot{r}_{ec}\right|^2 \tag{4.80}$$

In the materials that follow, the value σ_Σ^2 will be designated by $\Delta P_{j\Sigma}$. Accordingly, we obtain

$$\Delta P_{j\Sigma} = P_j(1 - |\dot{r}_{ec}|^2) \tag{4.81}$$

The relative degree of jamming attenuation η_c is determined by the ratio:

$$\eta_c = \frac{\Delta P_{j\Sigma}}{P_j} = 1 - |\dot{r}_{ec}|^2 \tag{4.82}$$

When determining the degree to which jamming is reduced, we should take into consideration attenuation of the level of the sidelobes as a result of the tapered illumination of the antenna aperture, as well as that occurring as a result of using antenna cancelers. For example, if the level of the sidelobe in a given direction is −32 dB, relative to $F^2(\Phi_j = 0)$, and the power attenuation, due to the canceler, of the jamming originating from battle formations is −10 dBW, then the total attenuation will be 42 dB and the range in jamming will increase by 21 dB(m). In particular, if $D_{s\min} = 32$ dB(m) (1.6 km), then the range in the direction of the lobe being analyzed will be 53 dB(m) (200 km).

The compensation scheme considered here is analog. Practically speaking, it can be implemented in a digital variant. In practice, multi-channel self-balancing cancelers with correlational feedback can be implemented based on phased or adaptive phased antenna arrays (PAA or APAA). Modern multielement PAA (or APAA) with high resolution can assure an attenuation of the jamming effect not only in the sidelobes, but in the main lobe. This can happen if the entropic nature of the jamming effect in the angular coordinates is not large. In order to exclude or impede the implementation of such a possibility, it is necessary to provide for a sufficiently high level of entropic power for angle noises. The dependency of η_c on the degree of cross-correlation of jamming in the main and cancellation channels indicates a possibility of increasing the effectiveness of jamming through developing suitable means for the decorrelation of jamming in the channels. In conclusion, it should be stressed once again that, in the case that it is necessary to overcome jamming cancellation channels, the number of jammers must exceed the number of cancellation channels by at least one.

The absence of information about the numerical values of a series of key parameters about the jamming target can make calculations using formulas (4.65), (4.66) or (4.69) (4.70) difficult. These parameters include: the ERPD of the radar $\frac{P_tG_t}{\Delta f_s}$, the pulse compression ratio $B = \Delta f_s\tau_s$, etc. Depending on the type of target being jammed, the determination of missing parameters and the performing of the corresponding calculations has its peculiarities. We consider them below. The issue of accounting for the absorption of radiation in the troposphere, evaluated using the coefficient α dB/km, is common for all targets and the calculations associated with them.

This issue is treated in detail in specialized literature [1, 2]. An analysis of this literature shows that it is possible to ignore the absorption of radiation in the troposphere in the range of waves $\lambda > 3$ cm with paths of limited length (approximately up to 100 km). Cases of significant precipitation can pose an exception. For example, when the rain rate is 25 mm/hr at wavelength $\lambda = 3$ cm, it is $\alpha = 0.2$ dB/km, and at wavelength $\lambda = 2$ cm, it is $\alpha = 0.6$ dB/km. According to (4.66), in the first case ($\lambda = 3$ cm), $D_{s\min}$ decreases by a factor of two if L_j is no less than 60 km. In the second case ($\lambda = 2$ cm), a decrease of $D_{s\min}$ by a factor of two takes place if $L_j = 20$ km. In the range $\lambda < 2$ cm, especially for millimeter wavelengths, it is mandatory to account for α. Particularly, at wavelength $\lambda = 0.8$ cm, $D_{s\min}$ is decreased by a factor of two if $L_j = 4$ km ($\alpha = 3$ dB/km) and jamming is generated from within the area. When jamming originates from within battle formations, L_j is only 2 km.

The antenna gain G_A is determined by its directive gain D_A and efficiency η_A

$$G_A = D_A \eta_A \tag{4.83}$$

In the material that follows, it is assumed that η_A is taken into consideration when determining the irradiated power P_s. This gives us a basis to assume that $G_A = D_A$. The gain G_A of an antenna with a flat aperture and an even distribution over the aperture of the amplitudes and phases of the elementary emitters can be found, provided the area of the antenna S_A and the wavelength λ are known:

$$G_A = \frac{4\pi S_A}{\lambda^2} \tag{4.84}$$

In antennas with an aperture in the shape of a rectangle with sides a and b, $S_A = ab$, whence it follows that

$$G_A = \frac{4\pi}{\frac{\lambda}{a}\frac{\lambda}{b}} \tag{4.85}$$

Using the results in Table 4.3, we obtain the well-known formula:

$$G_A = \frac{4\pi 0.88^2}{\Phi_{0.5}\Theta_{0.5}} \quad (4.86)$$

Here, $\Phi_{0.5}, \Theta_{0.5}$, are the values of the main lobe width in radians in orthogonal planes. If $\Phi_{0.5}$ and $\Theta_{0.5}$ are measured in degrees, then

$$G_A = \frac{37,700}{\Phi_{0.5}\Theta_{0.5}} \quad (4.87)$$

In the case of a round output antenna aperture and an even distribution of amplitudes and phases of the original emission sources,

$$G_A = \frac{32,000}{\Phi_{0.5}^2} \quad (4.88)$$

As was noted earlier, the distribution of amplitudes can be nonuniform. However, the value of the constant is (4.87) and (4.88) remain essentially the same regardless of illumination. The corresponding value of G_A for a reflector antenna, considering spillover and blockage, is

$$G_A = \frac{22,000}{\Phi_{0.5}\Theta_{0.5}} \quad (4.89)$$

4.4 Specifics of the Jamming Equation Using Active Jamming against Various Types of Radar

4.4.1 Noncoherent Radar Operating in Scan Mode

The structure of an optimal circuit for receiving pulse trains from noncoherent radar with a low duty cycle is given in Figure 4.16. It includes an optimum filter, matched to a single RF pulse, a linear envelope detector, a video pulse integrator and a threshold device. According to the threshold established (the adopted level of false alarms), a decision is made at the output of the threshold circuit with a probability P_{DET} as to the presence of a signal in the given sample function of n pulses.

The noncoherency of pulses and their low duty factor does not permit us to sum up energies in the linear part of the receiver. At the output to the envelope detector, the video pulse train energy is proportional to their number, if n is not too large. In the opposite case, it is necessary to take into

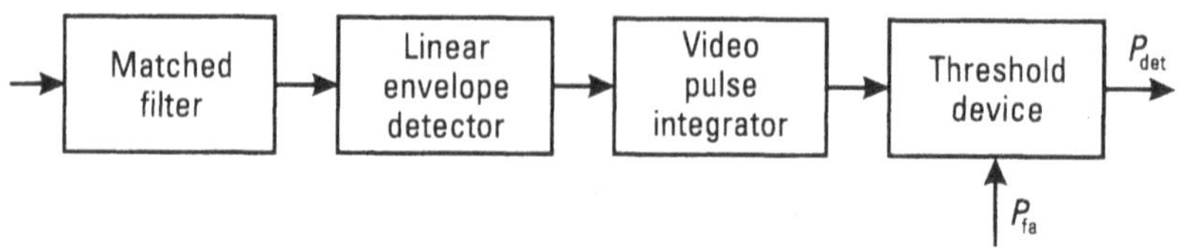

Figure 4.16 Structure of an optimal circuit for receiving pulse trains from noncoherent radar with a low duty cycle.

consideration losses in useful signal energy occurring as a result of noise jamming in the nonlinear element. For this reason, the total energy of the useful signal when n is large grows approximately proportionately to $\sqrt{n}$. Taking into consideration (3.68), (3.69) and (3.70) as well, the jamming coefficient for a radar using screening jamming, representing white Gaussian noise, is determined using the following approximate expressions:

$$K_j = K_{j0} n \quad n \leqslant n_0 \tag{4.90}$$

$$K_j = K_{j0}(n_0 + \sqrt{n - n_0}) \quad n > n_0 \tag{4.91}$$

where

$$K_{j0} = \frac{\lg P_{\text{det}}}{\lg P_{\text{fa}} - \lg P_{\text{det}}} \tag{4.92}$$

The ERPD of the radar is determined according to (4.58). We understand P_s to be the power in the pulse and G_s to be the maximum value of the antenna gain. $\Delta f_s = \frac{1}{\tau_s}$, where τ_s is the pulse length.

The number of pulses in train n when the antenna is rotating is determined by the beam width $\Phi_{0.5}$, the antenna rotation rate Ω_A and the pulse repetition rate F_p:

$$n = \frac{\Phi_{0.5}}{\Omega_A} F_p \tag{4.93}$$

For radars with circular rotation of the antenna, it is possible to substitute the quantity of revolutions of the antenna per minute N_A instead of Ω_A degrees/second. Then,

$$n = \frac{\Phi_{0.5} F_p}{6 N_A} \tag{4.94}$$

The jamming zone of a single radar using deception pulse jamming can be defined using the formulas given earlier. The jamming coefficient using deception pulse jamming can be assumed to be equal to one, $K_j = 1$.

4.4.2 Coherent Radar Operating in Scan Mode

The block diagram of a simplified form of a pulse-Doppler radar or, using a different terminology, a quasi-continuous wave (QCW) radar, is given in Figure 4.17 [5]. The coherence of radar signals is achieved by using stable (quartz) oscillators with frequency multipliers and a comparatively high pulse repetition rate F_p. The spectral width of the signal sample value is approximately hundreds of Hertz when the carrier frequency is of the order of 10 GHz. In a pulse sequence of n reflected signals, the number of coherent pulses n_c is quite large. A special feature of pulse-Doppler radar is the necessity of gating the receiving channels to select targets in range while gating out the transmission.

This task is solved by the strobing unit (SU) block. The length of the strobe and the transmitted pulses are identical and equal to τ_p. The optimum nature of receiving is assured sequentially at two levels. On the first level in the IFA matching the bandwidth with a pulse of length τ_p is provided for, as a result of which $\Delta f_{\text{rec}} \tau_p = 1$. In the second stage there occurs a matching of the spectrum of the sample function of the pulse-Doppler signal to the optimum linear filter, which, in the given case, is a

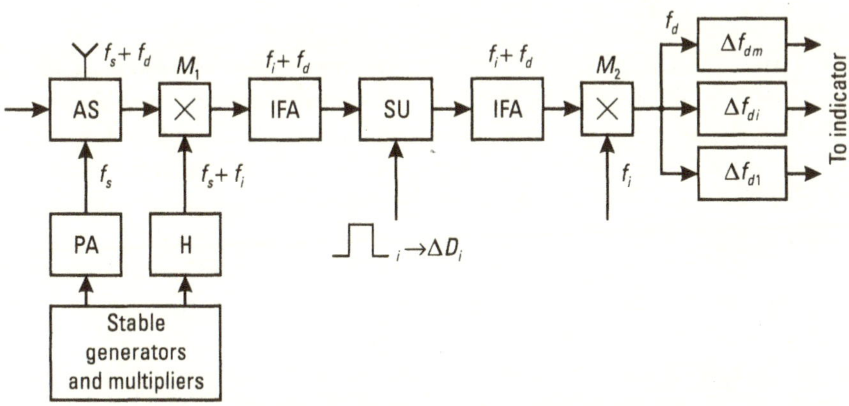

Figure 4.17 The block diagram of a simplified form of a pulse-Doppler radar (quasi-continuous wave radar).

comb filter, including a definite number of Doppler frequency filters, where signals arrive after the second mixer M_2. By definition, the bandwidth of an optimum comb filter is equal to the relative duration of the pulses, multiplied by the averaged passband of the filter of a separate comb. In the given case:

$$\Delta f_{QCW} = Q\Delta f_{\partial} \tag{4.95}$$

The number of coherent pulses in a train in the first approximation can be defined as the ratio of the pulse repetition frequency to the passband of the Doppler frequency filter:

$$n_c = \frac{F_n}{\Delta f_{\partial}} \tag{4.96}$$

From (4.95) and (4.96) it follows that the spectral width of the sample function of coherent pulses is equal to

$$\Delta f_{QCW} = \frac{\Delta f_{rec}}{n_c} \tag{4.97}$$

The ERPD of a radar with a pulse power of P_s and an antenna gain of G_s is determined using the expression:

$$\frac{P_s G_s}{\Delta f_{rec}} n_c \tag{4.98}$$

After the Doppler filters, forming the comb filter, the signals arrive at the envelope detectors and, further, at the smoothing filter, from the output of which the sequence of video pulses arrives at the threshold device, which, depending on the given false alarm level, gives a detection probability depending on the ratio of the useful signal (with an unknown initial phase) to white noise.

Recently, instead of a fairly complex system of analog Doppler frequency filters, digital ones are used. In this case, at the output to the IFA (Figure 4.16), instead of the mixer M_2, two phase detectors are included, forming two quadrature channels. In each channel an analog–digital converter (ADC) is included. The signals from the output of the ADC arrive at the processor that performs, using a fast Fourier transform (FFT) procedure, the formation of appropriate digital filters and the output of information to the user in digital form.

The jamming coefficient K_j of a pulse-Doppler radar using a screening jamming representing white Gaussian noise by analogy with (4.90) can be determined by the expression:

$$K_j = K_{j0} n_{cj} \tag{4.99}$$

where n_{cj} is the number of trains of coherently integrated pulses from the total number n, reflected from the target during its irradiation by the main lobe, K_{j0} is the jamming coefficient corresponding to the coherent train of n_c pulses, matched with the band of the optimum comb filter Δf_{QCW} (i.e., $n_c \tau_p \Delta f_{\mathrm{QCW}} = 1$). Accordingly, assuming the reflected signal to be fluctuating, K_{j0} by analogy with (4.92), it is possible to write

$$K_{j0} = \frac{\lg P_{\mathrm{det}}}{\lg P_{\mathrm{fa}} - \lg P_{\mathrm{det}}} \tag{4.100}$$

The value n_c, as follows from (4.93), in the first approximation is determined using the pulse repetition frequency F_n and Doppler filter passband Δf_∂ . The latter is seldom known to the jamming side. At the same time, an analysis of expressions (4.97), (4.98) and (4.99) indicate a possibility, when calculating the ratio of ERPDs shown, to manage without knowing Δf_∂. In reality,

$$\overline{K}_{\mathrm{REJ}} = \frac{P_s G_s}{P_j G_j} \frac{\Delta f_j}{\Delta f_{\mathrm{QCW}}} \frac{K_j \sigma_{\mathrm{BF}}}{4\pi \gamma_j F_j^2(\Phi_s, \Theta_s)}$$

Taking into consideration (4.94), (4.95) and (4.96), the expression for $\overline{K}_{\mathrm{REJ}}$ is transformed to the form:

$$\overline{K}_{\mathrm{REJ}} = \frac{P_s G_s}{P_j G_j} \frac{\Delta f_j}{\Delta f_{\mathrm{rec}}} \frac{K_{j0} n_{cj} n_c \sigma_{\mathrm{BF}}}{4\pi \gamma_j F_j^2(\Phi_s, \Theta_s)} \tag{4.101}$$

The calculation of $\overline{K}_{\mathrm{REJ}}$ using expression (4.101) does not require us to know Δf_∂. It is sufficient to simply know Δf_{rec}, which is defined by the length of the pulse τ_p. The ERPD of the radar already relates to the spectrum width Δf_{rec} (i.e., it decreases by n_c times), and the jamming coefficient K_j correspondingly increases by n_c times:

$$K_j = K_{j0} n \tag{4.102}$$

Here K_{j0} is already matched to one duration pulse τ_p and is determined, just as before, by the probability values of for accurate detection P_{det} and false alarm P_{fa} (cf. 3.69). The value $\overline{K}_{\text{REJ}}$ remains unchanged.

In conclusion we note that optimum methods of processing doubtlessly increase reception quality, however, the greatest degree of attenuation of the jamming effect is achieved by using a coherent narrowband signal in the radar, permitting us to perform effective filtering.

4.4.3 Pulse-Compression Radar Operating in Scan Mode

A variant of the block diagram for a pulse-compression radar with linear frequency modulation ramp is shown in Figure 4.18 [5]. The carrier frequency of the transmitted pulse of duration τ_s changes linearly in time from f_1 to f_2 (Figure 4.19). The latter is achieved by using a dispersive filter for modulation, having a signal propagation rate that increases with an increase in frequency. On the receiving side, the dispersive filter has a slope opposite that of the dispersion filter of the transmitting side. The parts of the signal received with lesser frequency (f_1) remain in the filter for a longer time than the high-frequency parts of the spectrum, which permits compressing the signal received at the output to the dispersion filter to the size τ_{sc}. The ratio of the signal lengths at the input (τ_s) and at the output of a dispersion filter (τ_{sc}) is call the compression ratio and is indicated using the letter B:

$$B = \frac{\tau_s}{\tau_{sc}} \tag{4.103}$$

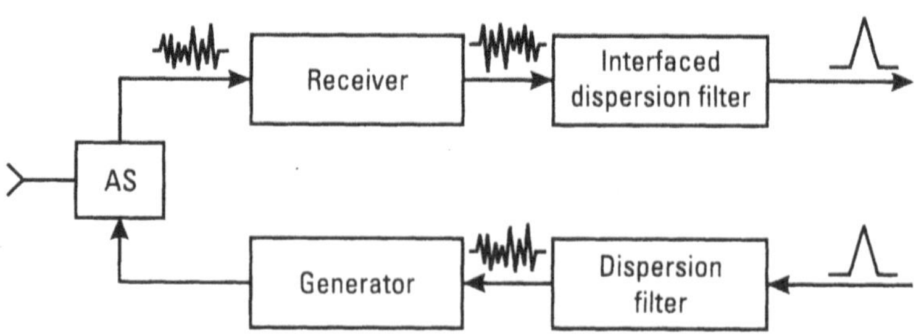

Figure 4.18 The block diagram for a pulse-compression radar with linear frequency modulation ramp.

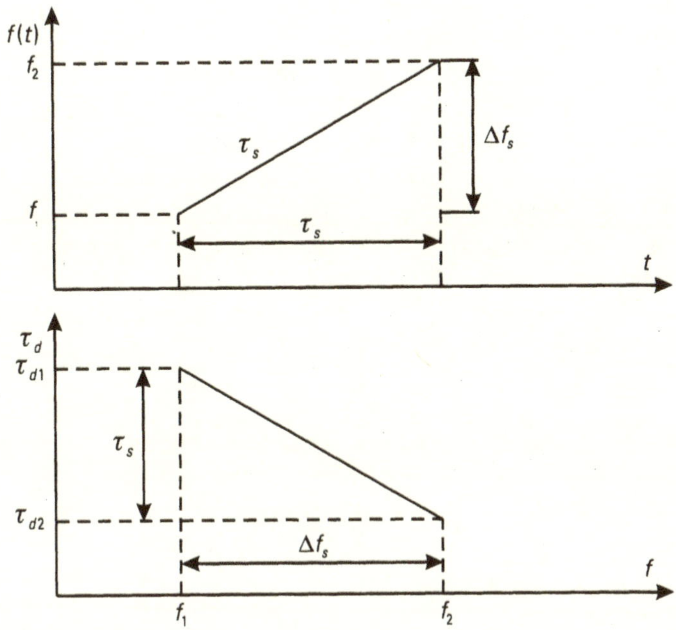

Figure 4.19 The carrier frequency of the transmitted pulse of duration τ_s changes linearly in time from f_1 to f_2.

The signal spectrum width Δf_s is determined by the length of the compressed pulse τ_{sc}:

$$\Delta f_s = \frac{1}{\tau_{sc}} \tag{4.104}$$

From (4.96) and (4.97) it follow that

$$\Delta f_s \tau_s = B \qquad B >> 1 \tag{4.105}$$

Pulse-compression radar differs from conventional radar in that the product of the signal length and its spectrum width is significantly greater than one. For conventional radar $B = 1$. Besides linear frequency modulation pulse-compression signals in radar can be formed using phase modulation [5, 6, 10]. In fields other than radar, the term time–bandwidth product is used rather than pulse compression ratio. Bearing in mind tendencies for radar signals to develop in the direction of increasing their pulse compression ratio and raising their security, it is convenient to use the single term "time–bandwidth product."

The signal compression process in a conjugate dispersion filter is optimum. The filter is matched with the pulse spectrum, which allows us to use the scheme for determining the jamming coefficient already considered before. A unit pulse is processed optimally, but the initial phase of the signal from pulse to pulse changes randomly. In view of the large pulsewidth of the signals from a pulse-compression radar with a large time–bandwidth product, the pulse repetition frequency in them is not large. Accordingly, the number of pulses in train (n) is small. By analogy to (4.90), (4.99), taking into consideration (3.68) and (4.105), can be written down:

$$K_j = K_{j0} n \tag{4.106}$$

$$K_{j0} = \tau_s \Delta f_s \frac{\lg P_{\text{det}}}{\lg P_{\text{fa}} - \lg P_{\text{det}}} \tag{4.107}$$

$$K_j = B \frac{\lg P_{\text{det}}}{\lg P_{\text{fa}} - \lg P_{\text{det}}} n \tag{4.108}$$

The specific ERPD of a pulse compression radar is defined by its ERPD $P_s G_s$ and spectrum width Δf_s (i.e., when ERPDs are identical, the specific ERPD of a pulse compression radar is B times less). At the same time, the jamming coefficient K_j, when all other conditions are equal, turns out to be B times greater because of matched filtering over the long pulse. Accordingly,

$$\overline{K}_{\text{REJ}} = \frac{P_s G_s}{P_j G_j} \frac{\Delta f_j}{\Delta f_{\text{rec}}} \frac{K_j \sigma_{\text{BF}}}{4\pi \gamma_j F_j^2(\Phi_s, \Theta_s)} \tag{4.109}$$

In the formulas given, losses taking place in the dispersion filters are not taken into consideration, which makes the value of information damage determined on their basis somewhat too low.

4.4.4 Synthetic Aperture Radar (SAR)

The principle of synthesizing a radar antenna aperture and the variant of an onboard system for recording a radio hologram of the target area, and subsequent optical processing of signals is explained in Figures 4.20 and 4.21

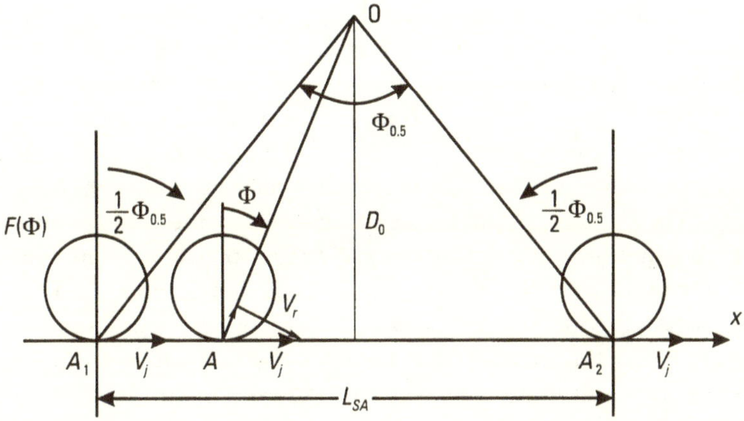

Figure 4.20 The principle of synthesizing a radar antenna aperture.

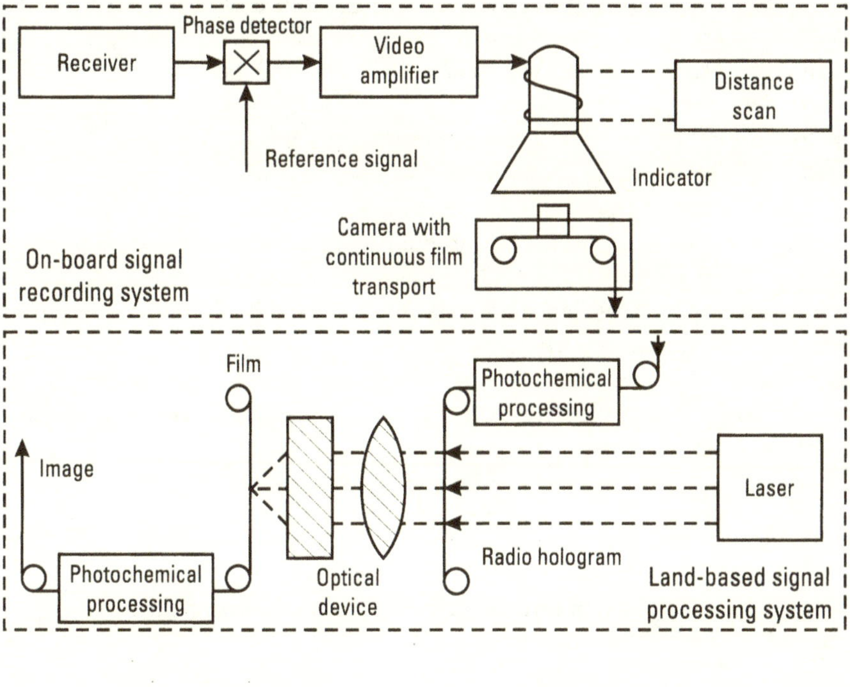

Figure 4.21 The variant of an onboard system for recording a radio hologram of the target area, and subsequent optical processing of signals.

[5, 11]. An airplane with side-looking radar makes a flight with a constant velocity V_j along axis x. The target area O is irradiated using coherent radar signals, beginning from point A_1 through point A_2. The target irradiation angle $\Phi_{0.5}$ corresponds to the 3-dB main lobe width of the physical aperture $F(\Phi)$. The coherent signals reflected from the target O from the output to the receiver, after transformation in the phase detector and amplification, are delivered (in the case of an analog processor) to the scope control grid, performing intensity modulation of the electron stream corresponding to the magnitude of the envelope and the phase of the signals received. The change in brightness on the scope screen is recorded by a photo camera with continuous film transport. This completes the first phase of forming the hologram onboard the airplane. The second phase, in the variant considered, is performed on the ground. It includes the photochemical processing of the film with its radio hologram and the subsequent reproduction of the image by irradiating the film using coherent radiation in the optical range (laser) and the capture of the image in visible light on another photo film. After photochemical processing of the latter, direct decoding of the information received can take place.

Instead of optical processing on the ground, digital processing of the radio hologram directly onboard the plane can be and, in practice, is performed. In this case, two quadrature channels are formed with two phase detectors at the receiver output, from the output of which, after analog–digital conversion, the discrete signals arrive at appropriate processors. Digital processing is more complex than optical.

Synthesizing of the radar antenna aperture permits us to significantly increase the resolution in one angular coordinate. Synthesizing of the antenna aperture can be considered as a procedure for compressing the signal received according to angle Φ, similar to the way that linear frequency modulation permits us to perform signal compression according to distance. The compression coefficient in angle B_Φ is determined as the ratio of angle $\Phi_{0.5}$ to the main lobe width Φ_{SA} of a synthesized antenna with aperture L_{SA}:

$$B_\Phi = \frac{\Phi_{0.5}}{\Phi_{SA}} \tag{4.110}$$

The 3-dB beam width of a synthesized aperture antenna L_{SA} is determined by the formula:

$$\Phi_{SA} = \frac{\lambda}{2L_{SA}} \tag{4.111}$$

The reduction of the beam width by a factor of two in a synthetic aperture antenna is caused by the doubled difference in the wave path during the process of beam formation. The latter increases by a factor of two the phase incursion when the change in the radiation incidence angle is identical.

In the first approximation, for large D_0 and small angles $\Phi_{0.5}$ (Figure 4.19),

$$L_{SA} = D_0 \Phi_{0.5} \tag{4.112}$$

After substituting (4.111) and (4.112) in (4.110), we obtain the final formula, defining the compression in angle Φ:

$$B_\Phi = \frac{2\Phi_{0.5}^2 D_0}{\lambda} \tag{4.113}$$

This same conclusion can be obtained by considering the synthesized length signal T_s to be linearly frequency modulated. The latter is caused by the linear change in Doppler signal frequencies during the time of the airplane from point A_1 to A_2. Indeed, at the intermediate point A (Figure 4.19), the Doppler frequency of the signal received is equal to

$$f_\partial(\Phi) = \frac{2V_r}{\lambda} = \frac{2V_j \sin\Phi}{\lambda}$$

where V_r is the radial component of the airplane velocity in the direction of target O. Normally the angle $\Phi_{0.5}$ is quite small, permitting us to write down the following expression for $f_\partial(\Phi)$:

$$f_\partial(\Phi) = \frac{2V_j}{\lambda}\Phi \tag{4.114}$$

During the time T_s, the function $f_\partial(\Phi)$ changes proportionately to Φ from the maximum value $f_{\partial\,\max} = \frac{V_j}{\lambda}\Phi_{0.5}$ to the minimum $f_{\partial\,\min} = -\frac{V_j}{\lambda}\Phi_{0.5}$. The spectrum width of the modulated wave is equal to

$$\Delta f_{m\partial} = f_{\partial\,\max} - f_{\partial\,\min} = 2\frac{V_j}{\lambda}\Phi_{0.5} \tag{4.115}$$

Accordingly, the signal reflected from target O turns out to be linearly modulated by frequency. By analogy with the broadband signal analyzed previously, which provides signal compression with regard to distance, in the given case it is possible to speak of signal compression by angle Φ.

The corresponding signal base by angle Φ, B_Φ, is determined using the equality:

$$B_\Phi = T_s \Delta f_{m\partial} = \frac{L_{SA}}{V_j} \Delta f_{m\partial} \tag{4.116}$$

Replacing L_{SA} and $\Delta f_{m\partial}$ by their values from (4.112) and (4.115), we obtain the expression for B_Φ, which corresponds, as was to be expected, to (4.113):

$$B_\Phi = \frac{2\Phi_{0.5}^2 D_0}{\lambda}$$

This expression is accepted as being initial for the analysis of the particulars of the jamming equation in the given case.

Besides compression by angle Φ in SAR, temporary pulse compression is performed using linear frequency modulation with a quite large deviation to improve the range resolution. The compression ratio with regard to τ is equal to B_τ. Thus, when determining the jamming coefficient K_j using screening jamming, as well as when determining the reduced ratio of the ERPDs $\overline{K}_{REJ}$, it is necessary to take into consideration two kinds of signal compression.

The value of the jamming coefficient K_j using screening jamming according to (4.91) and (4.108) is

$$K_{j0} = B_\tau \frac{\lg P_{det}}{\lg P_{fa} - \lg P_{det}} \quad K_j = K_{j0} \times \begin{cases} n, n \leqslant n_0 \\ n_0 + \sqrt{n - n_0} \end{cases} \tag{4.117}$$

$$B = B_\Phi B_\tau \quad n = T_s F_p \quad n_0 = 25 \cdots 30$$

$$\overline{K}_{REJ} = \frac{P_s G_s}{P_j G_j} \frac{\Delta f_j}{\Delta f_{rec}} B_\Phi \frac{K_j \sigma_0}{4\pi \gamma_j F_j^2(\Phi_s, \Theta_s)} \tag{4.118}$$

The probabilities P_{det} and P_{fa} are determined by the specific particulars of the problem being solved. In the given case, in particular, we may be speaking of the probability of recognizing the reconnaissance target and the probability of differentiating it among other similar ones.

The reduced ratios do not take into consideration energy losses in the useful signal during SAR processing. Rounding is performed in favor of the jamming target.

4.4.5 The Jamming Equation for Bistatic Radar Using Active Jamming

A variant for the implementation of a bistatic radar is given in Figure 4.22. The transmitter (F_1) and receiver (F_2) parts of the radar are separated by a distance of ΔD. Radar signals emitted, after being reflected from an aircraft in a battle formation BF, are received at point F_2, with is subjected to jamming. In accordance with (4.1), (4.8), (4.9) and Figure 4.22, the jamming equation is derived as a result of the following sequence of procedures:

$$(P_j)_{in} = \frac{P_j G_j}{4\pi D_j^2} \gamma_j A_{\text{rec}} F_{\text{rec}}^2(\Phi_j, \Theta_j) \frac{\Delta f_c}{\Delta f_j} \times F_j^2(\Phi_{\text{rec}}, \Theta_{\text{rec}})$$

$$(P_s)_{in} = \frac{P_s G_s}{4\pi D_{s1}^2} \sigma_{\text{BF}} \frac{1}{4\pi D_{s2}^2} A_{\text{rec}} \tag{4.119}$$

$$K = \left(\frac{P_j}{P_s}\right)_{in} = \frac{P_j G_j \Delta f_s}{P_s G_s \Delta f_j} \frac{4\pi\gamma_j F_j^2(\Phi_{\text{rec}}, \Theta_{\text{rec}})}{\sigma_{\text{BF}} D_j^2} \times D_{s1}^2 D_{s2}^2 F_{\text{rec}}^2(\Phi_j, \Theta_j)$$

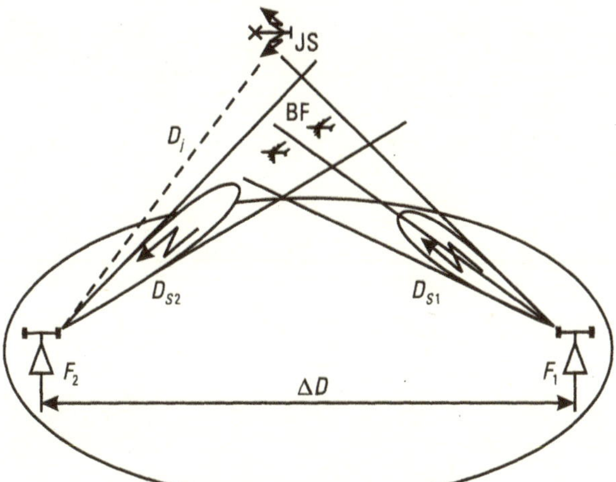

Figure 4.22 A variant for implementing bistatic radar.

Assuming $K = K_j$ and solving (4.119) with respect to the product $D_{s1}^2 D_{s2}^2$, we obtain

$$D_{s1}^2 D_{s2}^2 = \overline{K}_{\mathrm{REJ}} \frac{D_j^2}{F_{\mathrm{rec}}^2(\Phi_j, \Theta_j)} \tag{4.120}$$

where

$$\overline{K}_{\mathrm{REJ}} = \frac{P_s G_s}{P_j G_j} \frac{\Delta f_j}{\Delta f_{\mathrm{rec}}} \frac{K_j \sigma_{\mathrm{BF}}}{4\pi \gamma_j F_j^2(\Phi_{\mathrm{rec}}, \Theta_{\mathrm{rec}})} \tag{4.121}$$

From (4.120) it follows that the jamming zone of a bistatic radar can be represented using a Cassini oval [1, 12]. If we assume that points F_1 and F_2 are foci, then the corresponding equation for the oval will have the form:

$$D_{s1} D_{s2} = a^2 \tag{4.122}$$

where

$$a^2 = \frac{D_j}{F_{\mathrm{rec}}^2(\Phi_j, \Theta_j)} \sqrt{\overline{K}_{\mathrm{REJ}}} \tag{4.123}$$

The parameter a, having dimensionality of length, is compared to the value c, equal to half the distance between the foci F_1 and F_2 (the transmitter and receiver antennas):

$$c = \frac{\Delta D}{2}$$

The equation for a Cassini oval in the Descartes system of coordinates is written as

$$(x^2 + y^2)^2 - 2c^2(x^2 - y^2) = a^4 - c^4 \tag{4.124}$$

Depending on the value of the ratio:

$$\overline{a} = \frac{a}{c}$$

the form of the Cassini oval changes (Figure 4.23). When $\overline{a} > c\sqrt{2}$ ($\overline{a} > c\sqrt{2}$), the typical form of an oval is preserved (Figure 4.23(a)). If, however, $\overline{a} < 1$ (i.e., when the product of $D_j\sqrt{\overline{K}_{\mathrm{REJ}}}$ is small and the

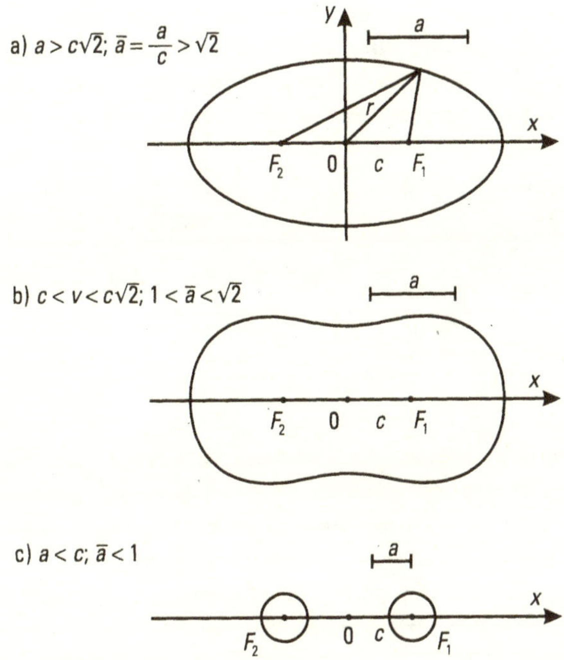

Figure 4.23 Changes in Cassini ovals: (a) oval preserved; (b) decreasing product and increasing distance; (c) oval divided into two circles with centers at the foci F_1 and F_2.

transmitter and receiver points are widely spaced), the oval divides into two circles with centers at the foci F_1 and F_2 (Figure 4.23(b),(c)). From (4.119) and (4.122) it follows that a Cassini oval in the given case represents the border of the jamming area that occurs for space–time and electric parameters specified. The equation for the Cassini oval in a polar system of coordinates (r, φ) follows directly from (4.124) and has the form [12]:

$$r^2 = c^2 \cos 2\varphi \pm \sqrt{c^4 \cos^2 \varphi + (a^4 - c^4)} \tag{4.125}$$

In the polar system of coordinates shown, in which $r = c\rho$ and $\bar{a} = \frac{a}{c}$, (4.125) is represented by the following expression [1]:

$$(\rho^2 + 1)^2 - 4\rho^2 \cos^2 \varphi = \bar{a}^4 \tag{4.126}$$

If jamming is generated using the individual devices of the airplane or a jammer from a fairly dense battle formation, then $D_{s2} = D_j$ and the form of the jamming area is determined according to (4.70). At the same time it is necessary to bear in mind that, when jamming missiles with semi-active seekers from dispersed battle formations, one must take into consideration the change in form of the jamming area, represented by a Cassini oval.

The minimal jamming range of a bistatic radar is determined using (4.120), if we put in it $F_{\text{rec}}(\Phi_j, \Theta_j) = 1$. Then, according to (4.122) and (4.123),

$$D_{s\,\text{min}} D_{s2} = D_j \sqrt{K_{\text{REJ}}} \tag{4.127}$$

An analysis of (4.127) and Figure 4.22 shows that, as a result of the distance between the transmitter and receiver parts of a bistatic radar, its jamming using active jamming from zones can be achieved only from certain directions, allowing for the possibility of observing a number of conditions, directly associated with the patterns on the transmitter and receiver sides. We are speaking of providing for jammer start-up conditions and attaining the required degree of jamming in the receiver.

Thus, before making a decision to implement a specific jamming technique for a bistatic radar from within the area, it is necessary to make sure that at all stages of implementing the selected route, the conditions mentioned are fulfilled. In practice, these conditions are fulfilled when generating jamming from within battle formations.

4.4.6 The Jamming Equation in the Case of Generating Active Jamming from Battle Formations of Aircraft against a Semiactive Homing Seeker

Figure 4.24 shows the variant of generating active jamming against a radar homing seeker of a surface-to-air missile (SAM). A target illumination radar irradiates the aircraft battle formation (BF) with its main lobe. The reflected signals arrive at the seeker which homes on the BF. It is also subjected to jamming, generated by the jammers of the battle formation. Let us determine the ratio of the powers of the interference and signal at the input to the seeker, as related to the passband in its linear part:

$$(P_j)_{in} = \frac{P_j G_j}{4\pi D_{\text{RHH}}^2} A_{\text{RHH}} F_{\text{RHH}}^2(\Phi_{j1}, \Theta_j) \gamma_j \frac{\Delta f_{\text{RHH}}}{\Delta f_j} \tag{4.128}$$

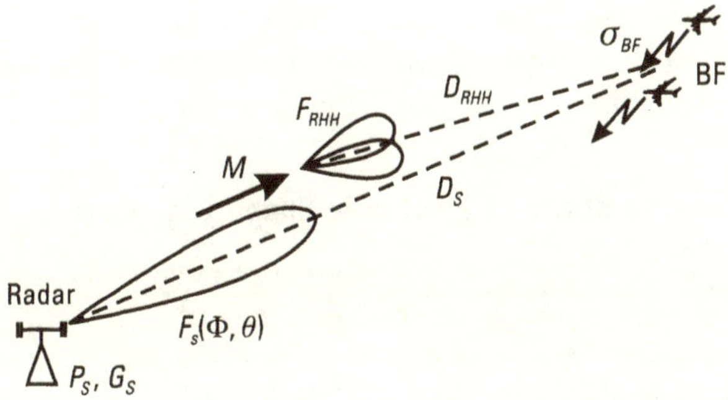

Figure 4.24 Variant of generating active jamming against a radar homing seeker of a surface-to-air missile (SAM).

$$(P_s)_{in} = \frac{P_s G_s}{4\pi D_s^2} \sigma_{\text{BF}} \frac{1}{4\pi D_{\text{RHH}}^2} A_{\text{RHH}}$$

It is assumed that the battle formation is reasonably dense. This permits us to suppose that $F_j^2(\Phi_{\text{R}}, \Theta_{\text{R}}) = 1$.

Then,

$$K = \left(\frac{P_j}{P_s}\right)_{in\text{RHH}} = \frac{P_j G_j}{P_s G_s} \frac{\Delta f_{\text{RHH}}}{\Delta f_j} \frac{4\pi\gamma_j}{\sigma_{\text{BF}}} \times D_s^2 F_{\text{RHH}}^2(\Phi_j, \Theta_j) \tag{4.129}$$

Assuming $K = K_j$, introducing the notation $\overline{K}_{\text{REJ}}$ and solving (4.129) relative to $D_s F_{\text{RHH}}(\Phi_j, \Theta_j)$, we obtain the formulation of the jamming equation in canonical form:

$$D_s F_{\text{RHH}}(\Phi_j, \Theta_j) = \sqrt{\overline{K}_{\text{REJ}}} \tag{4.130}$$

where

$$\overline{K}_{\text{REJ}} = \frac{P_s G_s}{P_j G_j} \frac{\Delta f_j}{\Delta f_{\text{rec}}} \frac{K_j \sigma_{\text{BF}}}{4\pi\gamma_j} \tag{4.131}$$

The minimum jamming range $D_{s\,\text{min}}$ is equal to

$$D_{s\,\text{min}} = \sqrt{\overline{K}_{\text{REJ}}} \tag{4.132}$$

Thus, in the case considered, the ratio of powers of the jamming and useful signal at the input to the seeker is determined, beyond all else, by the distance of the battle formation from the target illumination radar and does not depend on the distance from the BF to the seeker.

4.4.7 Range of Operations of Electronic Support Measures

A mandatory prerequisite for active jamming of radar is the detection of its radiation by the electronic support measures (ESM), which, as a rule, controls the jammer. In Figure 4.25 we show the ESM method using a special reconnaissance airplane. In the given case it is necessary to determine the maximum distance D_{ESM}, at which the airborne ESM can detect a signal emitted by the antenna of a radar in the direction $\Phi_{\text{R}}, \Theta_{\text{R}}$.

The power of the signal, arriving at the input to the ESM receiver, is determined by the expression:

$$P_{\text{ESM}}(\Phi_{\text{R}}, \Theta_{\text{R}}) = \frac{P_s G_s F_s^2(\Phi_{\text{R}}, \Theta_{\text{R}})}{4\pi D_{\text{ESM}}^2} A_{\text{ESM}} \gamma_j \times \Gamma_{s,\text{ESM}}^2 10^{-0,1\alpha L_{\text{SR}}} \tag{4.133}$$

where $A_{\text{ESM}} = \frac{\lambda^2 G_{\text{ESM}}}{4\pi}$ is the effective area of the ESM receiving antenna; γ_j is the polarization coefficient; $\Gamma_{s,\text{ESM}}$ is the propagation factor; α is the absorption coefficient in dB(km^{-1}) in a sector of length L_{SR}; λ is the wavelength; and G_{ESM} is the gain of the ESM receiving antenna.

The solution of the ESM problem is possible if

$$P_{\text{ESM}}(\Phi_{\text{R}}, \Theta_{\text{R}}) \geqslant P_{\text{rec min}} \tag{4.134}$$

where $P_{\text{rec min}}$ is the minimum required power at the input for the receiver, ensuring the solution of detection, determination and recognition tasks, or

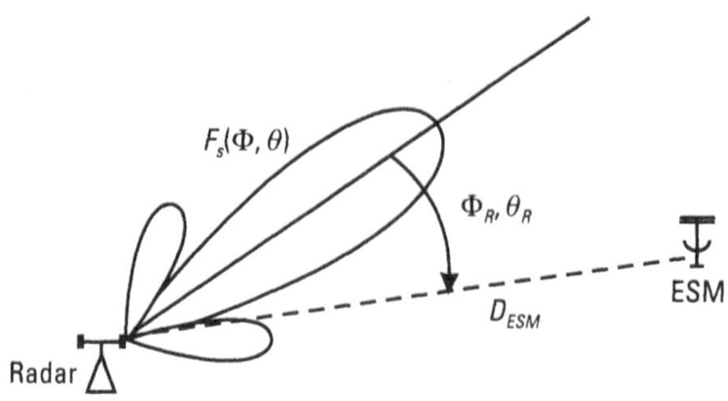

Figure 4.25 ESM method using a special reconnaissance airplane.

solving the problem of starting up an appropriate jamming station.

After substituting $P_{ESM}(\Phi_R, \Theta_R)$ in expression (4.133) by $P_{rec\,min}$ and solving it relative to D_{ESM}, we obtain

$$D_{ESM}(\Phi_R, \Theta_R) = \frac{\lambda}{4\pi}\Gamma_{s,ESM}F_s(\Phi_R, \Theta_R)10^{-\frac{1}{2}(0,1\alpha L_{SR})} \times \sqrt{\frac{P_s G_s G_{ESM} \gamma_j}{P_{rec\,min}}} \tag{4.135}$$

When converting to dB, expression (4.128) is written down in the following manner:

$$\begin{aligned} D_{ESM}\ \mathrm{dB(m)} = {} & D_{ESM}\ \mathrm{dB(m)} - 11 + \Gamma_{s,ESM}\ \mathrm{dB} + F_s(\Phi_R, \Theta_R)\ \mathrm{dB} \\ & - \frac{1}{2}\alpha L_{SR}\ \mathrm{dB} + \frac{1}{2}(P_s G_s)\ \mathrm{dBW} + \frac{1}{2}G_{ESM}\ \mathrm{dB} \\ & + \gamma_j\ \mathrm{dB} - P_{rec\,min}\ \mathrm{dBW} \end{aligned} \tag{4.136}$$

Here, $P_{rec\,min}$ is determined by the specific ESM conditions, the type of targets subject to reconnaissance, the required probability of recognition, etc.

A mandatory prerequisite of conducting ESM using airborne devices for detection of ground-based and other radars is verification that the line-of-sight condition for averaged conditions is met:

$$D_{rec\,max} = 4.12\left(\sqrt{h_{ESM}} + \sqrt{h_A}\right) \tag{4.137}$$

Here, $D_{rec\,max}$ is the maximum ESM range in km; h_{ESM} is the flight altitude of the airplane above sea level in meters; and h_A is the antenna height of the target subject to reconnaissance in meters.

4.5 Particulars of Jamming Radar Using Screening Jamming with Limited Information Quality Indicators — Use of the Jamming Equation for Analysis of the Electronic Environment

In conditions of information conflict, one of the key indicators is information stability, which is understood to be the ability of electronic targets to function in conditions of deliberate countermeasures taken by each of the conflicting sides. It was noted previously that the indicator for potential stability of jamming and corresponding devices is the information (entropy) quality coefficient η_H. Below, using specific examples, this position is expanded upon and clarified. In conclusion, the question of using the jamming equation for the comprehensive analysis of the radio electronic environment unfolding between the conflicting sides is discussed.

4.5.1 The Definition of the Radar Jamming Coefficient Using Narrowband Screening Jamming with a Limited Spectrum

In specific conditions, the limiting of the jamming spectrum leads to the reduction of its entropy power in the phase space of the concealed signal. Accordingly, the quality coefficients η_E and η_H are reduced and the jamming coefficient is increased. In the problem analyzed below, first the jamming coefficient K_j is determined and then the degree to which the entropy quality coefficient η_H effects it is established. In the given case, it is not possible in full measure to use the methodology of determining the radar jamming coefficient using jamming of the Gaussian white noise type, based on the likelihood function. The required threshold values of the signal/jamming ratio are determined directly from a likelihood ratio of type (1.1). In order to determine the ratio indicated, it is necessary to represent the sample function of the jamming in the form of a sum of uncorrelated random values of type (2.21). Such a model for the sample function of a random process is called a canonical expansion. Another name is also widespread — the Karhunen–Loeve expansion. At times people speak of a generalized Fourier series [13, 14]. The practical use of a canonical expansion in jamming is known only for narrowband Gaussian noise with a flat spectrum. In this case the random values u_i (sampling values) are distributed according to a normal law and their variances are the eigenvalues of the singular integral equation (2.32) or (2.33) [15]. The multidimensional probability density $p_n(u_1, \ldots, u_n)$ of the sample function of narrowband Gaussian noise of duration T is represented by expression (2.39). Bearing in mind first approximation estimates, we will assume the useful signal to be fully known. The only unknown fact remaining is whether the useful signal S is present or absent in the given sample function.

The probability density $p_n(u_0, \ldots, u_n, S)$ of the sample function, containing the useful signal S, can be written by analogy with (2.39) in the following way:

$$p_n(u_i, S) = \prod_{i=0}^{n} (2\pi P_j \lambda_j)^{-\frac{1}{2}} \times \exp\left(-\sum_{i=0}^{n} \frac{(u_i - S_i)^2}{2P_j\lambda_i} \right) \qquad i = \overline{0, n} \qquad (4.138)$$

The decision as to the presence of a useful signal in the given sample function is taken on the basis of comparing the likelihood ratio:

$$\Lambda = \frac{p_n(u_i, S_i)}{p_n(u_i)} \qquad i = \overline{0, n} \tag{4.139}$$

with a certain threshold value Λ_0. If $\Lambda < \Lambda_0$, it is assumed that there is no signal. The alternative decision occurs when

$$\Lambda \geqslant \Lambda_0$$

In the first instance, the hypothesis $H_0(\Lambda < \Lambda_0)$ occurs, and in the second $(\Lambda \geqslant \Lambda_0)$ — the hypothesis H_1.

Replacing $p_n(u_i, S)$ and $p_n(u_i)$ by their expressions in (4.138) and (2.39), we transform the likelihood ratio (4.139) to the following form:

$$\Lambda = \exp\left(-\sum_{i=0}^{n} \frac{(u_i - S_i)^2}{2\lambda_i'} + \sum_{i=0}^{n} \frac{(u_i^2)}{2\lambda_{i'}}\right) \tag{4.140}$$

where

$$\lambda_i' = \lambda_i P_j \quad \text{and} \quad i = \overline{0, n} \tag{4.141}$$

λ_i the eigenvalues, corresponding to (2.32), as given in tables [16].

After transformation (4.140) we write down the likelihood ratio in the following way:

$$\Lambda = \exp\left(-\sum_{i=0}^{n} \frac{u_i s_i}{\lambda_i'} - \sum_{i=0}^{n} \frac{s_i^2}{2\lambda_{i'}'}\right) \tag{4.142}$$

The latter formulation, taking into account the monotonic character of an exponential function, permits subsequent analysis to be made based on sufficient statistics of G, which carries all information related to the given problem represented by the expression:

$$G = \sum_{i=0}^{n} \frac{u_i s_i}{\lambda_i'} = \ln\Lambda + \sum_{i=0}^{n} \frac{s_i^2}{2\lambda_i'} \tag{4.143}$$

The threshold value G_0 is unambiguously determined by the threshold value of the likelihood ratio Λ_0:

$$G_0 = \ln \Lambda_0 + \sum_{i=0}^{n} \frac{s_i}{2\lambda'_i} \tag{4.144}$$

The random value G is a linear combination of random values u_i, $i = \overline{0, n}$, normally distributed. Accordingly, G is also a Gaussian random value. Its probability densities $P_0(G)$ and $P_1(G)$ are determined by the mathematical expectation $M[G]$ and the variance $D[G]$. $P_0(G)$ corresponds to the hypothesis H_0, and $P_1(G)$ to the hypothesis H_1.

From expression (4.144) it follows that

$$M[G] = \frac{a}{P_j} \sum_{i=0}^{n} \frac{1}{\lambda_i} \quad \text{(for hypothesis } H_1\text{)} \tag{4.145}$$

$$M[G] = 0 \quad \text{(for hypothesis } H_0\text{)} \tag{4.146}$$

$$D[G] = \frac{a^2}{P_j} \sum_{i=0}^{n} \frac{1}{\lambda_i} \tag{4.147}$$

$$\sigma_G = \sqrt{D[G]} \tag{4.148}$$

Here, a is the effective value of the voltage (current) of the useful signal in a resistance (resistor) of 1 ohm.

Knowing $M[G]$ and $D[G]$, it is possible to write an expression for the probabilities of false alarm P_{fa} and detection P_{det}, permitting us to directly move on to the solution of the problem posed: the determination of the jamming coefficient K_j for narrowband noise. We obtain

$$P_{\text{fa}} = \int_{G_0}^{\infty} p_0(G)\text{d}G = \frac{1}{2} - \Phi_0\left(\frac{G_0}{\sigma_G}\right) \tag{4.149}$$

$$P_{\text{det}} = \int_{G_0}^{\infty} p_1(G)\text{d}G = \frac{1}{2} - \Phi_0\left(\frac{G_0}{\sigma_G} - \frac{M[G]}{\sigma_G}\right) \tag{4.150}$$

Here, G_0 is determined by the threshold selected, in the given case, by the permissible value for the probability of a false alarm. The ratio $\frac{M[G]}{\sigma_G}$ is an analog of the parameter q (the signal/interference ratio) in formulas (1.3) and (1.4). Taking into consideration (4.146) and (4.147), we obtain

$$q = \sqrt{\frac{a^2}{P_j}\sum_{i=0}^{n}\frac{1}{\lambda}} \tag{4.151}$$

In the case of white noise ($n \to \infty$), (4.151) can be transformed into (1.3). Indeed, when $n \to \infty$, $\sum_{i=0}^{n}\frac{1}{\lambda} \to n$. According to the Kotelnikov theorem (the Sampling Theorem), if $n = 2\Delta f_j T$, then

$$q = \sqrt{\frac{a^2 2T}{P_j/\Delta f_j}} = \sqrt{\frac{2E}{N_0}} \tag{4.152}$$

where N_0 is the one-sided spectral density of white noise; and $E = a^2 T$ is the signal energy.

By definition, the jamming coefficient in the given case is equal to the power ratios of the useful and jamming, as related to the passband, when the probabilities P_{fa} and P_{det} do not exceed specified values. The latter define the threshold value $q = q_{thresh}$.

In the conditions being analyzed, it is necessary to distinguish two cases:

$$f_j = \Delta f_i \leqslant \Delta f_s \qquad \text{and} \qquad f_j = \Delta f_j > \Delta f_s \tag{4.153}$$

In the first case,

$$K_j = \frac{P_j}{a^2} \tag{4.154}$$

or, with consideration to (4.151),

$$K_j = \frac{\sum_{i=0}^{n}\frac{1}{\lambda_i}}{q^2_{thresh}} \tag{4.155}$$

In the second case, it is necessary to take into consideration the power reduction in the jamming, caused by extending of its spectrum:

$$K_j = \frac{P_j \frac{\Delta f_j}{\Delta f_s}}{a^2} \tag{4.156}$$

With consideration to (4.151), we obtain

$$K_j = \frac{1}{q_{\text{thresh}}^2 \Delta f_j T_s} \sum_{i=0}^{n} \frac{1}{\lambda_i} \tag{4.157}$$

Let us determine the energy quality coefficient η_E for narrowband screening jamming. According to (3.79),

$$\eta_E = \frac{\overline{K}_{j0}}{K_{j0}} \tag{4.158}$$

where $\overline{K}_{j0}$ is the jamming coefficient using white noise; and K_{j0} is the jamming coefficient using the noise being analyzed.

Both $\overline{K}_{j0}$ and K_{j0} are determined by the threshold value $q = q_{\text{thresh}}$. According to (3.66) and (3.67),

$$\overline{K}_{j0} = \frac{2}{q_{\text{thresh}}^2} \tag{4.159}$$

The threshold value q_{thresh} is determined only by the probabilities of false alarm P_{fa} and a detection P_{det}. In all cases considered it is the same. This permits us to represent the coefficient η_E in the following manner:

$$\Delta f_j \leqslant \Delta f_s \qquad \eta_E = \frac{2}{\sum_{i=0}^{n} \frac{1}{\lambda_i}} \tag{4.160}$$

$$\Delta f_j > \Delta f_s = \frac{1}{T} \qquad \eta_E = \frac{2 \Delta f_j T}{\sum_{i=0}^{n} \frac{1}{\lambda_i}} \tag{4.161}$$

when $\eta \to \infty$, $\eta_E \to 1$.

The results of calculating η_E depending on $\frac{\Delta f_j}{\Delta f_s}$ are given in Figure 4.26. Here we have shown the change of the information quality coefficient η_H as

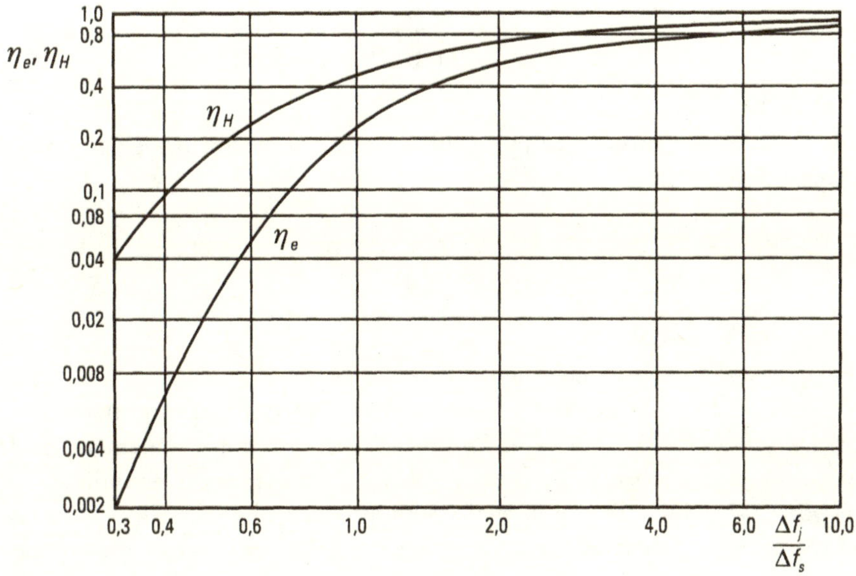

Figure 4.26 The results of calculating η_E depending on $\frac{\Delta f_j}{\Delta f_s}$.

well. The latter is determined in the following way. In accordance with (3.56), (3.39) and (3.57),

$$\eta_H = \frac{\overline{P}_j}{P_j}$$

where

$$\overline{P}_j = 2\frac{1}{2\pi e}\exp(2H')$$

In the given instance,

$$H' = \frac{1}{n+1}\sum_{i=0}^{n} H_i \tag{4.162}$$

where H_i is the entropy of a Gaussian random value with variance λ_i',

$$H_i = \frac{1}{2}\ln(2\pi e\lambda_{i'}')$$

With consideration to (4.141) and (4.162), we obtain

$$\overline{P}_j = P_j \left(\prod_{i=0}^{n} \lambda_i \right)^{\frac{1}{n+1}} \tag{4.163}$$

The information (entropy) quality coefficient sought η_H is determined using the formula:

$$\eta_H = \left(\prod_{i=0}^{n} \frac{1}{\lambda_i} \right)^{\frac{1}{n+1}} \tag{4.164}$$

As was to be expected, when $n \to \infty$, $\eta_H \to 1$.

An analysis of calculation results of how coefficients η_E and η_H depend on the values of the ratio $\frac{\Delta f_j}{\Delta f_s}$ (Figure 4.26) confirms the statement made earlier about the information quality indicator for jamming as an indicator of its potential stability to counteractions by the side being jammed. A decrease in the entropy power of the interference in the phase space of the jammed signal lower than the permissible level permits the conflicting side, by making the transition to optimum processing, to reduce jamming to a significant degree. This is illustrated by the following numbers, resulting from Figure 4.26. A reduction of η_H by -8, -9 dB, occurring as a result of expanding the passband of the receiver being jammed Δf_s by a factor of two in comparison to the spectrum width of the jamming Δf_j, leads to an increase of K_j by 16, 17 dB, and accordingly an increase in the ERPD of the jamming station of required by 16, 17 dB, which in a number of cases is problematic. The loss of entropy power as a result of a decrease in jamming spectrum width Δf_j to 0.3 of Δf_s requires an increase in the jammer ERPD to assure the specified degree of jamming at 27, 28 dB, which, practically speaking, in many cases is impossible. What has been discussed gives cause to mention the information (entropy) coefficient η_H in a broader sense than as a jamming quality indicator, and consider it as a jamming system information stability indicator, understanding by this its ability to successfully operate in conditions of information conflict. In particular, the assertion is confirmed that in a number of cases the loss of information stability cannot be compensated by an increase in ERPD. For example, in the instance analyzed above, when the jamming spectrum is decreased by a factor of three, its spectral density must increase by approximately 5 dB and, at the same time, the entropy power decreases by 15, 16 dB. The required energy expenditures at this time increase by 27, 28 dB. It is necessary, above all, to increase the entropy power of the jamming in the phase space of the target being jammed, and not just its spectral density.

The calculations shown for the jamming coefficient K_j using narrow-band noise permit us to determine the optimum jamming spectrum width, corresponding to its minimal power at the input to the receiver in the passband of its linear part, at which the required degree of jamming is achieved. The existence of an extremum is caused by the nature of the change in K_j as the ratio $\Delta f_j / \Delta f_s$ changes. As this ratio changes from one in the direction of smaller values, K_j increases and accordingly the required power $(P_j)_{in}$ increases. When the ratio changes in the direction of values larger than one, K_j decreases, but the total power of the jamming increases, since only part of its arrives at the passband of the receiver. The dependencies of $(P_j)_{in}$ and K_j on $\frac{\Delta f_j}{\Delta f_s}$ are shown in Figure 4.27. The jamming coefficient K_j is determined according to formula (4.158) in the assumption that $\overline{K}_{j0} = 1$. The energy quality coefficient η_E is determined using the formulas (4.160) and (4.161). The required input power was determined according to the following scheme:

$$(P_j)_{in} = \begin{cases} K_j \Delta f_j \leqslant \Delta f_s \\ K_j \dfrac{\Delta f_j}{\Delta f_s}, \Delta f_j > \Delta f_s \end{cases} \tag{4.165}$$

The power of the useful signal $(P_s)_{in}$ is assumed to be equal to unity.

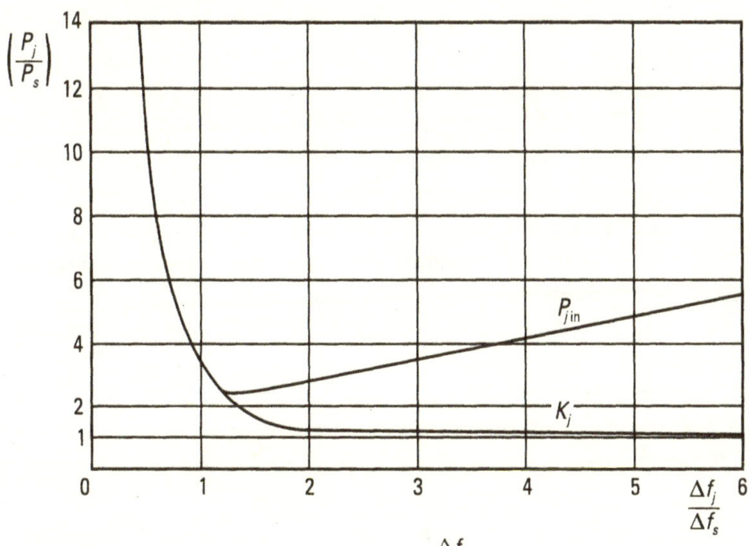

Figure 4.27 The dependencies of $(P_j)_{in}$ and K_j on $\frac{\Delta f_j}{\Delta f_s}$.

The notions mentioned above have a direct relationship to screening jamming of the "angle noise" type as well. In this case the passband of the target being jammed Δf_s should be understood to mean the 3-dB width of the main lobe of the antenna. By analogy to the preceding case of jamming a distance-measuring channel, it is possible to state that sources of angle noise must be distributed in a solid angle exceeding by several times the solid angle corresponding to the main lobe. Only when this condition is satisfied can sufficiently high information (entropy) quality coefficients be assured for angle noises $\eta_H(\Phi, \Theta)$.

Radar jamming of angular coordinates, generated by spot jammers with a high ERPD, but a very small $\eta_H(\Phi, \Theta)$, can potentially turn out to be informationally unstable. The side being jammed is presented with the potential possibility of implementing techniques providing for the information stability of its control system.

As a second example of a jamming with a limited information indicator, let us analyze the variant of jamming a radar using jamming formed by broadband frequency modulation of the carrier wave using noise. The effective modulation index is assumed to be quite large. The receiver band Δf_s, which is the jamming target, occupies only a relatively small part of the frequency band, corresponding to the effective deviation of the carrier frequency σ_f (i.e., $\sigma_f >> \Delta f_s$).

The jamming effect in a receiver takes place when the instantaneous frequency of the jamming is within the limits of the band Δf_s (Figure 4.28(a)). As a result, at the output to the receiver a sequence of radio pulses is formed with a random initial phase and a constant amplitude (Figure 4.28(b)). If the average interval between the jamming pules is much less that the IF filter time constant, then, according to the central limit theorem, there occurs a normalization of the jamming sequence when passing through the inertia narrowband linear system (IF) [17]. The minimal required quantity of summed radio pulses with a constant amplitude and a random initial phase required for normalization can be determined by estimating the degree of proximity of the characteristic functions of a Gaussian random value and the probability density of the sum of random values being analyzed. In the first approximation, five or six pulses are sufficient [18].

The average number of pulses summed $\bar{n}_j$ is determined by the average number of times the instantaneous frequency crosses the specified level f_s and by the duration of the pulse received τ_s (Figure 4.28(b)):

$$\bar{n}_j = 2\overline{N}_1 \tau_s \tag{4.166}$$

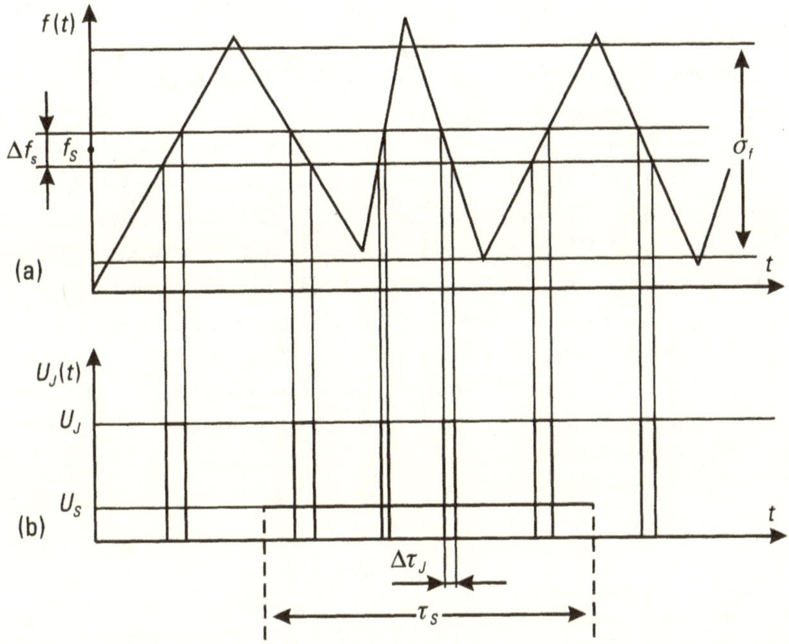

Figure 4.28 (a) The jamming effect in a receiver when the instantaneous frequency of the jamming is within the limits of the band Δf_s. (b) A sequence of radio pulses with a random initial phase and a constant amplitude at the output to the receiver.

Here, $\overline{N}_1$ is the average number of crossings of the specified level for a unit of time when the instantaneous frequency changes with a positive derivative. The doubling of the number of crossings is caused by additional crossings occurring when the instantaneous frequency changes in the opposite direction (with a negative derivative).

The average frequency of crossings of the specified level f_s is determined by the spectrum width of the modulating noise ΔF_{jm}. In the case of Gaussian noise with a flat spectrum [19],

$$\overline{N}_1 = \frac{\Delta F_{jm}}{\sqrt{3}} \exp\left(-\frac{f_s^2}{2\sigma_f^2}\right) \tag{4.167}$$

The characteristic feature of the jamming effect analyzed is the inclusion of an obviously expressed accompanying attribute, namely, the discrete nature of the influence and the significant difference in amplitudes of the jamming and useful signals. A limited degree of randomization in the jamming potentially permits the side being jammed to develop methods of significantly reducing the jamming effect even before it arrives at the victim receiver. One of the variants is a device of the "Dicke fix" type [8, 20]. It comprises a broadband amplifier (BA), a limiter (LIM), and a narrowband amplifier (NA). Normally the narrowband amplifier is an IFA. For this reason, after the limiter a mixer (MIX) and a heterodyne (H) are included (Figure 4.29). The level of limiting jamming pulses U_j approximately corresponds to the assumed level of the useful signal U_s. Such a device permits the reduction of the average power of the jamming effect (the noise variance at the output to the IF) by approximately $\overline{Q}_j$ times, where $\overline{Q}_j$ is the average relative duration of the jamming pulses:

$$\overline{Q}_j = \frac{\overline{T}_j}{\Delta\overline{\tau}_j} \tag{4.168}$$

Here, $\overline{T}_j$ is the average interval between jamming effects:

$$\overline{T}_j = \frac{1}{2\overline{N}_1} = \frac{\sqrt{3}}{2\Delta F_{jm}} \exp\left(\frac{f_C}{2\sigma_f^2}\right) \tag{4.169}$$

where $\Delta\overline{\tau}_j$ is the average duration of jamming pulses:

$$\Delta\overline{\tau}_j = \frac{1}{\Delta f_m} \tag{4.170}$$

Δf_m is the passband of the input broadband filter.

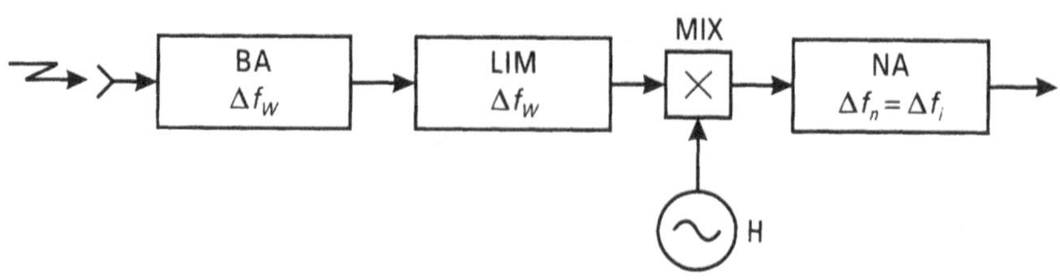

Figure 4.29 Inclusion of a mixer (MIX) and a heterodyne (H) after the limiter (LIM).

The power of the jamming at the output to the IF with a Dicke fix is equal to

$$(P_j)_{\text{out.IF}} = \frac{(P_j)_{in}}{\overline{Q}_j} \tag{4.171}$$

It is obvious that the Dicke fix circuit is effective if the value $\overline{Q}_j$ of is sufficiently large. The example analyzed is yet another confirmation of the thesis stated earlier about the reduction of jamming information stability, if the entropic nature of its parameters and attributes is insufficient.

4.5.2 Definition of the Optimum Limiting Level for Direct Noise Jamming

In real jammers, screening direct noise jamming is amplified a number of times, including in power amplifiers, the dynamic range of which is quite limited. In order to provide a maximum entropy power at the output to an amplifier with a fixed limiting level, it is necessary to select an optimum variance value for the jamming at the input to the limiter. The existence of an optimum (with respect to the criterion of maximum entropy) jamming variance value in an amplifier with fixed limiting directly follows from the analysis of probability densities $p(u)$ of the instantaneous values of the jamming at the output to the limiter shown in Figure 4.30. If the noise variance at the input to the limiter is small (Figure 4.30(a)), then only individual noise bursts significantly exceeding its effective value (σ) are limited.

The uncertainty of the random process is small, since basically it is concentrated in a comparatively narrow region (σ^2 is small). In the case of an excessively large input noise variance, the uncertainty at the output is also small, since the majority of the time the process is concentrated within the bounds of the limitation zone (Figure 4.30(b)). If the input noise variance is optimum, then the jamming is distributed more evenly over then entire limitation zone (Figure 4.30(c)) and the noise entropy at the output to the limiter is the greatest.

The probability density of instantaneous noise values at the output to the symmetric limiter $p_1(u)$ can be written down in the following fashion [19]:

$$p_1(u) = \begin{cases} p_1(u) + 2P(h)\delta(u \pm h), -h \leqslant u \leqslant h \\ 0, \quad u < -h, \quad u > h \end{cases} \tag{4.172}$$

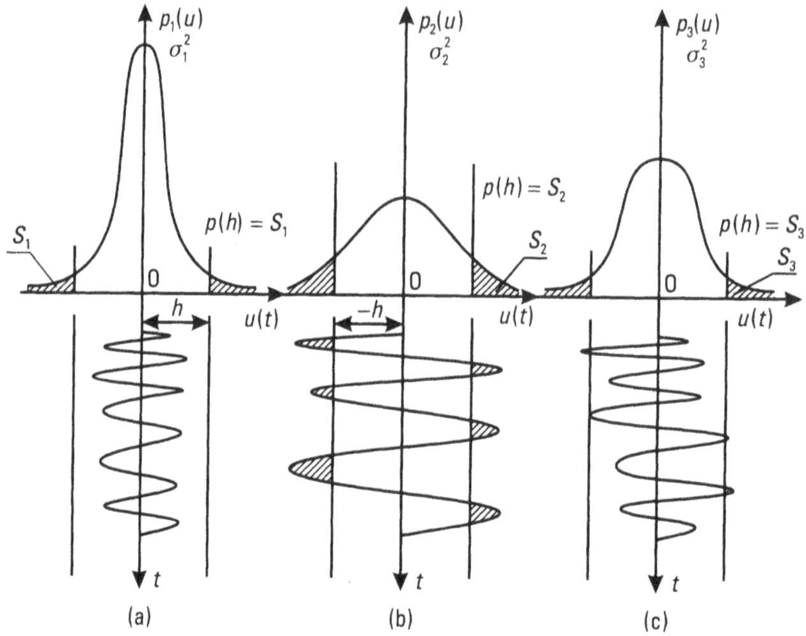

Figure 4.30 Instantaneous values of the jamming at the output to the limiter: (a) small noise variance; (b) an excessively large input noise variance; (c) the noise variance is optimum.

Here, $P(h)$ is the probability that the value u will exceed the limitation level (h or $-h$). It is equal to the corresponding area S (Figure 4.30); and $\delta(u \pm h)$ is the delta function.

The one-dimensional entropy of the random value u_c with a probability density of $p(u)$ (4.172) is determined by combining formulas (3.3) and (3.12) that determine the entropy of discrete and continuous distributions. If, at the input to the limiter, the probability density $p_1(u)$ is normal, then the one-dimensional entropy $H_1(u) = H_1(\sigma, h)$ is determined using the following formula:

$$H_1(\sigma, h) = \int_{-h}^{+h} \frac{1}{\sigma\sqrt{2\pi}} e^{-\frac{u^2}{2\sigma^2}} \ln \frac{1}{\sigma\sqrt{2\pi}} e^{-\frac{u^2}{2\sigma^2}} du - 2P(h) \ln P(h) \qquad (4.173)$$

After transformation, the expression (4.167) is reduced to the following form:

$$H_1(\sigma, h) = 2\Phi_0\left(\frac{h}{\sigma}\right)\ln\sqrt{2\pi e\sigma^2}$$
$$-\left(1-2\Phi_0\left(\frac{h}{\sigma}\right)\right)\ln\left(1-2\Phi_0\left(\frac{h}{\sigma}\right)\right)$$
$$-\frac{1}{\sigma\sqrt{2\pi}}\frac{h}{\sigma}e^{-\frac{h^2}{2\sigma^2}} \tag{4.174}$$

where the Gaussian probability integral $\Phi_0\left(\frac{h}{\sigma}\right)$ is determined by (3.128).

In Figure 4.31 we show the dependency of $H_1(\sigma, h)$ on σ when $h(|h| = 1)$ is fixed. The optimum value is $\sigma_{opt} = 1.43$. This value corresponds to a maximum entropy $H_1(\sigma, h)$ at the output to the limiter.

It is natural that limiting Gaussian noise leads to the reduction of its entropy power and to a corresponding reduction in the information quality coefficient η_H. Assuming that the average power of the jamming at the output to an ideal limiter is equal to the average power of the input signal (the variance of Gaussian noise at the input), using (3.39), (3.56) and Figure 4.31 we find, that in the given instance, the optimum value $\sigma_{opt} = 1.43$

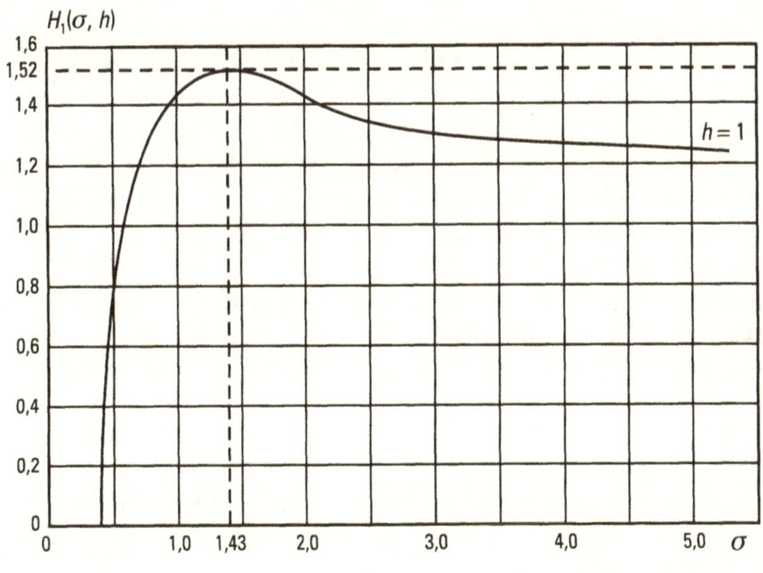

Figure 4.31 The dependency of $H_1(\sigma, h)$ on σ when $h(|h| = 1)$ is fixed.

corresponds to an entropy $H_1 = 1.52$, an entropy power $\overline{P}_j = 1.2$, a jamming power at the output to the limiter $P_j = \sigma^2_{\text{opt}} = 2$, and consequently $\eta_H = 0.6$. Let us note that it is possible in principle to provide a higher jamming quality at the output to the limiter if, for a fixed input noise variance, we increase the limitation threshold $(|h|)$. In particular, if $h \geqslant 1.7\sigma$, then it is practically speaking possible to assume $\eta_H = 1$. At the same time, it is necessary to bear in mind that then the energy possibilities of the output jammer amplifier are not completely utilized. For example, if in the case analyzed above $(|h| = 1)$ we set $\sigma = \frac{1}{1.7}$, then the jamming entropy power will be $\overline{P}_j = 0.35$ instead of 1.2 for the optimum ratio of σ and h.

In optimum mode, although a certain loss of jamming quality does occur, it is compensated for however by an increase in its power. Apparently, in individual specific instances, other conditional extreme co-relationships between σ and h can exist. The co-relationships derived above are optimum on the average.

4.5.3 Use of the Jamming Equation for Analysis of the Electronic Environment

In an information conflict, each of the sides conducts an analysis of the electronic environment as it develops, including the quantitative and qualitative indicators for jamming targets and jamming systems. Specifically, we are speaking of the indicators of individual devices (the number of jammers, their range and ERPDs, the range and ERPD of the victim radar, etc.). This can lead, in certain cases, to false conclusions. For example, comparing the ERPDs of jammers without taking into account the RCS of airplanes, covered by jamming, and the ERPDs of jammed radar, it is possible to assume that the side with greater jammer ERPDs, when compared to the corresponding jammer potentials of the enemy, possesses a superiority in an information conflict. However, in actuality, taking into consideration the RCS of airplanes and the ERPDs of radars, things turn out to be different.

The jamming equation, written in canonical form, permits us to perform a comprehensive analysis of the electronic targets of the sides participating in an information conflict. The reduced ratio of ERPDs $\overline{K}_{\text{REJ}}$, determined for each of the sides in the conflict, can serve as an overall indicator of the capabilities of the sides in a radar information conflict. In accordance with the definition (4.61),

$$\overline{K}_{\text{REJ}} = \frac{P_s G_s}{P_j G_j} \frac{\Delta f_j}{\Delta f_s} \frac{K_j \sigma_s}{4\pi\gamma_j F_j^2(\Phi_s, \Theta_s)}$$

The indicator $\overline{K}_{\text{REJ}}$ permits us to evaluate the degree of influence and quality of parameters related to the jammer $(P_j, G_j, \Delta f_j, F_j(\Phi_s, \Theta_s))$, the jamming target $(P_s, G_s, \Delta f_s)$, the object screened (σ_s) and both items — the jammer and the radar (K_j, γ_j).

The reduced ERPD ratio has a dimensionality of m^2. According to (4.64),

$$\sqrt{\overline{K}_{\text{REJ}}} = D_{s\min} \tag{4.175}$$

that is, the indicator analyzed directly stipulates a minimum radar jamming range when jamming is generated from battle formations.

A feature of the indicator $\overline{K}_{\text{REJ}}$ is the striving of each side in the conflict to make it as small as possible for himself, and significantly large for the enemy. The circumstance stated explains the co-relationship (4.175). Let us suppose that in the conflict two sides A and B are participating. Let us assume that side A has assured himself a lessor value for the indicator $\overline{K}_{\text{REJ}}$ than side B; that is,

$$\overline{K}_{\text{REJ}}(A) < \overline{K}_{\text{REJ}}(B) \tag{4.176}$$

This means, in accordance with (4.175), that

$$D_{s\min}(A) < D_{s\min}(B) \tag{4.177}$$

that is, the less $\overline{K}_{\text{REJ}}(A)$ is, the less $D_{s\min}(A)$ is too, and the deeper the airplanes of side A can potentially penetrate the AAD zone. Side B will also reason in a similar manner.

In practical operations $\overline{K}_{\text{REJ}}$ can be evaluated in range (m^2) and in decibels relative to m^2, using the value of the conditional minimum jamming range $D_{s\min}$ in km or in dBm. For example, $\overline{K}_{\text{REJ}} = 10^6$ m^2; $\overline{K}_{\text{REJ}} = 60$ dBm^2; $D_{s\min} = 10^3$ m $= 1$ km; $D_{s\min} = 30$ dBm.

Independently of the unit of measure selected for the overall indicator for the capabilities of the sides in a radar conflict, it without doubt influences decisions made by the conflicting sides in a most definitive manner.

References

[1] Skolnik, M., editor, *Spravochnik po radiolokatsii* (*Radar Handbook*). Moscow: *Sov. radio* (*Soviet Radio*), Vol. 4, 1976, pp. 193–215.

[2] Chernyy, F .B., *Rasprostranenie radiovoln* (*Propagation of Radio Waves*). Moscow: *Sov. radio* (*Soviet Radio*), 1962, pp. 89–147, 309–326.

[3] Fel'd, Ya. N. and L. S. Benenson, *Antenno-fidernye ustroystva* (*Antenna Feeder Devices*). Moscow: N. E. Zhukovsky Airforce Academy, 1959, pp. 510–520.

[4] Fel'd, Ya. N., and L. S. Benenson, *Antenny santimetrovykh i detsimetrovykh voln* (*Centimeter and Decimeter Wave Antennas*). Moscow: N. E. Zhukovsky Airforce Academy, 1955, pp. 45–108, 167–202.

[5] Dudnik, P. I. and Yu. I. Cheresov, *Aviatsionnye radiolokatsionnye ustroystva* (*Aircraft Radar Devices*). Moscow: N. E. Zhukovsky Airforce Academy, 1986, pp. 25–49, 69–112, 172–358.

[6] Shirman, Ya. D., et al., *Teoreticheskie osnovy radiolokatsii* (*The Theoretical Bases of Radar*). Moscow: *Sov. radio* (*Soviet Radio*), 1970, pp. 33–78, 365–440.

[7] Dulevich, V. E., et al., *Teoreticheskie osnovy radiolokatsii* (*The Theoretical Bases of Radar*). Moscow: *Sov. radio* (*Soviet Radio*), 1978, pp. 24–59.

[8] Maksimov, M. V., et al., *Zashchita ot radiopomekh* (*Safeguarding against Radio Interference*). Moscow: *Sov. radio* (*Soviet Radio*), 1976, pp. 214–234, 382–400. Translation: M. V. Maksimov, et al., *Radar Anti-jamming Techniques*, USA, Artech House, Inc., 1979, 421 pp.

[9] Shirman, Ya. D. and V. N. Manzhos, *Teoriya i tekhnika obrabotki radiolokatsionnoy informatsii na fone pomekh* (*The Theory and Technology of Processing Radar Information against an Interference Background*). Moscow: *Radio i svyaz'* (*Radio and Communications*), 1981, pp. 127–160, 312–370.

[10] Vakin, S. A., and L. N. Shustov, *Osnovy radioprotivodeystviya i radiotekhnicheskoy razvedki* (*The Basics of Radio Countermeasures and Radio Technical Intelligence*). Moscow: *Sov. radio* (*Soviet Radio*), 1968, pp. 9–122, 259–342.

[11] Kondratenkov, G. S., et al., *Radiolokatsionnye stantsii vozdushnoy razvedki* (*Airborne Intelligence Radar Stations*). Moscow: *Voenizdat* (*Military Publishing House*), 1983, pp. 28–101.

[12] Bronshteyn, I., and K. Semendyaev, *Spravochnik po matematike* (*Mathematics Handbook*). Moscow: *Nauka* (*Science*), 1986, pp. 72–73.

[13] Pugachev, V. S., *Teoriya sluchaynykh funktsiy i ee primenenie k zadacham avtomaticheskogo upravleniya* (*The Theory of Random Functions and its Application to Problems in Automatic Control*). Moscow: *Fizmatgiz* (*State Publishing House for Physics and Mathematics*), 1962, pp. 220–280.

[14] Tikhonov, V. I., *Statisticheskaya radiotekhnika* (*Statistical Radio Technology*). Moscow: *Radio i svyaz'* (*Radio and Communications*), 1982, pp. 427–477.

[15] VanTrees, H. L., *Teoriya obnaruzheniya, ocenok i modulyatsii* (*Detection, Estimation and Modulation Theory*). Moscow: *Sov. radio* (*Soviet Radio*), 1972, pp. 203–249.

[16] *Funktsii s dvoynoy ortogonal'nost'yu v radioe'lektronike i optike* (*Dually Orthogonal Functions in Radio Electronics and Optics*). Under the Editorship of M. K. Razmakhnin and V. P. Yakovlev. Moscow: *Sov. radio* (*Soviet Radio*), 1971, pp. 12–39, 232–237.

[17] Tikhonov, V. I., and Yu. N. Bakaev, *Statisticheskaya teoriya radiotekhnicheskikh ustroystv* (*The Statistical Theory of Radio Technical Devices*). Moscow: N.E. Zhukovsky Airforce Academy, 1976, pp. 5–119, 180–220, 263–416.

[18] Gonorovskiy, G. I., *Radiotekhnicheskie tsepi i signaly. Chast' II* (*Radio Technical Circuits and Signals. Part II*). Moscow: *Sov. radio* (*Soviet Radio*), 1967, pp. 181–204, 213–230, 245–319.

[19] Tikhonov, V. I., *Statisticheskaya radiotekhnika* (*Statistical Radio Technology*). Moscow: *Sov. radio* (*Soviet Radio*), 1966, pp. 356–380.

[20] Van Brunt, L. B., *Applied ECM.*, EW Engineering Inc., Dunn Loring, VA, USA, 1978, 973 pp.

5

Passive and Active–Passive Radar Jamming — The Jamming Equation

5.1 Types of Passive Jamming — Chaff

Passive jamming is understood to be signals formed at the input to the victim receivers as a result of the scattering of electromagnetic waves by objects employed in significant quantities. As a rule, the electromagnetic waves scattered are those emitted by victim antennas.

Normally jamming is generated by scatterers such as chaff that is employed on a massive scale. However, as a rule, the chaff cloud does not change the electrical properties of the environment, since the distance between the chaff dipoles in the cloud is many tens or hundreds of times greater than the wavelength.

Therefore, the effect of passive jamming is to form a masking background analogous to noise jamming. At the present time passive jamming is generated basically with antiradar chaff, dispersed in large quantities in the atmosphere.

Chaff is made from paper, glass fiber, or kapron, covered with a conductive layer. For this purpose, it is possible to use metallic foil. The length of the chaff and its thickness are selected to ensure effective scattering of radio waves over the widest range of frequencies possible. As a rule, its length is approximately equal to half the wavelength of the victim radar. However, chaff having lengths significantly exceeding the radar wavelength may be used.

Chaff is normally assembled in packets. Opening up after being launched from an aircraft, such a packet creates a chaff cloud, the reflected signal from which can be seen on the screen of a display such as a PPI in the form of a bright spot. If a fairly large number of packets are launched sequentially, then illuminated corridors of significant length are formed on the PPI.

At the present time, chaff is made of dielectrics or foil. The minimum thickness of the metallic covering is determined by the thickness of the working surface layer, formed due to the skin effect. The depth of penetration of current in the conductive layer depends on the frequency of the electromagnetic waves. In the centimeter band the depth of penetration can be very small ($d \approx 1$ μm). This permits us to make chaff in the form of very thin metalized strips or fibers several tens of microns in diameter. Practically speaking, the diameter used is controlled by the issues of durability and technology of manufacture.

The quantity of chaff in a packet, depending on the band, comprises tens of thousands or millions of units. As a result of the noncoherence of the fields dispersed by individual chaff dipoles, the RCS of a cloud consisting of dipoles of identical length will, on the average, be equal to the sum of the RCS of each chaff dipole; that is,

$$\overline{\sigma}_j = \sum_{i=1}^{N} \overline{\sigma}_i = N\overline{\sigma}_1 \tag{5.1}$$

where $\overline{\sigma}_j$ is the average RCS of the chaff cloud; $\overline{\sigma}_i = \overline{\sigma}_1$ is the average RCS of one chaff dipole; and N is the number of chaff dipoles in a packet.

The formula (5.1) is valid in the ideal case, when literally all chaff dipoles are used effectively. Practically, due to the clumping (sticking together) of chaff dipoles and their breakage, the RCS of the cloud will be less than $\overline{\sigma}_j$, determined using (5.1). Normally the RCS of a chaff cloud is calculated using a formula that takes into consideration the actual number of operational chaff dipoles in a packet:

$$\overline{\sigma}_j = \eta N\overline{\sigma}_1 \tag{5.2}$$

where η is the coefficient of the operational number of chaff dipoles.

The magnitude of the RCS of one half-wave chaff dipole ($\overline{\sigma}_1$) in the general case depends on its orientation relative to the electrical vector of the incident wave. As a result of atmospheric turbulence and aerodynamic properties, chaff dipoles are positioned in the cloud, as a rule, in a random fashion relative to one another. Moreover, to ensure a uniform distribution of orientations, an attempt is made in manufacture to shift the center of gravity of each dipole by a random interval from its center.

Therefore, the RCS of the entire cloud ($\overline{\sigma}_j$) is determined using the average value of the RCS of a single chaff dipole ($\overline{\sigma}_1$), positioned randomly in space. The value of ($\overline{\sigma}_1$) will be found below. Half-wave chaff is the most effective.

5.1.1 The Radar Cross Section of a Half-Wave Chaff Dipole Randomly Positioned in Space

According to definition (4.52) the RCS of a chaff dipole is equal to

$$\sigma_1 = S_1 G_1 \tag{5.3}$$

where $S_1 = P_2/p$ is the ratio of the power re-radiated by the chaff dipole (P_2) to the power density (p), of the plane wave incident on the chaff dipole; and G_1 is the orientation coefficient of the chaff dipole.

For a chaff dipole, positioned at an angle Θ with respect to the electrical vector $\overline{E}$ of the incident wave (Figure 5.1), the re-radiated power P_2 is equal to

$$P_2 = P_2 \cos^2 \Theta \tag{5.4}$$

where P_{20} is the power emitted by the chaff dipole in the equatorial plane (when $\Theta = 0$).

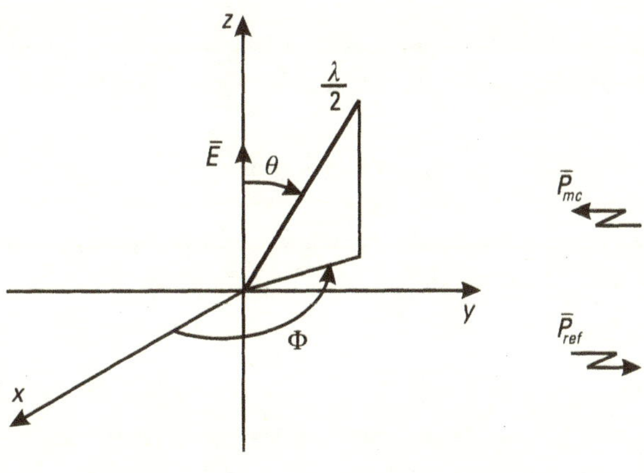

Figure 5.1 The re-radiated power P_2 for a chaff dipole positioned at an angle Θ with respect to the electrical vector $\overline{E}$.

As is known [1], the power value P_{20} can be found using the formula:

$$P_{20} = \frac{1}{2} i^2 R_\Sigma \tag{5.5}$$

where i is the current antinode amplitude; and R_Σ is the radiation impedance of the chaff dipole.

For a half-wave chaff dipole,

$$i = h_d \frac{E}{R_\Sigma} \tag{5.6}$$

where E is the amplitude of the electrical field of the plane wave received; $R_\Sigma = 73.3$ ohm is the radiation impedance of the half-wave chaff dipole; and $h_d = \lambda/\pi$ is the effective length of the half-wave chaff dipole.

From the co-relationships given above, we find the power, re-radiated by the half-wave chaff dipole:

$$P_2 = \frac{1}{2}\frac{\lambda^2}{\pi^2}\frac{E^2}{73.3}\cos^2\Theta \tag{5.7}$$

The power density of the incident wave (the absolute value of the Umov–Poynting vector) is determined using the formula:

$$p = \frac{E^2}{240\pi} \tag{5.8}$$

Thus, from (5.3), (5.7) and (5.8), taking into consideration that for a half-wave chaff dipole $G = 1.65\cos^2\Theta$, we finally obtain:

$$\sigma_1 = 0.86\lambda^2\cos^4\Theta \tag{5.9}$$

When polarizations of the chaff dipole and the incident wave correspond, the RCS of the half-wave chaff dipole will be maximum:

$$\sigma_{1\,\max} = 0.86\lambda^2 \tag{5.10}$$

In the general case, Θ is a random value. With accuracy sufficient for practical purposes and bearing in mind the considerations mentioned in the preceding paragraph, it is possible to consider the random value of Θ to be distributed with a uniform probability density within the limits of the entire solid angle Ω for all values of angle φ (Figure 5.1). This permits us to assume

$$\sigma_1(\Theta) = \sigma_1(\Omega) = \sigma_{1\,\max} \cos^4 \Theta \tag{5.11}$$

To determine the average RCS value of a chaff dipole $\overline{\sigma}_1$, it is necessary to find the parameters of the distribution law for a random value Ω.

The supposition of equiprobable positioning of chaff dipoles means that within the limits of any elementary solid angle $d\Omega_i$ (Figure 5.2), the number of chaff dipoles is approximately identical. The elementary solid angle $d\Omega_i$ can occupy any position with an equal probability within the limits of the entire solid angle 4π. Therefore, the probability density is equal to

$$p(\Omega) = \frac{1}{4\pi} \tag{5.12}$$

In the material that follows, we will consider the RCS of a chaff cloud to be independent of the co-relationship of the polarization of the receiving and transmitting antennas (i.e., we will consider their polarizations to be

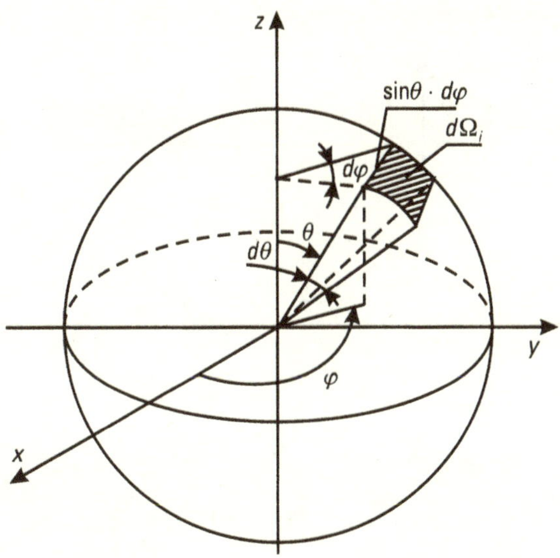

Figure 5.2 The supposition of equiprobable positioning of chaff dipoles.

identical). In the general case, it is necessary to consider any difference in polarizations.

The probability that a chaff dipole will be located within the limits of elementary solid angle $d\Omega$ is equal to

$$p(\Omega) = \frac{d\Omega}{4\pi} \tag{5.13}$$

In order to find the average RCS value of a chaff dipole (the mathematical expectation of $\overline{\sigma}_1$), it is necessary to average the value of σ_1, as determined using formula (5.9), in the space throughout the entire solid angle $\Omega = 4\pi$:

$$\sigma_1 = \int_{\Omega} \sigma_1(\Omega) p(\Omega) d\Omega = \int_{\Omega} \sigma_1(\Omega) \frac{d\Omega}{4\pi} \tag{5.14}$$

$$\sigma_1(\Omega) = \sigma_{1\,\max} \cos^4 \Theta$$

In a spherical system of coordinates, the surface element of the sphere with a single radius is equal to the element of the solid angle:

$$d\Omega = dS = \sin \Theta d\varphi d\Theta \tag{5.15}$$

Integrating (5.14), we obtain

$$\overline{\sigma}_1 = \int_0^{2\pi} d\varphi \int_0^{\pi} \sigma_{1\,\max} \cos^4 \Theta \frac{1}{4\pi} \sin \Theta d\Theta \tag{5.16}$$

whence

$$\overline{\sigma}_1 = \frac{\sigma_{1\,\max}}{5} = 0.17\lambda^2 \tag{5.17}$$

Thus, the average RCS of a chaff packet $\overline{\sigma}_j$ is equal to

$$\overline{\sigma}_j = N_{je}\overline{\sigma}_1 = 0.17\lambda^2 N_{je} \tag{5.18}$$

where $N_{je} = \eta N$ is the number of effectively operating chaff dipoles in the packet.

Above we derived formula (5.9) for determining the RCS of a chaff dipole, randomly positioned relative to the direction of the electrical vector of the incident wave. In the process of deriving formula (5.9) it was supposed that the points of emission and receiving are identical (monostatic radar), and the angle Θ between the electrical vector and the chaff dipole is a random value.

In other (bistatic radar) cases, it is important to know the RCS value in a direction not corresponding to the direction towards the emitter. Below we derive the formula for that case.

Let the angle, from which the signal reflected from a chaff element is received, be equal to ψ relative to the direction toward the emission source (Figure 5.3). We will designate by Θ the angle between the chaff dipole and the electrical vector of the incident field.

If $\sigma_{1\,max} = 0.86\lambda^2$ is the RCS of the chaff dipole when $\psi = 0$ and $\Theta = 0$, then for $\Theta \neq 0$ and $\psi \neq 0$,

$$\sigma(\Theta, \psi) = \sigma_{1\,max} \cos^2 \Theta \cos^2(\Theta + \psi) \tag{5.19}$$

The RCS value of interest to us for a single chaff dipole is determined as the mathematical expectation $\overline{\sigma(\Theta, \psi)}$; that is,

$$\sigma_\psi = \sigma(\Theta, \psi)p(\Omega)d\Omega \tag{5.20}$$

Integration is performed over the entire solid angle $\Omega = 4\pi$.

Replacing $\sigma(\Theta, \psi)$ and $p(\Omega)$ by their values from (5.19) and (5.12), as well as keeping in mind the well-known expression for the differential of a solid angle with a single radius $d\Omega = \sin\Theta d\Theta d\varphi$, we obtain

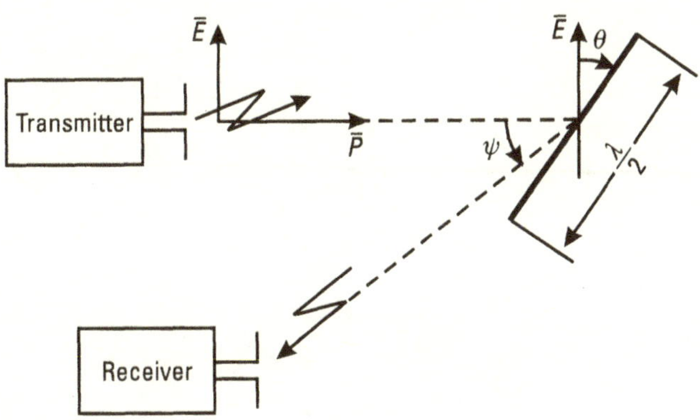

Figure 5.3 The case of bistatic radar.

$$\sigma_\psi = \overline{\sigma(\Theta, \psi)} = \int\limits_\Omega \sigma(\Theta, \psi) p(\Omega) d\Omega$$

$$= \frac{\sigma_{1\,max}}{4\pi} \int\limits_0^{2\pi} \int\limits_0^{\pi} \cos^2 \Theta \cos^2(\Theta + \psi) \sin \Theta d\varphi d\Theta d\varphi \qquad (5.21)$$

Thus, integrating (5.21), we finally obtain [2]

$$\sigma_\psi = \frac{\sigma_{1\,max}}{5} \cos^2 \psi + \frac{2}{15} \sigma_{1\,max} \sin^2 \psi \qquad (5.22)$$

or

$$\sigma_\psi = 0.17\lambda^2 \cos^2 \psi + 0.11\lambda^2 \sin^2 \psi \qquad (5.23)$$

When $\psi = 0$ we obtain the generally known formula (5.17).

The expression (5.23) permits us to come to the conclusion that the maximum scattering power corresponds to the angles $\psi = 0$ and $\psi = \pi$, and the minimum to the angles $\psi = \pi/2$ and $\psi = 3\pi/2$. The value σ_ψ in the direction of the minimum is approximately $0.65\sigma_{1\,max}$ (Figure 5.4).

The formulas derived are valid for an ideally conducting half-wave dipole. Real half-wave chaff as a result of finite conductivity and thickness has a broader bandwidth than an ideal half-wave dipole. An increase in the length of a chaff dipole to values greater than half a wave leads to a reduction

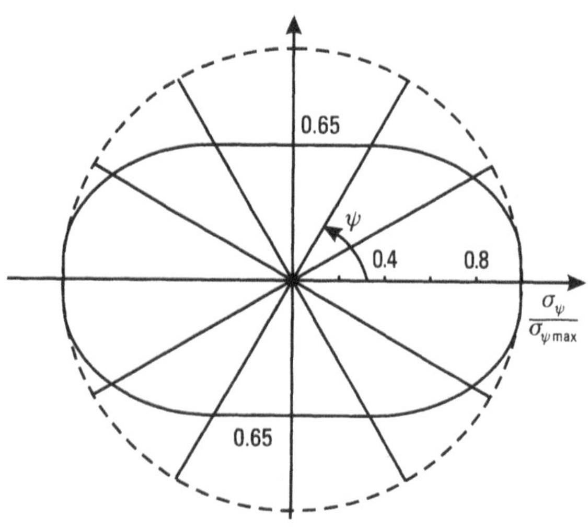

Figure 5.4 Maximum scattering power.

in RCS. However, for chaff length values that are multiples of the half-wavelength, its RCS once again increases and can be somewhat greater than that of a half-wave dipole (Figure 5.5). The latter does not mean that chaff, destined for the jamming of stations with longer waves, will also be effective against radar operating on shorter waves. The problem is that the quantity of chaff dipoles in the packets for longer waves decreases, since the weight of the packet is fixed. Accordingly, the RCS of a packet decreases for shorter waves.

The qualitative dependency of the RCS of chaff on its relative length is shown in Figure 5.5, where the ratio of the chaff dipole length to half the wavelength is offset along the axis of the abscissas.

5.1.2 The RCS of a Hertzian Dipole

The RCS of chaff with a length l that is significantly less than the wavelength λ (a Hertzian dipole), is tens of times less than the RCS of half-wave chaff. Quantitatively the RCS of a Hertzian dipole (σ_1) can be determined using the scheme analyzed earlier (cf. (5.3), (5.5), (5.8)), with an appropriate replacement of σ_1 with σ_l, h_d with l, and G_1 with G_l:

$$\sigma_l = \frac{P_2}{p} G_l \qquad P_2 = \frac{1}{2} R_\Sigma i^2$$

$$R_\Sigma = 80\pi^2 \left(\frac{l}{\lambda}\right)^2 \qquad i = \frac{El}{|z|}$$

$$|z| = W \cot \frac{2\pi}{\lambda} l \qquad p = \frac{E^2}{240\pi}$$

$$|z| >> R_\Sigma$$

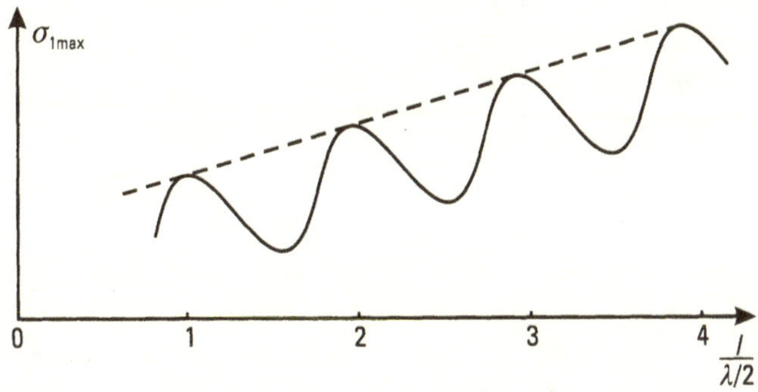

Figure 5.5 Qualitative dependency of the chaff RCS on its relative length.

Here, E is the intensity of the electrical field incident on the dipole. The vector of the incident field is assumed to be co-linear with the dipole; z is the complex input impedance of the dipole; and W is the wave impedance of the dipole.

According to conditions:

$$\frac{2\pi}{\lambda} l << l \quad \tan\frac{2\pi}{\lambda} l = \frac{2\pi}{\lambda} l \quad |z| = W\frac{\lambda}{2\pi l}$$

$$i^2 = \frac{4\pi^2 E^2}{W^2}\left(\frac{l^2}{\lambda}\right)^2$$

$$P_2 = \frac{1}{2} 80\pi^2 \left(\frac{l}{\lambda}\right)^2 \frac{4\pi^2 E^2}{W^2}\left(\frac{l^2}{\lambda}\right)^2$$

$$\frac{P_2}{p} = \frac{240\pi 80\pi^2 2\pi^2}{W^2}\frac{l^6}{\lambda^4}$$

For a Hertzian dipole $W = 1000$ ohm [2], and the orientation coefficient $G_l = 1.5$. Accordingly, the maximum RCS value for a Hertzian dipole is determined using the expression:

$$\sigma_l = 0.57\pi^3 \frac{l^6}{\lambda_4} \tag{5.24}$$

which corresponds to the results obtained by Rayleigh when researching the scattering of waves by small bodies.

In order to estimate the order of σ_l, let us assume that $l/\lambda = 0.1$. Then, $\sigma_l = 1.8 \times 10^{-3}\ l^2$. If $l = 0.01$m, then $\sigma_l = 1.8 \times 10^{-7}$m^2. The RCS of a half-wave chaff dipole at wavelength $\lambda = 0.1$m is equal to $\sigma_1 = 0.86\lambda^2 = 0.0086$ m^2:

$$\frac{\sigma_1}{\sigma_l} = \frac{8.6 \times 10^{-3}}{1.8 \times 10^{-7}} = 4.8 \times 10^4$$

Therefore, a half-wave chaff dipole with $l = 5$ cm and a wavelength of $\lambda = 10$ cm has an RCS σ_1 approximately 50,000 times larger than a chaff dipole 1 cm in length.

5.2 Formation Dynamics and Statistical Characteristics of Chaff Clouds

5.2.1 Space and Time Parameters of a Chaff Cloud

In order to determine the information damage caused by passive jamming, it is necessary to know the RCS of that part of the chaff cloud entering the resolution element of the victim radar at a given moment in time. The latter requires a knowledge of the dynamics of chaff cloud development after the opening of the packet launched by the jammer.

The mass of a chaff dipole and its aerodynamic characteristics ensure free movement under the influence of air currents. Practically speaking, the majority of operating chaff dipoles in the packet wander randomly according to the laws of turbulent diffusion in the atmosphere. Quantitative evaluations of the first approximation, permitting the approximate evaluation chaff cloud parameters, can be obtained by considering the simplest problem in the theory of one-dimensional walks [4, 5].

In the case analyzed, the diffusing inertialess dipole is a chaff dipole, performing random walks (fluctuations) under the influence of forces caused by the turbulent diffusion of the atmosphere. It is assumed that every time interval τ, the chaff dipole performs a discrete movement of distance h to the right or left of the axis OX. Movement begins from the origin of the coordinates ($OX = 0$). The movements are performed according to a random law to the right with probability p, and to the left with probability q.

The parameters τ, h, p and q are assumed to be constant during the observation time:

$$p + q = 1 \tag{5.25}$$

Let us determine the probability that after n steps a the moment of time $t = n\tau$ the chaff dipole has moved away from the origin of the coordinates by a distance of x. The latter occurs if during time t the chaff dipole moves to the right m times and to the left $(n - m)$ times. Accordingly,

$$x = mh - (n - m)h = (2m - n)h \tag{5.26}$$

The probability sought $P_n(m)$ is determined using the Bernoulli formula [6, 7]:

$$P_n = C_n^m P^m (1 - P)^{n-m} \tag{5.27}$$

where

$$C_n^m = \frac{n!}{m!(n-m)!}$$

At the limit when $\tau \to 0$ and $h \to 0$ for probability $P_n(m)$, it is possible to convert to probability $P(x, t)$. The walk process is Markovian. Based on the laws of Markovian processes, we determine the probability $P(x, t+\tau)$ that the chaff dipole at the moment of time $t+\tau$ will be at the point with coordinate x. According to the Markovian Theorem, which in the given case is reduced to the formula for full probability, we obtain

$$P(x, t+\tau) = P(x-h, t)p + (x+h, t)q \tag{5.28}$$

$$P(0,0) = 1 \quad P(x,0) = 0$$

Using the equality (5.25), the expression (5.28) can be written in the following way:

$$\begin{aligned} P(x, t+\tau) - P(x, t) = {} & (P(x-h, t) - P(x, t))p \\ & + (P(x+h, t) - P(x, t))q \end{aligned} \tag{5.29}$$

For sufficiently small τ and h in (5.29), the probability differences are represented in the form of the first terms of the corresponding series:

$$P(x, t+\tau) - P(x, t) = \frac{\partial P(x, t)}{\partial t}\tau + O(\tau)$$

$$P(x-h, t) - P(x, t) = \frac{\partial P(x, t)}{\partial x}h + \frac{1}{2}\frac{\partial^2 P(x, t)}{\partial x^2}h^2 + O(h^2)$$

$$P(x+h, t) - P(x, t) = \frac{\partial P(x, t)}{\partial x}h + \frac{1}{2}\frac{\partial^2 P(x, t)}{\partial x^2}h^2 + O(h^2)$$

where $O(\tau)$ and $O(h^2)$ are the residuals of the corresponding orders of smallness. Substituting the expressions obtained for the probability differences into (5.28) and omitting the sufficiently small values of $O(\tau)$ and $O(h^2)$, we obtain

$$\frac{\partial P(x, t)}{\partial t}\tau = -(p-q)h\frac{\partial P(x, t)}{\partial x} + \frac{1}{2}h^2\frac{\partial^2 P(x, t)}{\partial x^2} \tag{5.30}$$

Converting to the limit $\tau \to 0$ and $h \to 0$, we introduce limitations not permitting conversion of τ and h, arbitrary and independent of one another, to the limit. The limitations assume the distribution speeds of fluctuations and their dispersions are finite. Specifically, this comes down to the existence of finite limits:

$$\lim_{\tau \to 0} \frac{(p-q)h}{\tau} = A_x \qquad \lim_{\tau \to 0} \frac{h^2}{\tau} = B_x \tag{5.31}$$

With consideration to the limitations made, (5.30) is converted to the form:

$$\frac{\partial P(x,t)}{\partial t}\tau = -A_x \frac{\partial P(x,t)}{\partial x} + \frac{1}{2} B_x \frac{\partial^2 P(x,t)}{\partial x^2} \tag{5.32}$$

The solution of the equation must satisfy the following initial condition:

$$P(0,0) = 1 \quad P(x,0) = 0 \tag{5.33}$$

Equation (5.32) is a classical diffusion equation. It was obtained at the beginning of the century by the physicists Fokker and Planck. It was derived by A. N. Kolmogorov in the general form, and quite rigorously. The most widely used name for it is the Fokker–Planck–Kolmogorov (FPK) equation.

If the function $P(x,t)$ is differentiable, then the FPK equation can also be written for the probability density $p(x,t)$:

$$\frac{\partial p}{\partial t} = -A_x \frac{\partial p}{\partial x} + \frac{1}{2} B_x \frac{\partial^2 p}{\partial x^2} \tag{5.34}$$

The initial condition:

$$p(x, t_0 | x_0, t_0) = \delta(x - x_0) \tag{5.35}$$

where $\delta(x - x_0)$ $\delta(x - x_0)$ is the delta function. When $x_0 = 0$ and $t_0 = 0$, which is the case in this instance, we obtain $p(x, 0|0, 0) = p(x, 0) = \delta(x)$.

The solution of (5.34) can be obtained by using the method for separation of variables [8]. If the coefficients for drift A_x and diffusion B_x are constant, the solution of (5.34), given initial conditions (5.35), is written down in the following form:

$$p(x,t) = \frac{1}{\sqrt{2\pi B_x t}} \exp\left(-\frac{(x - A_x t)^2}{2B_x t}\right) \tag{5.36}$$

In the expression derived, $A_x t$ is equal to the average value of the chaff dipole displacement from the origin of the coordinates ($x = 0$) during the time t, which in the materials that follow is designated by x_0:

$$A_x t = x_0 \tag{5.37}$$

The mean square deviation of the chaff dipole to the axis x is defined by the equality:

$$\sigma_x = \sqrt{B_x t} \tag{5.38}$$

The averaging sign over σ_x, x_0 and other values in this case and in the materials that follow is omitted.

The characteristic special feature of the probability density $p(x,t)$ is its unsteadiness. The parameters σ_x and x_0 depend on t.

In Figure 5.6 we show the dependence of $p(x,t)$ for two moments of

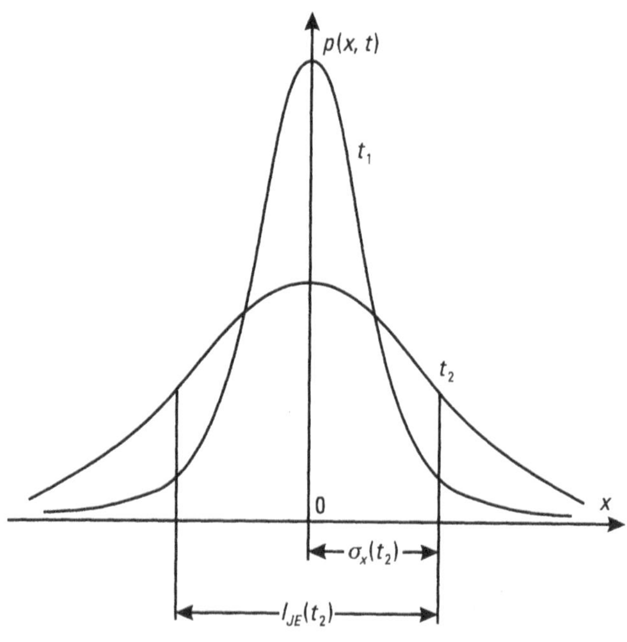

Figure 5.6 The dependence of $p(x,t)$ on ASD of σ_{x1} and σ_{x2}; effective chaff bandwidth $l_{JE,x(t_2)}$.

time t_1 and t_2 ($t_1 < t_2$), corresponding to the ASD of σ_{x1} and σ_{x2}. It is assumed that $A_x = x_0 = 0$. Here too is shown the effective chaff band-width $l_{JE,x}(t_2)$, defined by the standard deviation $\sigma_x(t_2)$ and equal to

$$l_{JE,x}(t_2) = 2\sigma_x(t_2) \tag{5.39}$$

In the future, we will designate the chaff distribution variance in a turbulent environment by $\sigma_x^2(\sigma_y^2, \sigma_x^2)$:

$$B_x = \sigma_x^2 \tag{5.40}$$

Assuming the independence of turbulent atmospheric diffusion along axes OX, OY and OZ, which is totally admissible in the analysis of this problem, it is possible to write down the FPK equation for probability densities $p(y, t)$ and $P(z, t)$, with the corresponding initial conditions, and obtain solutions analogous to (5.36). As is known, processes of turbulent atmospheric diffusion differ from the classical case described by the FPK equation with constant coefficients.

Depending on the variance value σ^2 for chaff distribution in a turbulent environment, the turbulent dispersion coefficient B changes in various ways over time, where

$$B = \frac{\partial}{\partial t}\sigma^2$$

A. A. Zagorodnikov's research [3, 9] showed that the distribution of chaff in a turbulent atmosphere, when the effective value of the cloud width is $l_{JE} \leqslant 50$m, is determined by the turbulent diffusion coefficient B, which depends on the time according to the quadratic law (the Kolmogorov–Obukhov Law of Turbulent Diffusion). Furthermore, the ASD of σ_x depends on the time to the 3/2 power:

$$\sigma_x = \eta_x t^{3/2}$$

where η_x is the proportionality coefficient.

If, however, $50\text{m} \leqslant l_{JE} \leqslant 1000\text{m}$, then the turbulent diffusion coefficient varies proportionately with time

$$B = \eta_1 t \text{ and } \sigma_x = \eta_2 t$$

Accordingly, for large values of effective cloud width ($l_{JE} > 1000$m), normal diffusion occurs $B = \text{const}$, $\sigma = \eta\sqrt{t}$.

In practice, the second case presents the greatest interest $25\text{m} \leqslant \sigma \leqslant 500\text{m}$. For the conditions indicated, the solution of the turbulent diffusion equation leads to the following formula for the cubic probability density of $p(v, t)$:

$$p(v,t) = p(x,y,z,t) = \frac{1}{(\sqrt{2\pi})^3 \sigma_{x0}\sigma_{y0}\sigma_{z0}t^3}$$
$$\exp\left[-\frac{1}{2t^2}\left(\frac{(x-x_0)^2}{\sigma_{x0}^2} + \frac{(y-y_0)^2}{\sigma_{y0}^2} + \frac{(z-z_0)^2}{\sigma_{z0}^2}\right)\right] \tag{5.41}$$

Here, $x_0 = A_x t$; $y_0 = A_y t$; $z_0 = A_z t$; C_x, C_y, C_z are the air movement velocities (current velocities) along axes x, y, z; $\sigma_{x0} = 0.0168A'_x$; $\sigma_{y0} = 0.0168A'_y$; $\sigma_{z0} = 0.0168A'_z$; and A'_x, A'_y, A'_z are the averaged wind speeds along axes x, y, z.

5.2.2 The Radar Cross Section Density of a Chaff Cloud

The probability density $p(v, t)$ permits us also to determine the average volume density (concentration) of operating chaff dipoles $n(v, t)$, and therefore, the averaged RCS density (i.e., related to a unit of volume σ_v):

$$n(v,t) = N_\Sigma p(v,t) \text{ chaff dipoles/m}^3 \tag{5.42}$$

where N_Σ is the average number of operating chaff dipoles, launched at the same time at a given local area point in space:

$$N_\Sigma = n_j N_1$$

n_j is the number of launched chaff packets, and N_1 is the average number of chaff dipoles in each packet.

The RCS density of chaff at the point with coordinates x, y, z is equal to $\sigma_v = n(v)\sigma_1$, or

$$\sigma_v(t) = n_j \sigma_j p(v,t) \text{ m}^2/\text{m}^3 \tag{5.43}$$

Of practical interest is the RCS density for a given moment of time t σ_v, determined within the limits of the central part of the cloud, within the bounds of a parallelepiped with sides $l_{\text{JE}x}$, $l_{\text{JE}y}$, $l_{\text{JE}z}$:

$$\sigma_v(t) = \frac{n_j \sigma_j}{l_{\text{JE}x}(t) l_{\text{JE}y}(t) l_{\text{JE}z}(t)} \text{ m}^2/\text{m}^3 \tag{5.44}$$

Here, $l_{\mathrm{JE}x}(t) = \sigma_{x0}(t)$, $l_{\mathrm{JE}y}(t) = \sigma_{y0}(t)$, $l_{\mathrm{JE}z}(t) = \sigma_{z0}(t)$.

In many cases the effective width along l_y and l_z is much less than the angular sizes of the radar resolution element (l_φ and l_Θ) (Figures 5.7 and 5.8). If $l_\varphi << l_{\mathrm{JE}y}$ and $l_\Theta << l_{\mathrm{JE}z}$, where $l_\varphi = D\varphi_{0.5}$ and $l_\Theta = D\Theta_{0.5}$, then instead of the volume RCS density σ_v, it is possible to speak of a linear RCS density σ_l, which is understood to be

$$\sigma_v = \sigma_l = \frac{n_j \sigma_j}{l_{\mathrm{JE}l}(t)} \tag{5.45}$$

If the launching of chaff packets is performed by the jammer with an interval of τ_j and it is possible to consider the chaff density along the jamming band to be constant for a limited time, then

$$\sigma_v = \sigma_l = \frac{n_j \sigma_j}{V_{\mathrm{JS}} \tau_j} \tag{5.46}$$

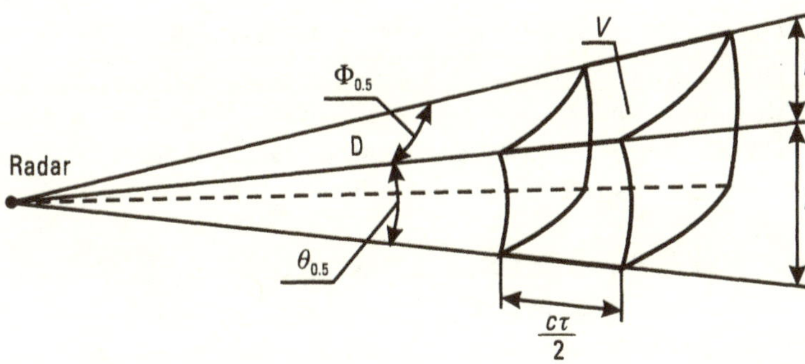

Figure 5.7 Effective width along l_y and l_z.

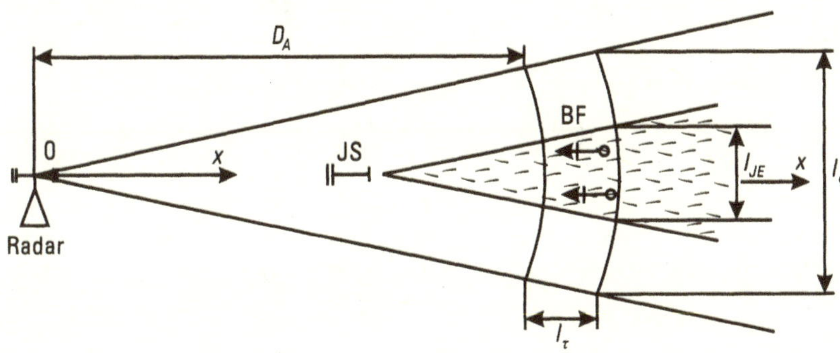

Figure 5.8 The averaged velocity of chaff dipoles in the direction *OX*.

Here, V_{JS} is the jammer velocity; and n_j is the quantity of chaff packets launched at one time.

The jamming signal affecting the victim radar receiver is formed as a result of reflections from a part of the chaff dipoles irradiated, which enter the resolution element (the pulse volume). In the material that follows, the radar cross section caused by the chaff located in the resolution element (the pulse volume) will be denoted by σ_{j0} and σ_{pr}.

5.2.3 The Fluctuation Spectrum of Signals Reflected by Chaff

Atmospheric turbulence, chaff aerodynamics and other factors influence the speed with which a chaff cloud disperses in space. As a result of this, the amplitude of the resulting reflected signal changes randomly in time and, besides, a widening of the frequency spectrum of the total reflected signal occurs. The widening of the spectrum of the total reflected signal takes place as a result of the Doppler components of the spectrum caused by the following reasons:

- The diffusion per se of the chaff in the atmosphere;
- The diffusion in space under the influence of atmospheric turbulence;
- Movement under the influence of wind;
- The descent of chaff under the influence of the force of gravity;
- Diffusion in spaces under the influence of airplane jet turbulence.

Additional reasons for the widening of the spectrum of fluctuations of the reflected signal amplitude are: chaff rotation per se, nonuniformity of the radar antenna radiation pattern, etc.

The normalized function of the spectral density for amplitude fluctuation of the reflected signal due to the movement of chaff dipoles, according to experimental research, can be represented in the form of a Gaussian curve [9, 10]:

$$S(F) = \exp\left(-0.7\left(\frac{F}{F_{0.5}}\right)^2\right) \tag{5.47}$$

where F is the frequency; $F_{0.5}$ is the half-power (−3-dB) spectrum width of the envelope:

$$F_{0.5} = \frac{4\overline{V}_x}{\lambda} \tag{5.48}$$

$\overline{V}_x$ is the averaged velocity of chaff dipoles in the direction OX (Figure 5.8); and λ radar wavelength; $\overline{V_x^2} = \overline{(V_x^2)_a} + \overline{(V_x^2)_b} + \overline{(V_x^2)_c}$ V_x is the velocity component of the chaff dipole along radar axis OX (Figure 5.9); $(V_x)_a$ is the velocity component of the descent of chaff dipoles due to their own weight; $(V_x)_b$ is the velocity component of chaff dipole movement due to the influence of the wind; $(V_x)_c$ is the velocity component of chaff dipoles from movement due to the effect of atmospheric turbulence.

From (5.47), we find the half-power spectrum width for the envelope (at the output to the amplitude detector):

$$F_{0.5} = \frac{4}{\lambda}\sqrt{\overline{V_x^2}} = \sqrt{F_a^2 + F_b^2 + F_c^2}$$

where

$$F_a^2 = \frac{16}{\lambda^2}\overline{(V_x^2)_a}$$
$$F_b^2 = \frac{16}{\lambda^2}\overline{(V_x^2)_b}$$
$$F_c^2 = \frac{16}{\lambda^2}\overline{(V_x^2)_c}$$

Spectrum widening F_a is caused by the presence in the cloud of chaff dipoles, descending at various speeds. As an example, in Figure 5.10 we show the probability density of the velocity of descending dipoles. The values, given on the axis of ordinates, show the relative quantity of cloud chaff dipoles having descent velocity V. The drawing shows that there are two stable groups of chaff dipoles: "slow" and "fast." The presence of "slow" chaff dipoles is explained by the fact that they tend to be oriented

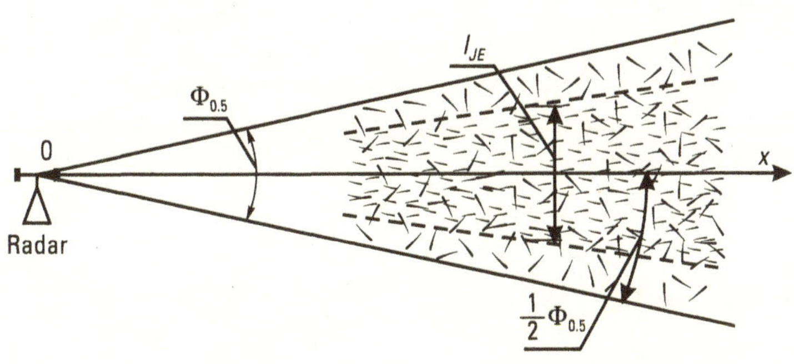

Figure 5.9 The velocity component of the chaff dipole along radar axis OX.

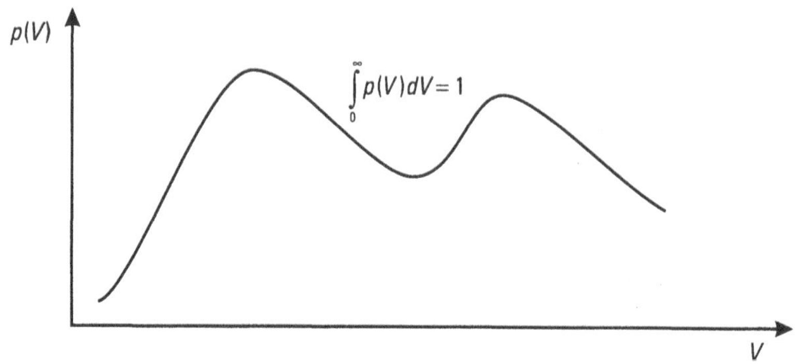

Figure 5.10 The probability density of the velocity of descending dipoles.

horizontally. A part of the chaff dipoles due to small nicks, deformations etc. are similar to small aerodynamic rudders, which provide for a stable descent with a primarily vertical orientation. These chaff dipoles form the "fast" group. The nature of movement of the "slow" and "fast" chaff dipoles is shown in Figure 5.11. Experimental data indicates a primarily horizontal distribution [11].

The spectrum component F_a for the spread of descent velocities is maximum when the cloud size is small compared to the sizes of the radar pulse volume. As the cloud size grows, this component gradually becomes

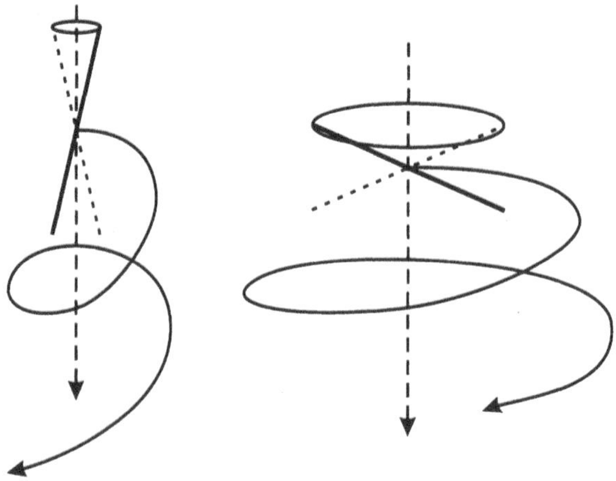

Figure 5.11 The nature of movement of "slow" and "fast" chaff dipoles.

smaller to a certain constant value, the magnitude of which increases together with atmospheric turbulence. The value of the component F_a to a great extent depends on the elevation angle of the cloud. For radar, operating with small elevation angles, it can be ignored.

The spectrum component F_b for the movement of chaff dipoles due to the influence of the wind is proportional to the spread of wind velocities along the vertical (the wind shear) and grows as the cloud progressively increases in size. The average wind velocity, carrying the cloud as a whole, causes a shift in the spectrum and only an insignificant widening of it. For small clouds, the component F_b does not depend on the size of the cloud. This component can reach a significant magnitude when jamming a radar, if the chaff cloud has a large vertical size within the limits of $\Theta_{0.5}$.

The component due to atmospheric turbulence practically speaking does not depend either on velocity or on the angular coordinates of the cloud, but is determined basically by the meteorological parameters of the atmosphere. It is possible to assume approximately:

$$F_{0.5w} = \frac{\overline{V}_w}{4\lambda} \tag{5.49}$$

where $\overline{V}_w$ is the averaged wind velocity, and not the velocity of the chaff dipoles; and λ is the wavelength of the victim radar.

The meteorological parameters of the atmosphere exert a significant influence on the spectrum width of reflected signals:

- The average wind velocity;
- The vertical gradient of the wind velocity (wind shear);
- The average temperature of the air layer;
- The vertical gradient of temperatures and atmospheric turbulence.

These meteorological parameters change with altitude. Therefore, the spectrum width of reflected signals for a given distribution of meteorological parameters depends on the altitude. In Figure 5.12 we show the dependency of the normalized spectral density for fluctuations of signals reflected by a chaff cloud on the product $F\lambda$ (F is the fluctuation frequency in Hertz, and λ is the wavelength in centimeters) [10]. The dependency has been experimentally recorded for several fixed meteorological conditions.

Measurements were made for two radar wavelength values $\lambda = 3.2$ cm and $\lambda = 9.2$ cm. The half-power spectrum width of the jamming signal envelope is $F_{0.5} = 27$ Hz ($\lambda = 3.2$ cm) and $F_{0.5} = 8.5$ Hz ($\lambda = 9.2$ cm). The

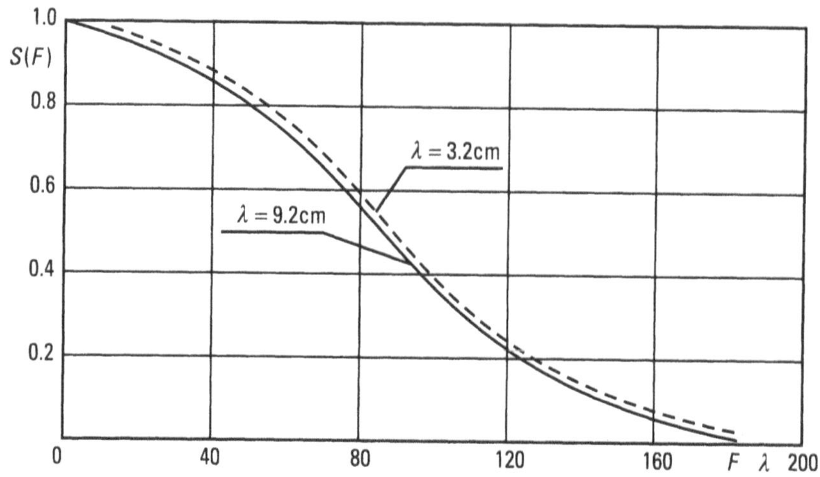

Figure 5.12 The dependency of the normalized spectral density for fluctuations of signals reflected by a chaff cloud on the product $F\lambda$.

averaged chaff dipole velocity according to measurement data lies within the limits $0.18 \leqslant \overline{V}_x \leqslant 0.6$ m/sec, which confirms the validity of formula (5.48).

Analogous results for calculating $F_{0.5}$ are obtained if, instead of (5.48), we use formula (5.49), placing in it not the averaged velocity of the chaff dipoles, but of the wind $\overline{V}_w$. In the example analyzed [10], $\overline{V}_w = 11$ m/sec.

5.3 The Equation for Radar Jamming Using Passive Jamming — The Jamming Coefficient for Noncoherent Radar

According to the concept formulated earlier, masking jamming signals that have a sufficient degree of randomness (i.e., that have a sufficiently entropic nature), can be stable with respect to countermeasures. In the case analyzed, the jamming signal is formed by chaff reaching the resolution element of the victim radar. The jamming signal will be proportional to the RCS of the chaff dipoles in the element σ_{pv}. In turn, σ_{pv} is determined by the pulse volume (by the resolution element volume) V_{pv} and the RCS density of the chaff dipoles σ_v. The latter, according to (5.43), is proportional to the probability density $p(v, t)$, which is the solution to the FPK equation.

Thus, the RCS of chaff, located in any arbitrarily selected resolution element of the victim radar σ_{pv}, is a random value obtained as a result of the linear transformation of the initial random value with the probability density $P(v,t)$, determined using the expression (5.41). Therefore, the probability density of the random value σ_{pv} is Gaussian.

The notions cited provide a basis to consider, at a fixed moment of time, the RCS of the resolution element σ_{pv} of the radar being jammed, in the given case chaff dipoles, as a set of uncorrelated Gaussian random values. In other words, potentially the chaff cloud has informationally stable masking attributes with respect to countermeasures.

The required quantity of chaff to provide a specified degree of radar jamming is determined based on the jamming equation, which is understood to be dependent on the power ratio of the jamming and useful signals on criterial norms, as well as the parameters of the radar of the protected target and the jamming devices.

In analogy to active jamming the jamming equation can be written in the following way:

$$\frac{(P_j)_{\text{inp}}}{(P_s)_{\text{inp}}} \geqslant K_j \tag{5.50}$$

Here, $(P_j)_{\text{inp}}$ and $(P_s)_{\text{inp}}$ are the powers of the jamming and useful signals at the input to the receiver of the victim radar, and K_j is the jamming coefficient.

According to Figure 5.13,

$$(P_j)_{\text{inp}} = \frac{P_s G_s}{4\pi D_s^2}\sigma_{pv}\frac{1}{4\pi D_s^2}A_{\text{rec}}$$

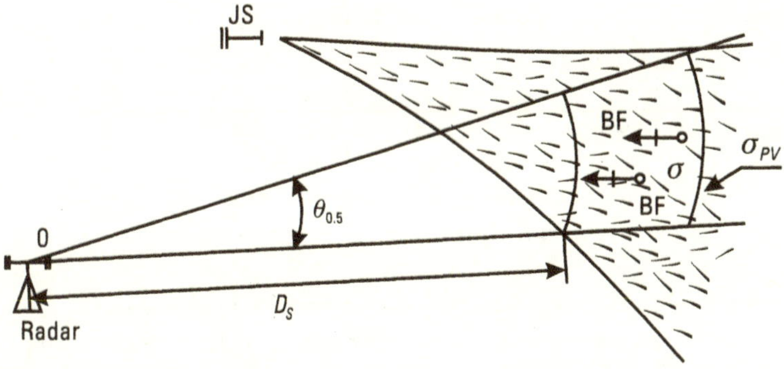

Figure 5.13 The powers of the jamming and useful signals.

$$(P_s)_{\text{inp}} = \frac{P_s G_s}{4\pi D_s^2} \sigma_s \frac{1}{4\pi D_s^2} A_{\text{rec}}$$

Thus, in the case of passive jamming, the jamming equation can be written in the following form:

$$\sigma_{pr} \geqslant K_j \sigma_s \tag{5.51}$$

As an explanatory example, let us write the jamming equation for the case when $l_\varphi >> l_{\text{JE}y}$ and $l_\Theta >> l_{\text{JE}z}$ (Figure 5.8) and the jamming target is a radar, located in a sector with jammer JS and protected battle formation BF. As follows from (5.46), the RCS density σ_{r}, in the conditions considered, is determined by the resolution element $l_\tau = c\tau_s/2$, where τ_s is the pulse length. Accordingly,

$$\sigma_{pr} = \sigma_1 \cdot V_{pr} = \sigma_l l_\tau \tag{5.52}$$

$$\sigma_{pr} = \frac{n_j \sigma_j}{V_{\text{JS}} \tau_j} l_\tau \tag{5.53}$$

Equation (5.51) is transformed to the form:

$$\frac{n_j \sigma_j}{V_{\text{JS}} \tau_j} l_\tau \geqslant K_j \sigma_{\delta_j} \tag{5.54}$$

where σ_{δ_j} is the average RCS of the aircraft battle formation protected by jamming.

Knowing σ_{δ_j} and K_j, from (5.54) it is possible to determine, with what interval τ_j and what quantity of chaff packets it is necessary to launch to assure the jamming conditions for the radar under consideration.

Ratio (5.54), in particular, shows that the required quantity of chaff is inversely proportional to the pulse length of the radar jammed. It follows from this that, if all other conditions are equal, the required degree of jamming for a broadband radar, the signal bandwidth of which is equal to B, is reached if the specific RCS of the initial chaff band σ_l is increased by no less than B times. Initial in the given case is understood to mean the chaff band, the RCS density of which is sufficient to jam a signal with a bandwidth $B = 1$.

In the first approximation the jamming coefficient K_j can be defined on the basis of energy ratios only, without regard to the fine structure of the signal reflected from the aircraft. Such an approach is valid for noncoherent radars that do not have moving target indication cancelers. Based on results of the first approximation, it is also possible to determine the jamming coefficient K_j in the more general case.

Based on energy ratios, the detection decision can be made according to the Neyman–Pearson criterion by comparing the likelihood ratio with a certain threshold value Λ_0. The basis for such a formulation of the question is the possibility established to represent the resolution element RCS σ_{p0} in the form of a Gaussian random value, the probability density of which $p(\sigma_0)$ depends on which of two alternative hypotheses occurs, to be more precise:

- Hypothesis H_0: there is only chaff in the resolution element;
- Hypothesis H_1: in the resolution element besides the chaff there is a certain target with the RCS σ_{BF}.

The hypothesis H_0 corresponds to a probability density of

$$P(\sigma_{p0}) = \frac{1}{\sqrt{2\pi D_\sigma}} \exp\left(-\frac{\sigma_{p0}^2}{2D_\sigma}\right) \tag{5.55}$$

where D_σ is the variance of the random value σ_{p0}.

The hypothesis H_1 corresponds to the probability density:

$$p_1(\sigma_{p0}) = \frac{1}{\sqrt{2\pi D_\sigma}} \exp\left(-\frac{(\sigma_{p0} - \sigma_{BF})^2}{2D_\sigma}\right) \tag{5.56}$$

In the materials that follow, it is assumed that σ_{p0} is a random value with a zero mathematical expectation $M[\sigma_{p0}] = 0$, if hypothesis H_0 occurs and $M[\sigma_{p0}] = \sigma_{BF}$; if hypothesis H_1 is valid, then the RCS of the battle formation σ_{BF} is assumed to be a constant and known value.

The likelihood ratio:

$$\Lambda = \frac{p_1(\sigma_{p0})}{p_0(\sigma_{p0})} \tag{5.57}$$

is transformed to the following form:

$$\Lambda = \exp\frac{2\sigma_{p0}\sigma_{BF} + \sigma_{BF}^2}{2D_\sigma} \tag{5.58}$$

The detection decision in the given resolution element of battle formation BF is taken by comparing Λ to a certain threshold value of the likelihood ratio Λ_0.

If $\Lambda > \Lambda_0$, then the decision is made that a BF has been detected, and in the opposite case $\Lambda < \Lambda_0$ it is assumed that in the resolution element there is only chaff. The expression (5.58) is quite complex. Taking into consideration the attributes of an exponential function, instead of the likelihood ratio Λ, it is possible to use sufficient statistics, mutually and unambiguously linked with it:

$$G = \sigma_{p0}\sigma_{\mathrm{BF}} \tag{5.59}$$

and take a decision based on the comparison of G with a certain threshold statistical value:

$$G_0 = D_\sigma \ln \Lambda_0 - \frac{1}{2}\sigma_{\mathrm{BF}}^2 \tag{5.60}$$

As follows from (5.59), the statistical value G is a Gaussian random value and is fully determined by the first two distribution moments $M[G]$ and D_G.

In the case of hypothesis H_0,

$$M[G] = 0 \quad D_G = \sigma_{\mathrm{BF}}^2 D_\sigma \tag{5.61}$$

If H_1 is valid, then

$$M[G] = \sigma_{\mathrm{BF}}^2 \quad D_G = \sigma_{\mathrm{BF}}^2 D_\sigma \tag{5.62}$$

The probability densities that correspond to hypotheses H_0 and H_1 are written in the following manner:

$$p_0(G) = \frac{1}{\sqrt{2\pi D_G}} \exp\left(-\frac{G^2}{2\sigma_{\mathrm{BF}}^2 D_\sigma}\right) \tag{5.63}$$

$$p_1(G) = \frac{1}{\sqrt{2\pi D_G}} \exp\left(-\frac{(G - \sigma_{\mathrm{BF}}^2)^2}{2\sigma_{\mathrm{BF}}^2 D_\sigma}\right) \tag{5.64}$$

Specifying the threshold value G_0, it is possible to determine the probabilities of a false alarm P_{fa} and correct detection P_{det}:

$$P_{\text{fa}} = \frac{1}{2} - \Phi_0(X_0) \quad X_0 = \frac{G_0}{\sigma_{\text{fa}}\sigma_{p0}} \tag{5.65}$$

$$P_{\text{det}} = \frac{1}{2} - \Phi_0\left(X_0 - \frac{\sigma_{\text{BF}}}{\sigma_{p0}}\right) \tag{5.66}$$

For a fixed level of false alarm (P_{fa}) and using expression (5.66), we determine the jamming coefficient K_j equal to the ratio $\sigma_{p0}/\sigma_{\text{BF}}$ at which the probability of correct detection P_{det} does not exceed the specified value.

As an example, let us determine the radar jamming coefficient using passive jamming, where the probability of correct detection does not exceed 0.001. The probability of a false alarm is fixed at the level 10^{-3}. Specifying $P_{\text{fa}} = 10^{-3}$, from (5.65) we find $X_0 = 3.1$. Assuming in (5.66) $P_{\text{det}} = 0.10$, we obtain $\sigma_{\text{BF}}/\sigma_{p0} = 0.77$. The jamming coefficient is $K_j = \sigma_{p0}/\sigma_{\text{BF}} = 1.3$.

If we take the probability of detection $P_{\text{det}} = 0.1$ to be sufficient, then at the same level for a false alarm $P_{\text{fa}} = 10^{-3}$, $K_j = 0.55$. The estimations for K_j obtained correspond to those of the criterial norm established by years of experience: $\sigma_{p0} \approx \sigma_{\text{BF}}$ (i.e., $K_j \approx 1$).

The energy criterial norms obtained in the first approximation for jamming of noncoherent radar using passive jamming does not take into consideration the possibility of a reduction of the jamming effect as a result of the appearance of attendant attributes, to be more specific, determinate changes in the phase time of the reflected signal, caused by the movement of the aircraft. Signal discrimination from a moving target against a background of reflections from slow-moving chaff occurs in pulse-coherent radar.

5.4 The Jamming Coefficient Using Passive Jamming for Coherent Pulse-Radar

There are three basic types of coherent pulse-radar known and, to be more specific, with low, high and average pulse repetition rates.

Coherent pulse-radars with a low pulse repetition rate are broadband. Moving target indication (MTI) in them is performed using pulse-to-pulse cancellation of jamming signals reflected from stationary or slow-moving targets.

Coherent pulse-radars with medium or high pulse repetition rates are narrowband. Target discrimination is performed in them using Doppler frequency filtering of signals reflected from moving targets. The latter provides a basis to call these radars pulse-Doppler radar or quasicontinuous-wave radar (QCW). By filtering Doppler frequencies of useful signals, the jamming effect generated by chaff is reduced by many tens of decibels. This, in essence, excludes the possibility of jamming pulse-Doppler radar with passive jamming using traditional methods. Practically speaking, it is possible to jam broadband coherent pulse-radar using passive jamming. The required degree of increase in the volume density of chaff can be determined using a quantitative estimate of the degree of reduction of the jamming effect in the MTI cancellation circuit.

In Figure 5.14 we give a simplified schematic for a coherent pulse-radar with a single-delay canceler. Here *M* is a modulator, forming short pulses for the modulation of the high-frequency generator (transmitter) *G* and long pulses for enabling the coherent oscillator CH during time *T*, which is approximately equal to the pulse repetition period. Normally this time is chosen somewhat smaller, in order to give an opportunity for inherent oscillations in circuits of the coherent oscillator to die down by the beginning of the next pulse repetition interval. The coherent oscillator is synchronized in phase with the high-frequency generator pulse, in this

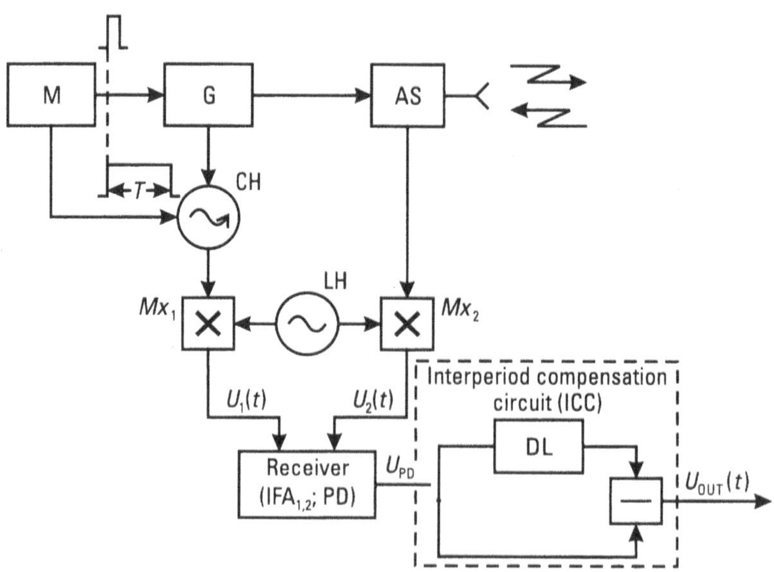

Figure 5.14 A simplified schematic for a coherent pulse radar with a single-delay canceler.

manner providing for the coherency of its signal and the emitted signal for a time approximately equal to T.

The next new high-frequency pulse once again performs (for itself) phasing of the coherent oscillator. As a result of the high pulse rate, wide bandwidth and the small time of coherency, this variant of a pulse coherent radar is called a pseudo- or quasi-coherent radar.

The reflected (useful) signal and the signal of the coherent oscillator after transformation in mixers Mx_1 and Mx_2, which are also fed signals from the local oscillator LH, arrive at the intermediate frequency amplifiers IFA_1 and IFA_2 and then on to the phase detector PD. From the output of the phase detector, the signals arrive at the canceler circuit including a delay line DL with delay time T and a subtractor (cancellation) circuit.

The resulting signal at the output to the ideally operating quasi-coherent radar $U_{out}(t)$ is equal to zero if the reflecting target is stationary, and unequal to zero if the target is moving with a radial velocity that does not correspond to the so-called blind speed. The latter is understood to be the target (airplane, helicopter) velocity, at which, during the pulse repetition period T, the target moves in a direction towards the radar equal to half the wavelength of the radar. Below the definition is illustrated using specific examples.

Let us assume that both the useful and the jamming signals effect the input to the victim receiver. We will denote the ratio of their powers at the input by K:

$$K = \left(\frac{P_j}{P_s}\right)_{\text{inp}} \tag{5.67}$$

Let us define the ratio of the powers of the jamming and useful signals at the output to the receiver of a coherent pulse radar with IFA.

The jamming signal at the input to the receiver in the given case for a limited time interval can be represented in the form of a quasi-sinusoidal wave of the following type:

$$u_n(t) = U_j(t)\cos(\omega_0 t + \varphi_j(t)) \tag{5.68}$$

where $U_j(t)$ and $\varphi_j(t)$ are the envelope and initial phase representing in comparison with $u_n(t)$ slowly changing random processes; and ω_0 is the carrier frequency.

The average power (variance) of the jamming signal is by definition:

$$(P_j)_{\text{inp}} = \overline{(U_j(t))^2} \tag{5.69}$$

Accordingly, the useful signal at the input is represented in the form of a sinusoidal wave, the initial phase of which varies with time as a result of the forward movement of the aircraft (or other target):

$$u_s(t) = U_s \cos(\omega_0 t + \varphi_s(t)) \tag{5.70}$$

Here,

$$\varphi_s(t) = \omega_{\text{D}}(t) \tag{5.71}$$

ω_{D} is the Doppler frequency:

$$\omega_{\text{D}} = 2\pi \frac{2V_r}{\lambda} \tag{5.72}$$

V_r is the radial component of the aircraft velocity in the direction towards the victim radar; λ is the radar wavelength; and U_s is the envelope of the useful signal.

By definition, the power of the useful signal at the input and the amplitude of the useful signal are linked by the relationship:

$$U_s = \sqrt{2(P_s)_{\text{inp}}} \tag{5.73}$$

Assuming the channels of the coherent heterodyne and the receiver to be identical, it is possible to write the following expressions for voltages of the useful and jamming signals at the output to the phase detector. The useful signal:

$$u_{\Phi D,s}(t) = K_{\text{rec}} U_s \cos \varphi_s(t) \tag{5.74}$$

where K_{rec} is the transmission coefficient of the entire processing circuit.

The jamming signal:

$$u_{\Phi Dj} = u_j^{(1)}(t) = K_{\text{rec}} U_j \cos \varphi_j(t) \tag{5.75}$$

Both the useful and the jamming signals are pulsed. An observed sample function of n pulses corresponds to a time interval:

$$t = nT \tag{5.76}$$

where T is the pulse repetition period.

Let us determine the voltage $\Delta u_s(t)$ of the useful signal at the output of the IFA, occurring during the passage of the $(n-1)$th and nth radar pulses:

$$\Delta u_s(t) = K_{\text{rec}} U_s(\cos(\omega_{\text{D}}(n-1)T) - \cos(\omega_{\text{D}} nT))$$

or

$$\Delta u_s(t) = K_{\text{rec}} 2U_s \sin\frac{\omega_{\text{D}} T}{2} \sin\left(\omega_{\text{D}} nT - \frac{1}{2}\omega_{\text{D}} T\right) \tag{5.77}$$

Thus, the voltage of the useful signal at the output to the IFA is a sinusoidal wave with frequency ω_{D} and an envelope equal to $2U_s \sin(\omega_{\text{D}} T/2)$. Accordingly, the power of the useful signal at the output to IFA $(P_s)_{\text{out}}$, taking into consideration (5.73) and (5.77), is written in the following way:

$$(P_s)_{\text{out}} K_{\text{rec}}^2 4(P_s)_{\text{inp}} \sin^2\frac{\omega_{\text{D}} T}{2} \tag{5.78}$$

With the help of (5.77), it is possible to clarify the definition of the blind speed of a moving target relative to a specific radar with an IFA. Let us define the conditions, in which the envelope in (5.77) or the power in (5.78) is equal to zero. Apparently, this occurs when

$$\frac{\omega_{\text{D}} T}{2} = m\pi \quad m = 0, 1, 2 \ldots \tag{5.79}$$

Taking into consideration (5.72) from (5.79), we find

$$V_r T = \frac{\lambda}{2} m \quad m = 0, 1, 2 \ldots \tag{5.80}$$

The voltage of the useful signal at the output to the IFA is equal to zero if, during the pulse repetition period T, the target being detected covers a path $V_r T$ in the direction of the radar that is a multiple of a whole number of half-wavelengths.

The turbulent atmospheric diffusion process develops comparatively slowly over a time equal to the radar pulse repetition period and the chaff cloud can be considered to be stationary. This permits us, with consideration to (5.75), to write the jamming signal at the output to a one-time IFA in the following manner:

$$\Delta u_j^{(1)}(t) = K_{\text{rec}}\left(u_j^1(t) - u_j^1(t+T)\right) \tag{5.81}$$

The power of the jamming signal at the output to the IFA when there is single-delay cancellation is defined by the expression:

$$(P_j)_{\text{out}} = \overline{\left(\Delta u_j^{(1)}(t)\right)^2} = K_{\text{rec}}^2 2(P_j)_{\text{inp}}(1 - r(T))_{\Phi} \tag{5.82}$$

Here, $(P_j)_{\text{inp}}$ is the power of the jamming signal at the input to the receiver, $r(T)$ is the normalized correlation function of the jamming signal envelope, calculated for the value $\tau = T$. In doing this, it was taken into consideration that

$$\overline{(u_j(t))^2} = \overline{\left(u_j^1(t)\right)^2}$$

As a result of the assumed stationary nature of the jamming signal, the normalized correlation function can be represented in the following fashion:

$$r(\tau) = \frac{\overline{u_j(t)u_j(t+\tau)}}{\overline{(u_j(t))^2}} \tag{5.83}$$

The overbar denotes the averaging operation with respect to time. With the help of (5.78) and (5.82), let us define the ratio sought for the powers of the jamming and useful signals:

$$K_{\text{out}}^{(1)} = \frac{(P_j)_{\text{out}}}{(P_s)_{\text{out}}} = \frac{(P_j)_{\text{inp}}}{(P_s)_{\text{inp}}} \frac{1 - r(T)}{2\sin^2 \frac{\omega_D T}{2}} \tag{5.84}$$

Let $K_{\text{out}}^{(1)} = K_{B\min}$ be the minimum necessary ratio of the powers of the jamming and useful signals at the input to the indicator device, at which the required degree of radar jamming is attained. This value $K_{B\min}$ corresponds to a certain value of the jamming/signal ratio at the input $K = (P_j/P_s)_{\text{inp}}$, which in the given conditions is also the minimum required to jam the radar. Accordingly, for a radar with single-delay cancellation, the jamming coefficient is determined using the expression:

$$K_j = K_{B\min} \frac{2\sin^2 \frac{\omega_D T}{2}}{1 - r(T)} \tag{5.85}$$

The minimum necessary jamming/signal ratio at the output to the victim receiver in the case of a noncoherent radar is practically equal to its jamming coefficient using passive jamming K_{j0}, determined using specified values for the probabilities of a false alarm and a correct detection (5.65) and (5.66). What has been said permits us to write (5.85) in the following form:

$$K_j = K_{j0} \frac{2 \sin^2 \dfrac{\omega_D T}{2}}{1 - r(T)} \tag{5.86}$$

The use in the radar of double-delay cancellation increases somewhat the required norms for expending chaff. The power of the jamming signal at the output to the double-delay canceler circuit is equal to

$$P_j^{(2)} = \overline{\left[\Delta^{(2)} u_j(t)\right]^2}$$

Due to the principle of operations of a double-delay canceler circuit,

$$\Delta^{(2)} u_j(t) = \Delta^{(1)} u_j(t) - \Delta^{(1)} u_j(t + T) \tag{5.87}$$

Therefore,

$$\left(P_j^{(2)}\right)_{\text{out}} = 2P'_{\text{jout}} - \overline{2\Delta^{(1)} u_j(t) \Delta^{(1)} u_j(t + T)} \tag{5.88}$$

For this, by analogy with (5.82), we find

$$\left(P_j^{(2)}\right)_{\text{out}} = 4P_{\text{jinp}}[1 - r(T)]\left[1 - r^{(2)}(T)\right] \tag{5.89}$$

Here, $r^{(2)}(T)$ is the normalized correlation function of random process $\Delta^{(2)} u(t)$, calculated for the value $\tau = T$.

Accordingly, for m-delay cancellation, the output power $\left(P_j^{(m)}\right)_{\text{out}}$ is equal to

$$P_{\text{jout}}^{(m)} = \overline{\left[\Delta^{(m)} u(t)\right]^2} = 2^m P_{\text{jinp}} \prod_{j=1}^{m} \left[1 - r^{(j)}(T)\right] \tag{5.90}$$

where $\prod$ is the product sign.

The normalized correlation function in circuits with double-delay cancellation $r^{(2)}(T)$ can be calculated using comparison (5.88) and (5.89). The direct calculation of the correlation function of the random process $\Delta^{(1)}u_j(t)$ yields

$$\overline{\Delta^{(1)}u_j(t)\Delta^{(1)}u_j(t+\tau)} = P_{\text{jinp}}[2r(\tau) - r(\tau - T) + r(\tau + T)] \tag{5.91}$$

From here it follows that

$$r^{(2)}(T) = \frac{2r(T) - r(2T) - 1}{2[1 - r(T)]} \tag{5.92}$$

Substituting (5.92) in (5.88), we find that

$$P_{\text{out}}^{(2)} = 2P_{\text{jinp}}[3 - 4r(T) + r(2T)] \tag{5.93}$$

In the MTD schematic with double-delay cancellation of the useful signal, the output voltage is defined as a difference:

$$\Delta^2 u_s(n) = \Delta u_s(n) - \Delta u_s(n+1) \tag{5.94}$$

Using (5.77), we obtain

$$\Delta^{(2)}u_s(n) = K_{\text{rec}} 4u_s \sin^2 \frac{\omega_D T}{2} \cos n\omega_D T \tag{5.95}$$

The amplitude of the wave at the output of the double-delay cancellation circuit is equal to

$$\Delta^{(2)}u_s(n) = K_{\text{rec}} 4u_s \sin^2 \frac{\omega_D T}{2} \tag{5.96}$$

Accordingly, for *m*-delay cancellation, the wave amplitude at the output of the IFA is

$$\Delta^{(m)}u_s = K_{\text{rec}} u_s 2^m \sin^m \frac{\omega_D T}{2} \tag{5.97}$$

The ratio of the powers of the jamming and useful signals at the output of the double-delay canceler circuit is equal to

$$\left(\frac{P_j}{P_s}\right)^{(2)}_{out} = K^{(2)}_{out} = \left(\frac{P_j}{P_s}\right)_{inp} \frac{3 - 4r(T) + r(2T)}{8 \sin^4 \frac{\omega_D T}{2}} \tag{5.98}$$

The radar jamming coefficient with double-delay cancellation is equal to:

$$\frac{K_j = K_{p0} 8 \sin^4 \frac{\omega_D T}{2}}{3 - 4r(T) + r(2T)} \tag{5.99}$$

The normalized correlation function, that enters into formulas (5.86) and (5.99), can be determined using the well-known spectral density $S(F)$ and the Khinchin–Wiener transformation [6]:

$$R(\tau) = \int_0^\infty S(F) \cos(2\pi F \tau) dF . F > 0 \tag{5.100}$$

According to (5.47),

$$S(F) = \exp\left(-\left(\frac{0.84F}{F_{0.5}}\right)^2\right) \tag{5.101}$$

After replacing (5.101) in (5.100), we obtain

$$R(\tau) = \frac{\sqrt{\pi} F_{0.5}}{1.68} \exp\left(-\frac{1}{0.7}\pi^2 \tau^2 F_{0.5}^2\right) \tag{5.102}$$

We are interested in the normalized correlation function, defined at moment $\tau = T$. After normalization and an appropriate substitution, we have

$$r(T) = \exp\left(-14(F_{0.5} T)^2\right) \tag{5.103}$$

The formula (5.86) defines the value of the jamming coefficient supposing the ideal operation of the moving target indication circuit. In reality, to a significant degree, instabilities in the frequency of the coherent and local oscillators, pulse repetition frequencies, and the inherent movement of the antenna have an effect on the operating quality of an MTI circuit. All of this, when taken together, can be, to a certain degree, accounted for by the equivalent widening of the jamming signal spectrum by approximately a factor of 1.5 to 2. Considering what has been said, for estimates of the first approximation, we assume

$$r(T) = \exp(-50(F_{0.5}T)^2)$$

In modern coherent pulse-radars the pulse repetition rate and, consequently, the repetition period are changed within the limits of one pulse train to eliminate blind speeds. When jamming a radar system that is a part of an ACS for AAD forces, the course angles of the jammed targets are changed within broad limits. This permits, with the exception of certain special cases, when determining K_j in the formula (5.86) to orient ourselves towards the averaged value $\sin^2(\omega_D T/2)$, equal to 0.5. As was already noted earlier, it is possible to assume that K_{p0} is equal to 1. What has been said makes it possible to write expression (5.86) in the following form:

$$K_j = \frac{1}{1 - \exp\left[-50(F_{0.5}T)^2\right]} \tag{5.104}$$

If the jamming signal spectrum width $F_{0.5}$ is small and $50(F_{0.5}T)^2$, then K_j can be determined using the approximate formula:

$$K_j = \frac{1}{50(F_{0.5}T)^2} \tag{5.105}$$

For example, if $T = (1/200)$ s, $F_{0.5} = 25$ Hz, then $K_j = 8$.

Let us note that the formulas (5.104) and (5.105) are valid if the spectral density of fluctuations in the jamming signal envelope can be approximated with a Gaussian curve (5.101).

One of the basic methods of deliberately reducing the jamming coefficient using passive jamming is the broadening of the jamming signal spectrum by increasing the effective width of the chaff band in the vertical plane (by the elevation angle with the limits of $\Theta_{0.5}$), when assuring at the same time a sufficient level of RCS density in the chaff cloud σ_p. In this case the broadening of the jamming signal spectrum occurs as a result of the wind shear. In anticipation of its linear change within the limits of the given altitude range ΔH, the maximum difference in chaff dipole speeds, borne by the wind as a single whole, ΔV_H is equal to

$$\Delta V_H = \beta \frac{\Delta H}{100} \text{ m} \tag{5.106}$$

where β is the wind shear coefficient. Quantitatively it is equal to the change in the wind speed in the sector $\Delta H = 100\text{m}$: $\beta = (\Delta V\ \text{m/s})/100\text{m}$ The spectrum of the jamming signal envelope is determined by the Doppler frequency:

$$\Delta F_{\mathrm{D}} = \frac{2\Delta \overline{V}_{\mathrm{HX}}}{\lambda} \tag{5.107}$$

where $\Delta \overline{V}_{\mathrm{HX}}$ is the averaged component of the difference in wind speeds in the direction of X (Figure 5.8).

The broadening of the spectrum of the jamming signal envelope in the given case is caused not by the turbulent atmospheric diffusion, but by the presence of movements with varying speeds of compact groupings of chaff dipoles with quite high RCS within the resolution element (of the pulse volume) of the radar. In order to determine the jamming coefficient, in the given case it is necessary to use formula (5.86) and find the value $r(T)$ with the help of (5.81) and (5.82), understanding the voltage at the output to the phase detector to be the value:

$$u_{\Phi \mathrm{D},j} = u_j'(t) = K_{\mathrm{rec}} U_j(t) \cos\big(\omega_{\mathrm{D}} + \varphi_j(t)\big)$$

Accordingly, by analogy with (5.83), we obtain

$$r(T) = \frac{\overline{u_j(t)\cos\big(\omega_{\mathrm{D}} t + \varphi_j(t)\big) u_j(t+T)\cos\big(\omega_{\mathrm{D}}(t+T) + \varphi_j(t+T)\big)}}{\overline{\big(u_j(t)\cos\big(\omega_{\mathrm{D}} t + \varphi_j(t)\big)\big)^2}}$$

Assuming $\varphi_j(t) \approx \varphi_j(t+T)$, we determine the values sought:

$$r(T) = \cos\left(\frac{1}{2}\omega_{\mathrm{D}} T\right) \tag{5.108}$$

$$K_j = \frac{1}{1 - \cos\left(\frac{1}{2}\omega_{\mathrm{D}} T\right)} \qquad \omega_{\mathrm{D}} = 2\pi \frac{2\overline{\Delta V}_{\mathrm{HX}}}{\lambda} \tag{5.109}$$

The spectrum width of the envelope of the reflected signal, caused exclusively by the turbulent atmospheric diffusion $F_{0.5w}$ according to measurements performed [9, 10] is determined using the formula (5.49):

$$F_{0.5w} = \frac{\overline{V}_w}{4\lambda}$$

where $\overline{V}_w$ is the averaged component of the wind speed within the radar resolution element in the direction of its antenna beam.

In Figure 5.15 we show the normalized spectral density of the jamming signal envelope for various values of averaged wind speeds in the resolution element of the radar jammed.

The methods obtained for calculating the jamming coefficient are approximate. All the same, for practical purposes their accuracy can be sufficient, since source data is not always reliable. The RCS of chaff packets can noticeably differ from the nominal value. The RCS of an airplane is significantly different depending on the bearings of the radar being jammed and meteorological data is approximate. The formulas offered permit estimation in the first approximation. In many cases this is enough to make a decision.

5.5 Effectiveness of Radar Jamming Using Passive Jamming — Determination of the Required Quantity of Chaff

5.5.1 The Width of the Concealment Area for a Standalone Radar — Determination of the Required Quantity of Chaff

The basic effectiveness indicators for radar jamming using passive jamming are the information damage they cause to the radar system, as well as the

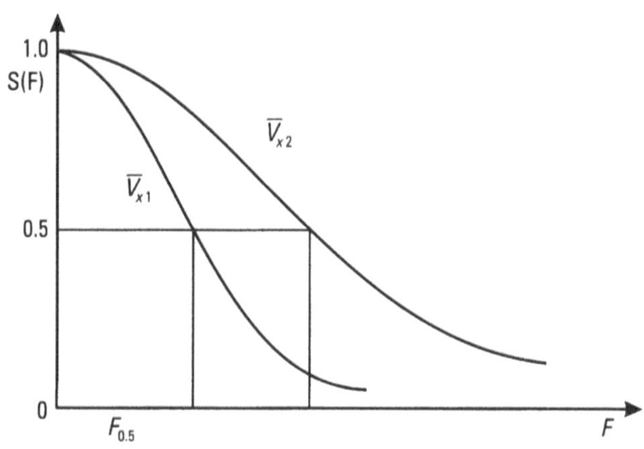

Figure 5.15 The normalized spectral density of the jamming signal envelope for various values of averaged wind speeds.

required quantity of chaff to assure that the needed degree of jamming is achieved, based on operational and tactical considerations.

Depending on the problems being solved, and also the techniques and tactics used by the air force, the degree of jamming and the required amount of chaff can vary. One of the basic ways of employing passive jamming is the creation of chaff bands of significant length that exclude the possibility of a radar detecting aircraft (helicopter) battle formations.

In order to determine the required quantity of chaff, it is necessary to know the path length that must be covered using passive jamming with the required chaff corridor width according to navigational support conditions, as well as the required number of jammers. We will solve this problem in two stages.

In the first stage we determine the width of the concealment area provided by the chaff corridor, located in the range of the jammed radar and protecting the battle formation BF (Figure 5.8). The selected variant is more difficult for the jamming side, since in this case the resolution element of the radar jammed has the smallest dimensions. It is determined by the pulse length l_τ, which accordingly requires an increase in the chaff density. We are interested in the concealment area width L_M at a distance of D_A from the radar (Figure 5.16). Within the limits L_M, the radar cannot discriminate the

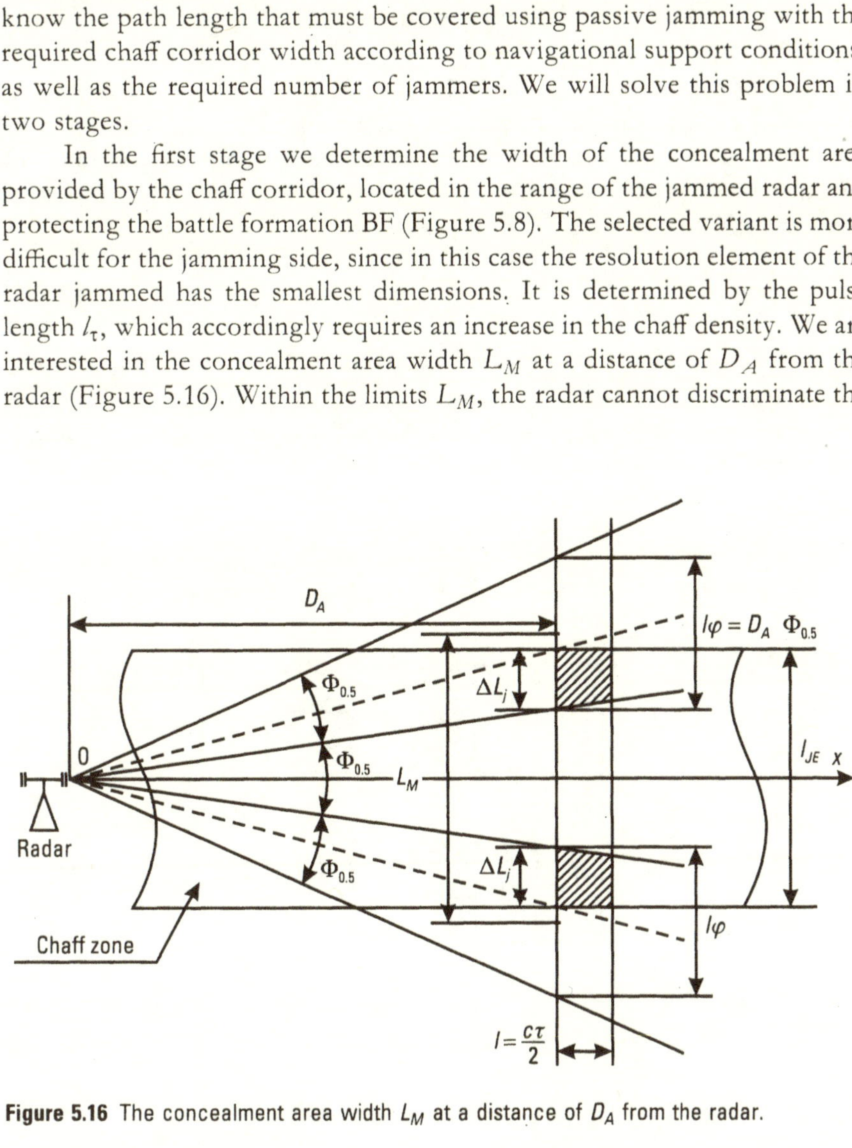

Figure 5.16 The concealment area width L_M at a distance of D_A from the radar.

battle formation BF against the passive jamming background, if the jamming conditions are ensured by that part of the chaff corridor that is in the radar resolution element. In Figure 5.16 this part is crosshatched. Its length is equal to L_τ, and its width is Δl_j.

In Figure 5.17, in accordance with Figure 5.16, we have shown an extended model of the chaff corridor section being analyzed, from which it directly follows that

$$L_M = l_\varphi + (l_{\text{JE}} - 2\Delta l_j) \tag{5.110}$$

Let us consider specific particular cases of determining the width of the masking area L_M using (5.110). Let us assume that $\Delta l_j = l_{\text{JE}}$ (i.e., the battle formation is concealed in the case where all chaff dipoles determining the effective width of the corridor ($l_\varphi \geqslant l_{\text{JE}}$) enter into the radar resolution element l_φ). Then, according to (5.110),

$$L_M = l_\varphi - l_{\text{JE}} \tag{5.111}$$

In particular, such a variant is assumed when deriving the jamming equation represented by (5.69). In order to increase the width of the masking area, it is necessary to increase the RCS density of the area σ_V accordingly. Increasing σ_V approximately four times, in the variant being analyzed it is

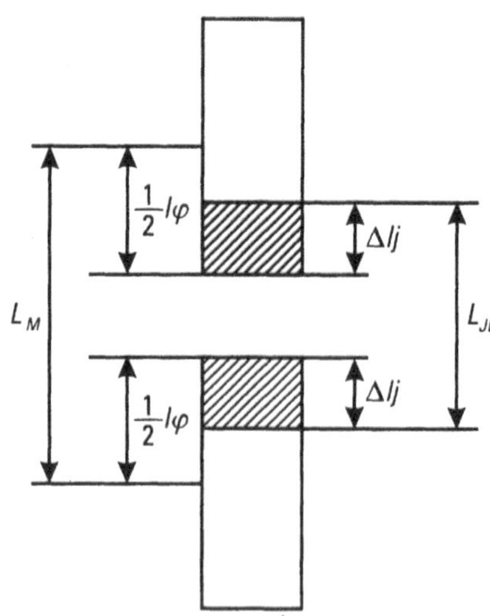

Figure 5.17 An extended model of a chaff corridor section.

possible to assure the required degree of jamming. When $\Delta l_j = l_{JE}/4$, we have

$$L_M = l_\varphi + \frac{1}{2} l_{JE} \tag{5.112}$$

The width of the masking area in the latter case is increased by $1.5 l_{JE}$ in comparison with L_M, as determined from (5.111).

As follows from (5.38) and (5.39), the effective width of the chaff corridor increases with time. In the first moments of time, the increase l_{JE} takes place proportionally to the time to the power 3/2. Subsequently, the effective width grows proportionally to the time. For large values of l_{JE} ($l_{JE} > 1000$m), widening of the corridor occurs according to the law of diffusion, described by the FPK equation with constant coefficients. Qualitatively, the nature of the change over time of the effective width of the corridor is shown in Figure 5.18. The nature of the change in the width of the corridor over time was experimentally established for the first time by V. T. Borovik.

The size of the resolution element $l_\varphi = D_A \Phi_{0.5}$ grows proportionally to the distance. For small distances ($<$ 50–60 km), radar jamming conditions may not be fulfilled due to the fact that the size of the resolution element l_φ is smaller than the effective chaff corridor width $l_\varphi < l_{JE}$. So that this does not happen, it is necessary to correspondingly increase the RCS density of the chaff corridor σ_v, m^2/m^3.

The required width of the masking area L_{mj} and the needed quantity of passive jammers to achieve it, as well as of chaff, can be determined, if we

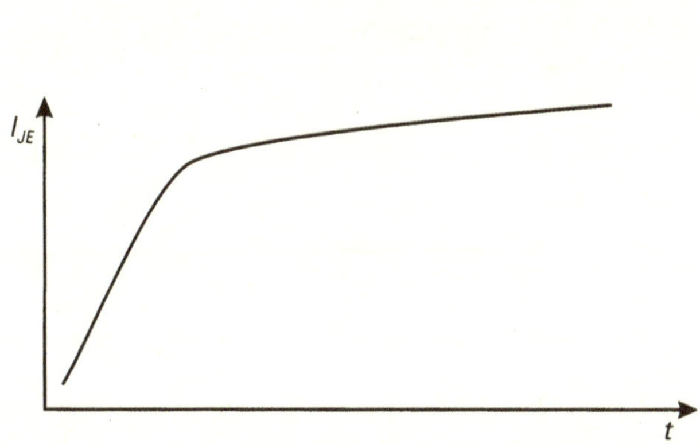

Figure 5.18 The nature of the quantitative change over time of the effective width of a chaff corridor.

know the capabilities of the navigational support system and the required probability that the battle formation will enter the given bound. In anticipation of a Gaussian law for the distribution of navigational errors, the probability that a random error X will occur in the interval $(-0.5L_{mj}, 0.5L_{mj})$ is determined using the following expression:

$$P\left(-\frac{L_{mj}}{2} \leqslant X \leqslant \frac{L_{mj}}{2}\right) = 2\Phi_0\left(\frac{L_{mj}}{2\sigma_n}\right) \tag{5.113}$$

where σ_n is the navigation rms error, and $\Phi_0(X)$ is the probability integral (3.128).

Knowing $P(-0.5L_{mj} \leqslant X \leqslant 0.5L_{mj})$ and σ_n, it is possible using tables to determine the required value for the width of the masking area L_{mj} [12]. If L_{m1} is the width of the masking area, generated by one jammer, then the required quantity of passive jammer aircraft N_{JS} is equal to:

$$N_{JS} = \frac{L_{mj}}{L_{m1}} \tag{5.114}$$

The definition of the required quantity of chaff can be performed in the following sequence. First of all, the specific chaff expenditure is determined by measuring the number of packets expended for a kilometer of the path (n_{pac} packets/km) when jamming a radar requiring the greatest expenditure of chaff, when all other conditions are equal, to achieve the specified degree of jamming. As a rule, such jamming targets are radars located within the range of a jammer and covered by a battle formation (Figure 5.8).

According to (5.52), the dimensions of the radar resolution element, in the given case, are determined by the pulse length. If n_j is the quantity of chaff packets, launched within the resolution element $l_\tau = mc\tau/2$, then the specific quantity of packs is determined using the expression:

$$n_{pac} = \frac{1000}{l_\tau} \text{ packets/km} \tag{5.115}$$

The number of packets n_{j1} expended by one jammer over then entire path length D_m is equal to

$$n_{j1} = D_m n_{pac} \text{ packets} \tag{5.116}$$

The total chaff expenditure is

$$n_{j\Sigma} = N_{JS}n_{n1} = N_{JS}n_{pac}D_m \tag{5.117}$$

The increase in information stability (the level of radio electronic protection) of modern coherent pulse-radar requires a significant increase in the specific expenditure of chaff n_{pac}, which is far from always being possible. The use of passive jamming using traditional methods in the interests of jamming pulse-Doppler radar is practically speaking ineffective. Up to the present time, three trends have been established, ensuring the maintenance of and, in a number of cases, an increase in the effectiveness of both passive and active jamming. They include: the discrete launching of chaff; the creation of large volume chaff clouds with high RCS density σ_v so as to attenuate the radar radiation power density due to its diffusion in the chaff corridor; and the combining of active and passive jamming.

5.5.2 The Discrete Launching of Chaff

Priority jamming targets, in this case, are coherent pulse-radars with MTI cancellation of the jamming signal. The basic idea of the discrete method of launching comes down to ensuring radar jamming conditions using a limited amount of chaff by significantly increasing the density of the chaff dipoles, launched in separate discrete areas in space with simultaneous randomization of the time intervals between launches. A certain reduction in the masking properties of a discrete chaff corridor is compensated by an increase in jammers. The quantity of concurrently launched chaff is selected so as to ensure overcoming the IFA circuit by the jamming signal reflected from the chaff cloud. In the conditions analyzed, the greatest uncertainty (entropy) in the jamming environment occurs when the probability density of time intervals between chaff launches is uniform. The average launch interval T_{ave} in this case is equal to (Figure 5.19)

$$T_{ave} = \frac{T_{max} + T_{min}}{2} \tag{5.118}$$

where T_{max} and T_{min} are the maximum and minimum launch intervals.

When the quantity of jammers JS is sufficiently large, discrete launching can turn out to be no less effective than continuous launching and lead to the disruption of centralized control of AAD forces as a result of the loss of radar tracking for trajectories of both battle formations and individual aircraft.

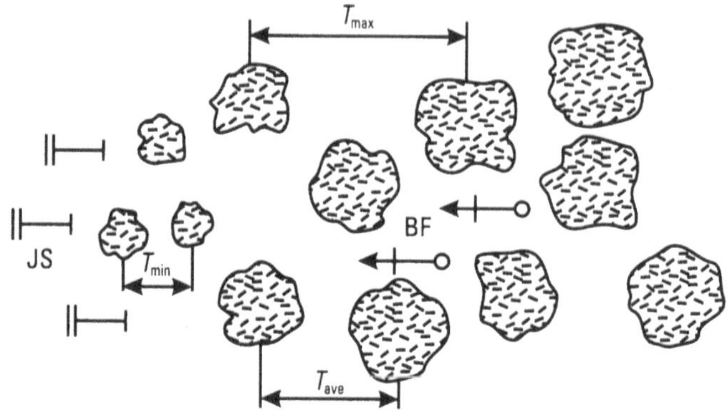

Figure 5.19 The chaff launch interval.

The required quantity of chaff to be launched concurrently in a single resolution element (pulse volume) of a radar being jammed can be determined from the jamming equation, written in the form of expression (5.53). The radar jamming coefficient with single-delay cancellation depending on the technique implemented for creating discrete chaff corridors is determined using (5.105), if separation of discrete corridors by altitude has not been provided for, or using (5.109), if such separation has been provided for.

Certain peculiarities occur in the case of discrete launching of chaff to protect individual aircraft [13, 14]. The basic problem here is to ensure that the chaff cloud launched is formed at a sufficiently small distance from the object being protected. This distance must be commensurable with the resolution element of the jamming target.

5.5.3 Attenuation of Radar Signals in a Chaff Band

Let us determine the degree of attenuation of radar signals in a chaff cloud, caused by radiation being scattered by chaff (Figure 5.20). We assume that the radiation scattering by chaff dipoles is isotropic and proportional to its averaged RCS. The fields diffused by the chaff dipoles are noncoherent and the power density of the total diffused field is equal to the sum of the power densities of the individual scatterers.

According to what has been mentioned, the value of Δp is proportional to the power density, if the incident wave $p(x)$, as well as the total RCS of the chaff dipoles of elementary volume with a thickness of Δx, are equal to

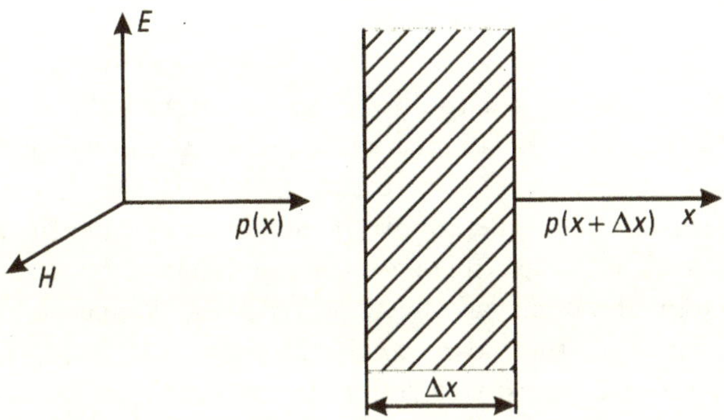

Figure 5.20 The degree of attenuation of radar signals in a chaff cloud caused by the radiation scattered by chaff.

$$n\overline{\sigma}_1\Delta x = \sigma_v\Delta x$$

where σ_v is the averaged RCS density of the chaff cloud, and $\overline{\sigma}_1$ is the averaged RCS of an arbitrarily positioned chaff dipole.

Accordingly,

$$\Delta p = -p(x)\sigma_v\Delta x \tag{5.119}$$

After converting to the limit, we obtain the following differential equation for determining the sought degree of attenuation in the power density in a chaff corridor of length L_j:

$$\frac{\mathrm{d}p}{p} = -\sigma_v \mathrm{d}x \tag{5.120}$$

The boundary condition: $x = 0$, $p(0) = p_0$. Integrating (5.120), we obtain

$$p = p_0 \exp(-\sigma_v L_j) \tag{5.121}$$

The degree of attenuation of the power density in decibels is determined using a expression resulting directly from (5.121):

$$p = p_0 10^{-0.1\alpha_v L_j} \tag{5.122}$$

where

$$\alpha_j = 4.3\sigma_v \text{ dB/m} \tag{5.123}$$

If passive jamming is used directly to reduce the radar detection range, then the required level of attenuation in dB of the power density can be determined using (5.122) after increasing L_j by a factor of two.

As an example, let us determine the required cubic density and the chaff corridor length, at which there occurs a reduction by 10 times in the operating range of a 3-cm radar. Accordingly, the degree of attenuation of the power density of the radar signal must be 40 dB. Using (5.122), with consideration to the necessity to double L_j, we obtain the equation defining the required value of the product $\sigma_v L_j$:

$$40 = 8.6\sigma_v L_j$$

or

$$\sigma_v L_j = 4.6 \tag{5.124}$$

If $L_j = 10^4$m, then the required specific RCS is

$$\sigma_v = 4.6 \cdot 10^{-4}\text{m}^2/\text{m}^3$$

Such an RCS density in the 3-cm band occurs if the cubic chaff density is three chaff dipoles/m^3. This is a quite high concentration that is difficult to attain. To jam a 10-cm band radar to the same degree, a significantly lesser cubic density (by approximately one order of magnitude) is required, but this density is also difficult to achieve. The possibilities of radar power jamming increase when active and passive jamming are combined.

5.6 Active–Passive Jamming

The combined use of active and passive jamming in the interests of increasing the effectiveness of radar jamming is possible in different variants [15]. Particularly, passive jamming can be used for the following purposes:

- Increasing the effectiveness of active jamming operating directly on the radar;
- Deliberate modification of the arrival angle of the jamming by illuminating the chaff cloud;

- Increasing the power ratio of the jamming and useful signal at the input to the receiver being jammed as a result of the placement of chaff corridors of great density in the radar scan area in the range of the active jammer.

In principle, other variants of the combined use of active and passive jamming are possible. Let us examine the three variants mentioned in greater depth.

In the first case (Figure 5.21) the jamming target is a coherent pulse (pseudo-coherent) radar with MTI. The placement of a passive jamming corridor (PJC) with a comparatively high RCS density (σ_v) compels the radar operator to turn on the MTI circuit, which leads to the reduction by 3–5 dB of the of the radar jamming coefficient using active jamming. The latter occurs as a result of the summing of the variances of the noncorrelated jamming signals and the frequency cancellation of useful signals at the output to the MTI. In the end, the target detection range of the radar being jammed is reduced by 2–2.5 dB(m), whereas the energy potential of the jammer remains the same.

If, due to the effect of active jamming generated by jammer JS, the radar detection bound of a battle formation of airplanes BF is $D_{A,1}$ (Figure 5.21) then, after setting up a passive jamming corridor (PJC), the detection bound is reduced to a value of $D_{A,2}$, less than $D_{A,1}$ by 2–2.5 dB(m).

In the second variant, the jamming target is directly acted upon by a jamming signal reflected from a chaff cloud. This significantly attenuates (by tens of dB or more) the active jamming power at the input to the jamming

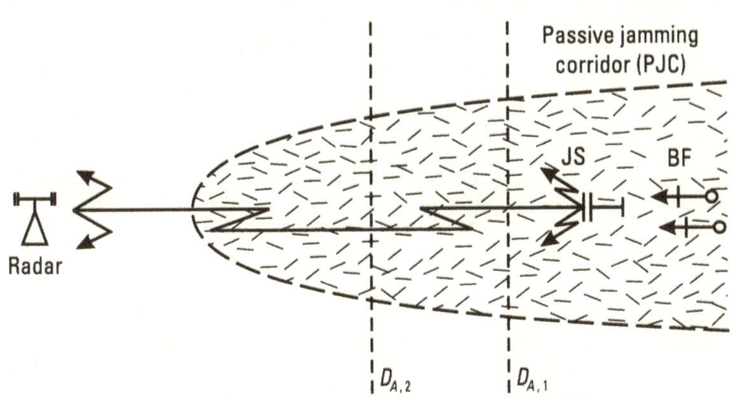

Figure 5.21 A coherent pulse (pseudo-coherent) radar with MTI as the jamming target.

target. This limits the sphere of use of the given jamming variant, in the main, to continuous and quasi-continuous radiation radar, the useful signal power of which is commensurable with the average power of the jamming signal. The corresponding variant of jamming the radar homing head (RHH) of a missile with semiactive homing when the target is continuously illuminated is given in Figure 5.22. In the case considered, the RHH of the missile is acted upon by the jamming signal reflected from chaff cloud CC.

The jamming equation using active–passive jamming of an RHH can be written in the following form:

$$K = \left(\frac{P_j}{P_s}\right)_{\text{inp,RHH}} = \frac{P_j G_j}{P_s G_s} \frac{\Delta f_{\text{RHH}}}{\Delta f_j} \gamma_j \frac{\sigma_{p0}}{\sigma_A} \frac{D_A^2}{L_{\text{JS}}^2} \frac{D_{\text{AM}}^2}{D_{\text{JM}}^2} \tag{5.125}$$

Here P_jG_j is the effective radiated power of the jammer JSA; P_sG_s is the effective radiated power of the radar illuminating the target; Δf_{RHH} is the passband of the linear part of the RHH receiver; Δf_j is the bandwidth of the jamming signal; γ_j is the polarization coefficient; σ_{p0} is the RCS of the part of the cloud CC, entering into the RHH resolution element, as determined by its antenna pattern, and occurring when scattering is bistatic (cf. (5.22) and (5.23)); σ_A is the aircraft RCS JSA; D_A is the distance from the aircraft to

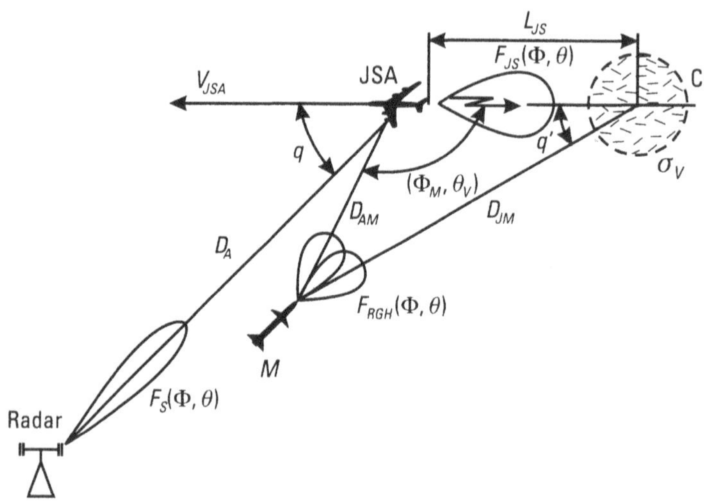

Figure 5.22 The variant of jamming a radar homing head (RHH) of a missile with semiactive homing.

the illuminating radar; L_{JS} is the distance of the chaff cloud with RCS density σ_p from the aircraft JSA; D_{AM} and D_{JM} are the distances from the missile to the aircraft JSA and to the chaff cloud CC; K is the ratio of the powers of the jamming and useful signal, as related to the passband of the linear part of the RHH receiver, implemented at a certain moment in time t (Figure 5.22).

The value of L_{JS} is selected with consideration to the following circumstance. The ratio $\sigma_{p0}/4\pi L_{JS}^2$ can not be smaller than $F_j^2(\Phi_p\Theta_p)$, the level of the sidelobe of the jammer (JSA) antenna pattern in the direction of missile P. The RCS of the chaff cloud σ_{p0} at small distances is not great, and accordingly the value of L_{JS} is also limited. The value of L_{JS} is also limited by the RHH jamming condition, in accordance with which the value of K, determined by the REJ equation (5.125) must be greater than or equal to the jamming coefficient K_j. The latter, in the given instance, can be determined using the relationships (3.83) and (3.84). Let us note that the value $L_j = 100$m corresponds to reduction in K by -40 dB. The presence of limitations on the value of L_{JS} on one side and the existence, when guiding the missile using the proportional navigation method, of a critical distance D_{Mcr}, after which the missile is no longer controlled [16] (furthermore, $D_{Mcr} >> L_j$), permits us to write down the following approximate formula for D_{JM}:

$$D_{JM} = D_{AM}\left(1 - \frac{L_{JS}}{D_{AM}}\cos(\pi - q)\right) \tag{5.126}$$

In many cases when performing practical calculations, it is possible to assume $D_{JM} = D_{AM}$. Then the jamming equation in canonical form, solved relative to the minimal radar jamming distance, is written down in the following manner:

$$D_{A\min} = L_{JS}\sqrt{\overline{K}_{REJ}} \tag{5.127}$$

where

$$\overline{K}_{REJ} = \frac{P_sG_s}{P_jG_j}\frac{\Delta f_j}{\Delta f_{RHH}}\frac{K_j}{\gamma_j}\frac{\sigma_A}{\sigma_{p0}} \tag{5.128}$$

In the case where the third variant is employed using active–passive jamming, the increase in effectiveness of jamming is achieved because the degree of attenuation of the useful signal is two times greater, since it passes through the chaff cloud twice, compared with a jamming signal that passes through this same cloud only once (Figure 5.23).

The required amount of chaff to ensure the reduction of the minimal jamming range by the available active jammer to the required limit $D_{A,2}$ can be determined from the jamming equation, written in logarithmic form for the two cases according to formulas (4.67) and (4.68).

Active jamming is generated from an area, located at a distance of D_j from the target:

$$D_{A\min} = \frac{1}{2} D_j + \frac{1}{4} \overline{K}_{\mathrm{REJ}} - \frac{1}{4} \alpha_j L_j$$

Here,

$$\overline{K}_{\mathrm{REJ}} = 10 \log \left(\frac{P_s}{P_j} \frac{G_s}{G_j} \frac{\Delta f_j}{\Delta f_s} \frac{K_{\Pi} \sigma_{\mathrm{BF}}}{4\pi \gamma_j F_j^2(\Phi_s, \Theta_s)} \right) \ \mathrm{dBm}^2$$

$\alpha_j = 4.3\sigma_v$, dB/m; L_j is the length of the chaff corridor with RCS density σ_v $\mathrm{m}^2/\mathrm{m}^3$.

Active jamming is generated from battle formations:

$$D_{A\min} = \frac{1}{2} \overline{K}_{\mathrm{REJ}} - \frac{1}{2} \alpha_j L_j$$

For example, if in the latter case it is required to reduce the minimum jamming range by a factor of two (i.e., to ensure conditions), in which $D_{A,2} = D_{A,1}/2$, where $D_{A,2}$ and $D_{A,1}$ are the minimum jamming ranges occurring correspondingly when active and passive jamming are

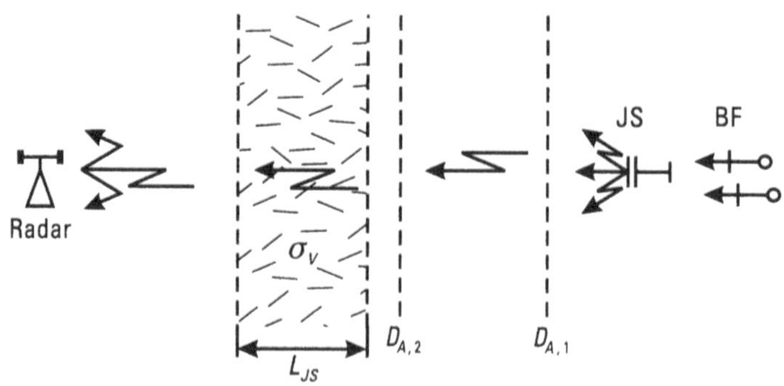

Figure 5.23 The third variant using active–passive jamming.

combined and without passive jamming, then the following equality must occur:

$$\frac{1}{2}\alpha_j L_j = 3 \text{ dB}$$

From this it follows that the RCS density of chaff cloud must be $\sigma_v = 1.4L_j^{-1}$.

In the particular case where the cloud is a cube with a side of 10 km, the mass of operating chaff dipoles, forming a cloud with an RCS of 1000 m^2, is 1 kg. The overall mass of chaff dipoles, required to solve the problem posed is $M_\Sigma = 140$ tons. In the case where jamming is generated from within the area, the required amount of chaff is doubled. The reduction of the cloud volume by an order of magnitude, when reducing σ_v, permits us to limit ourselves to only 14 tons of chaff.

5.6.1 Quantitative and Qualitative Indicators of the Effectiveness of Passive Jamming

The basic effectiveness indicators of passive jamming implemented using chaff are the averaged RCS of a unit of mass of chaff dipoles in a single standard bundle (packet), and the width of the frequency band they block out. Besides this, the important indicators should include:

- The effective chaff diffusion time;
- The average time of effective operation on the jamming target;
- The averaged RCS for a unit of mass of chaff σ_m is determined by the ratio of the RCS of the operating chaff dipoles after packet diffusion σ_j to their mass in packaged form m_j:

$$\sigma_m = \frac{\sigma_j}{m_j} \text{ m}^2/\text{g} \tag{5.129}$$

The required bandwidth of chaff in a single packet is correspondingly achieved by including half-wave chaff dipoles of varied length in it. An example can be the chaff in shell RR-170B/AL for chaff and thermal trap launching device AN/ALE-40, installed on U.S. Airforce tactical aircraft [17, 18]. In each shell there is chaff from glass fiber, covered with aluminum, of six various sizes (0.96, 1.57, 1.78, 2.24, 2.80 and 5.10 cm), which assures the masking of targets in the band from 2.3 to 18 GHz (wavelengths of jammed radar $\lambda = 1.7$–13 cm). The dependency of the RCS of the packet analyzed for frequency is shown in Figure 5.24.

The average diffusion time depends on the turbulent atmospheric diffusion parameters, as well as the specific launching techniques and devices. A significant number of variants for launching devices and technologies for the manufacture of chaff are known [14, 19]. At the present time, in the first approximation, it is possible to orient ourselves towards an average descent speed of chaff dipoles of 1 m/s, which, in many respects, determines the operating time of a cloud on a radar. The average RCS of operating chaff dipoles 1 m^2 corresponds to their mass 1g.

Basically, the indicators of active–passive jamming are determined by the capabilities of chaff and its launching devices. Requirements for the antenna pattern of jammers, performing illumination of chaff clouds, should be mentioned separately. The main lobe of the pattern must be quite broad in order to ensure illumination of the cloud, taking into consideration its random deviations from the beam axis. Besides, the level of the sidelobe radiation in the direction of the jamming target must be significantly reduced (Figure 5.22).

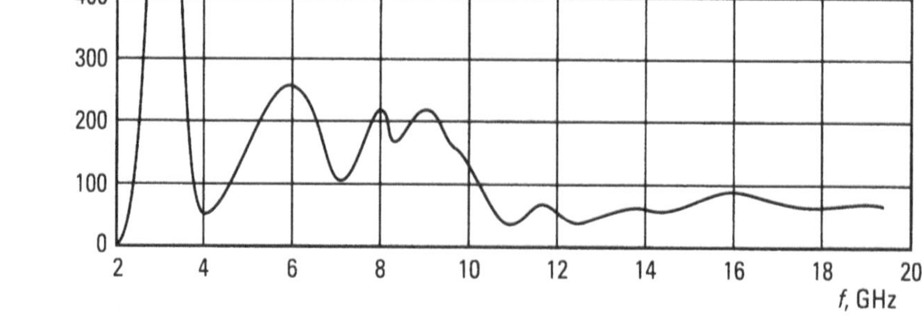

Figure 5.24 The dependency of the RCS of the packet analyzed for frequency.

References

[1] Fel'd, Ya. N., and L. S. Benenson, *Antenny santimetrovykh i detsimetrovykh voln* (*Centimeter and Decimeter Wave Antennas*). Moscow: N. E. Zhukovsky Airforce Academy, 1955, pp. 45–108, 167–202.

[2] Pistol'kors, A. A., *Antenny* (*Antennas*). Moscow: *Gosizdat* (*State Publishing House*), 1947, p. 102.

[3] Vakin, S. A., and L. N. Shustov, *Osnovy radioprotivodeystviya i radiotekhnicheskoy razvedki* (*The Basics of Radio Countermeasures and Radio Technical Intelligence*). Moscow: *Sov. radio* (*Soviet Radio*), 1968, pp. 9–122, 259–342.

[4] Gnedenko, B. V., *Kurs teorii veroyatnostey* (*A Course in the Theory of Probability*). Moscow: *Nauka* (*Science*), 1965, pp. 293–301.

[5] Rytov, S. M., *Vvedenie v statisticheskuyu radiofiziku* (*An Introduction to Statistical Radio Physics*). Moscow: *Nauka* (*Science*), 1966, pp. 105–116.

[6] Tikhonov, V. I., and Yu. N. Bakaev, *Statisticheskaya teoriya radiotekhnicheskikh ustroystv* (*The Statistical Theory of Radio Technical Devices*). Moscow: N.E. Zhukovsky Airforce Academy, 1976, pp. 5–119, 180–220, 263–416.

[7] Venttsel', E. S., *Teoriya veroyatnostey* (*The Theory of Probability*). Moscow: *Fiz.-mat.literat.* (*Physics and Mathematics Literature*), 1962, pp. 504–547.

[8] Tikhonov, A. N., and A. A. Samarskiy, *Uravneniya matematicheskoy fiziki* (*The Equations of Mathematical Physics*). Moscow: *Nauka* (*Science*), 1966, pp. 181–203.

[9] Zagorodnikov, A. A., *Nekotorye rezul'taty izmereniy turbulentnosti v svobodnoy chistoy atmosfere radiolokatsionnym sposobom. Doklady AN SSSR* (*Certain Results of Measuring Turbulence in a Free Pure Atmosphere Using Radar. Reports of the USSR Academy of Sciences*), Vol. 156, 1964, No. 6.

[10] *Rasprostranenie ul'trakorotkikh radiovoln* (*The Propagation of Ultra Short Radio Waves*). Under the Editorship of B. A. Sillerov. Moscow: *Sov. radio* (*Soviet Radio*), 1954, pp. 539–580, 691–697.

[11] Palermo, C. J., and L. H. Brauer, *Dvuxpozitsionnoe sechenie rasseyaniya dipol'nykh otrazhateley primenitel'no k zadacham radiosvyazi* (*Bistatic Scattering Cross Section of Chaff Dipoles with Application to Communications*). Proc. IEEE 53, Vol. 8, August 1965, pp. 1119–1121.

[12] Bronshteyn, I., and K. Semendyaev, *Spravochnik po matematike* (*Mathematics Handbook*). Moscow: *Nauka* (*Science*), 1986, pp. 72–73.

[13] Atrazhev, M. P., V. A. Il'yin, and N. P. Mar'in, *Bor'ba s radioe'lektronnymi sredstvami* (*Warfare against Radio Electronic Systems*). Moscow: *Voenizdat* (*Military Publishing House*), 1972, pp. 98–180.

[14] Plokhikh, A. P., and D. S. Shabanov, *Radiolokatsionnye otrazhateli i ikh primenenie* (*Radar Reflectors and their Application*). Moscow: *Zarubezhnaya radioe'lektronika* (*Foreign Radio Electronics*), Vol. 8, 1992, pp. 77–109.

[15] Van Brunt, L. B., *Applied ECM.*, EW Engineering Inc., Dunn Loring, VA, USA, 1978, 973 pp.

[16] Maksimov, M. V., G. I. Gorgonov, and V. S. Chernov, *Aviatsionnye sistemy radioupravleniya* (*Aircraft Radio Control Systems*). Moscow: N. E. Zhukovsky Airforce Academy, 1984, pp. 11–46, 50–197.

[17] Zaporozhets, G. V., *Standartnoe ustroystvo vybrosa sredstv radioe'lektronnoy bor'by samoletov takticheskoy aviatsii VVS SHAA* (*Standard Device for Launching Radio Electronic Warfare

Systems from US Airforce Tactical Aircraft). Moscow: N. E. Zhukovsky Airforce Academy, 1983, 46 pp.

[18] Zaporozhets, G. V., *Sredstva radioe'lektronnoy bor'by samoleta neposredstvennoy aviatsionnoy podderzhki A-10 (SSHA)* (*Electronic Warfare Systems on Board the A-10 Direct Aircraft Support Plane (USA)*). Moscow: N. E. Zhukovsky Airforce Academy, 1982, 28 pp.

[19] Paliy, A. I., *Radioe'lektronnaya bor'ba* (*Radio Electronic Warfare*). Moscow: *Voenizdat* (*Military Publishing House*), 1989, pp. 58–98.

6

False Radar Targets and Decoys

6.1 Types of False Radar Targets, Decoys and Disposable EW Devices

The use of:

- False radar targets,
- Radar decoys,
- Thermal decoys,
- Expendable sources of jamming emissions,
- Other expendable devices

are EW means of disrupting the information stability of electronic and optoelectronic systems [1–4].

Decoy radar targets are used to counteract AAD radar for detection, guidance and target designation (DGTD radar). By simulating real aircraft, they increase the stream density λ of targets entering the antiaircraft defense area. Given a constant stream density μ of targets to be serviced, the probability of a target being serviced by the AAD automated control system (ACS) for that area is thus decreased. In particular, when centralized control is disrupted, the service probability P'_{serv} according to (1.65) is equal to

$$P'_{serv} = \frac{1}{1 + \frac{\alpha}{n}}$$

where $\alpha = \frac{\lambda}{\mu}$, and n is the number of fire channels being serviced by the given ACS.

False radar targets can be used successfully to protect airplanes and other airforce targets at airports, in particular, by attracting high-precision weapon attacks, without having to use aircraft models.

Radar and thermal decoys are used to counteract the operation of the tracking sensors of weapon control systems. When expendable, a decoy should capture the tracking sensor of, for example, the radar or thermal homing head of the missile. The probability of capture is determined by the simulation quality and the probability that the decoy will be recognized.

Recently, jammers launched from airplanes and ships are being used as disposable devices. By simulating a jammer, they are able both to capture tracking sensors and to deliberately affect DGTD radar forming a part of AAD ACS.

Therefore, the problem is to provide a quality simulation of the basic attributes of the targets being simulated for all cases considered.

6.2 Parameters Simulated by False Radar Targets and Radar Decoys

In the given case, we are directly emulating the reflected radar signals from the target being simulated. Accordingly, it is necessary to simulate the characteristics of both the information and attendant parameters of the radar signals (including their statistical characteristics as well). In addition, it may be necessary to emulate the characteristic attributes inherent in the target being simulated (i.e., the turbine effect) [5].

Target simulation on first-generation radar displays was done by deliberate reproduction of the envelope of the reflected pulses. In practice, this was achieved with repeater pulse jammers [6] or through the use of chaff with the same RCS value as the target being simulated.

As the radar information stability increased, it became necessary to take into account the fluctuation characteristics of the signal reflected from an airplane, in particular, the probability and spectral densities of the envelope [7]. This required appropriate adjustments to the signals being simulated. In radar trackers, a coherent signal is formed as a collection of reflections from a series of scattering sources. As a result of this, angle noise is produced with specific statistical characteristics [7, 8]. The spectrum of signals reflected from the airplane in the Doppler frequency is typical.

A most informative attribute used in radar recognition is the polarization scattering matrix [5]. In radar signals reflected from the doubly curved surface segments of an airplane, depolarization of the electrical vector occurs due to the change in direction of electrical current when passing along the curved surface. The degree of depolarization can vary depending on the arrival angle of the wave and the radius of curvature of the surface. Quantitatively, polarization characteristics are represented in the form of appropriate scattering matrices for the airplane (missile, helicopter) received for various aspect angles [5, 7, 9, 10]. Other polarization characteristics can also be determined, in particular, the parameters of the polarization ellipse [10]. The elements of the polarization scattering matrix are complex reflection coefficients that occur when an aircraft (or other target) is illuminated by a plane wave. Whereas plane wave polarization is determined by two orthogonal electrical vectors (i.e., E_Φ and E_Θ), that form the polarization basis in a spherical system of coordinates, the polarization scattering matrix consists of four complex reflection coefficients S, defined for the orthogonal polarization components of the incident wave E_r and the reflected wave E_s.

In the general case, the orthogonal polarization components of the reflected plane wave $E_{s\Phi}$ and $E_{s\Theta}$ are expressed using the orthogonal components of the incident wave $E_{r\Phi}$ and $E_{r\Theta}$ in the following linear combinations:

$$\begin{aligned} E_{s\Phi} &= S_{\Phi\Phi} E_{r\Phi} + S_{\Theta\Phi} E_{r\Theta} \\ E_{s\Theta} &= S_{\Phi\Theta} E_{r\Phi} + S_{\Theta\Theta} E_{r\Theta} \end{aligned} \tag{6.1}$$

The indices Φ and Θ determine the pair of orthogonal polarization components corresponding to each of the reflection coefficients. Each of them in turn is determined by the modulus and phase of the reflection coefficient $(\sqrt{\sigma}, \psi)$, together with the required indices. The polarization diffusion matrix S in the given case is defined by the following expression:

$$S = \begin{Vmatrix} \sqrt{\sigma_{\Phi\Phi}} e^{j\psi_{\Phi\Phi}} & \sqrt{\sigma_{\Theta\Phi}} e^{j\psi_{\Theta\Phi}} \\ \sqrt{\sigma_{\Phi\Theta}} e^{j\psi_{\Phi\Theta}} & \sqrt{\sigma_{\Theta\Theta}} e^{j\psi_{\Theta\Theta}} \end{Vmatrix} \tag{6.2}$$

Here, $\sigma_{\Phi\Phi}$ is the RCS of the target. The polarization of its electrical vector $E_{r\Phi}$ is collinear with the vector of the polarization basis $E_{r\Phi}$, on the condition that the reflected signal is received by an antenna of the same polarization; and $\psi_{\Phi\Phi}$ is the phase of the reflection coefficient corresponding to the conditions which define $\sigma_{\Phi\Phi}$.

The meanings of the other reflection coefficients with various combinations of the indices Φ and Θ are analogous.

Using matrix (6.2), the linear form (6.1) can be expressed in the following way:

$$E_s = SE_r \tag{6.3}$$

The polarization scattering matrix of a false radar target or decoy must be determined in accordance with results of processing statistical measurement data for the scattering matrices of quite a large group of simulated aircraft. The measurements must be made in various frequency bands corresponding to the victim AAD radars, and for various bearing angles.

Along with the simulation of characteristic parameters of reflected radar signals, it is necessary to take into consideration changes in the signal generated by a given target. For example, the turbine effect [5], produces amplitude and frequency modulation in signals reflected from airplanes with turbojet engines. Other effects also occur.

These phenomena introduce significant difficulties in developing false radar targets and radar decoys that permit aircraft to overcome antiaircraft defense. In practice, the problem can be solved if the jamming energy is sufficient to provide simulation. This is achieved if, as a false target, we use an aircraft with an expendable active jammer operating on a radar station of the detection, guidance and target designation type. An analogous situation occurs when we launch expendable jammers and thermal decoys to protect individual aircraft.

The task becomes simpler when we deliberately simulate aircraft as false targets at an airport or other airbase. In this case, problems associated with motion are removed and quality results may be achieved by simulating only key attributes. This is determined for each case according to the radio environment and the problems to be solved. As a rule, it is necessary to assure a correspondence between the statistical RCS characteristics of the simulator and the averaged RCS of the simulated target. An indicator of the simulation quality for statistical characteristics is the difference in entropies of the probability distributions determined in the sample functions of random processes and representing the changes over time of parameters in the real and false targets (cf. p. 3.2).

6.3 Methods of Increasing the Radar Cross Sections of False Targets and Decoys

6.3.1 Repeaters

The implementation of repeaters does not present any real problems. In Figure 6.1 we show the block diagram of a simple repeater. The signals from the victim radar are received by antenna A_1, amplified by the preamplifier PA and fed to the final power amplifier FA. The signals are modulated here using noise voltage (current) generated by modulator M, ensuring a quality simulation. After amplification in the final amplifier FA, the signal is emitted by antenna A_2. Normally traveling-wave tubes are used as amplifiers with a broad passband and a high gain.

Taking into account the isolation being implemented, let us determine the gain of a repeater that can provide the required jamming signal power at the input to a given radar (Figure 6.2).

The signal power of the victim radar at the input to the repeater receiving antenna is equal to

$$P_{\text{inp}} = \frac{P_s G_s}{4\pi D_m^2} F_s^2(\Theta_m, \Phi_m) F_m^2(\Theta_s^1, \Phi_s^1) A_{r1} \tag{6.4}$$

where $P_s G_s$ is the ERP of the radar being jammed; $F_s(\Theta, \Phi)$ is the normalized antenna pattern of the radar being jammed; $F_m(\Theta, \Phi)$ is the normalized pattern of the repeater receiving antenna; Θ_m, Φ_m are the angular coordinates of the repeater (decoy), calculated from the beam axis of the victim radar; Θ^1, Φ^1 are the angular coordinates of the victim radar, calculated from the beam axis of the repeater; A_{r1} is the maximum effective area of the repeater receiving antenna; and D_m is the distance between the repeater (decoy) and the victim radar.

The ERP of the repeater transmitting antenna in the direction defined by angles Θ and Φ, is equal to

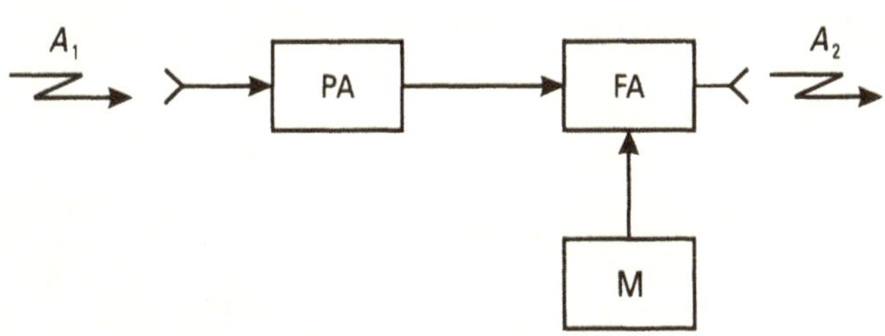

Figure 6.1 The block diagram of a simple repeater.

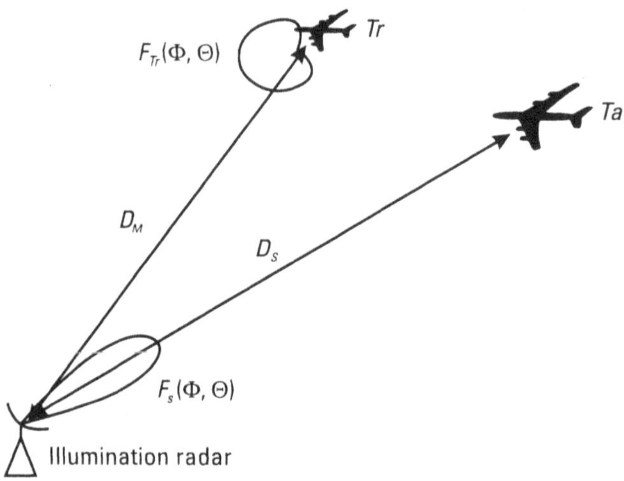

Figure 6.2 Determining the gain of a repeater that can provide the required jamming signal power.

$$P_{\text{outp } m} = P_{\text{inp } m} K_m G_2 F_{m2}^2(\Theta, \Phi) \tag{6.5}$$

where K_m is the power gain of the repeater; G_2 is the gain of the repeater transmitting antenna; and $F_{m2}^2(\Theta, \Phi)$ is the normalized pattern of the repeater transmitting antenna. In the material that follows, for the sake of simplicity, we let

$$F_{m1}(\Theta, \Phi) = F_{m2}(\Theta, \Phi) \quad G_2 = G_n$$

Between the receiving and transmitting antennas of the repeater it is necessary to ensure appropriate isolation, at least no less than K_m as a power ratio. In any case, K_m can not exceed the K of the isolation.

The powers of the jamming and useful signals at the input to the receiver of the victim radar are accordingly equal to

$$P_{n \text{ inp}} = \frac{P_s G_s}{4\pi D_m^2} \frac{A_{r1} A_{r2}}{4\pi D_m^2} G_n F_s^2(\Theta_m, \Phi_m) F_m^4(\Theta, \Phi) K_m \tag{6.6}$$

$$P_{s \text{ inp}} = \frac{P_s G_s}{4\pi D_s^2} \frac{\sigma_{tr}}{4\pi D_s^2} A_{r2} F_s^4(\Theta_s', \Phi_s') \tag{6.7}$$

where σ_{tr} is the RCS of the airplane being protected (Tr); D_s, Θ'_s, Φ'_s are the polar coordinates of the airplane being protected Tr (the angles Θ'_s, Φ'_s are calculated from the antenna axis of the radar being jammed); and A_{r2} is the maximum effective area of the antenna of the radar being jammed.

From (6.6) and (6.7), we have the jamming/signal ratio at the input to the victim receiver:

$$K = \left(\frac{P_n}{P_s}\right)_{\text{inp}} = \frac{A_{r1} G_n F_s^2(\Theta_m, \Phi_m) F_m^4(\Theta, \Phi) D_s^4 K_m}{\sigma_{tr} F_s^4(\Theta'_s, \Phi'_s) D_m^4} \tag{6.8}$$

Hence, assuming $K = K_j$, we easily obtain the expression for the required value of the repeater gain:

$$K_m = \frac{K_j \sigma_{tr} F_s^4(\Theta_s^1, \Phi_s^1) D_m^4}{A_{r1} G_n F_s^2(\Theta_m, \Phi_m) F_m^4(\Theta, \Phi) D_s^4} \tag{6.9}$$

In the particular case where the distance between the decoy and the airplane being protected is small in comparison to the distance between the airplane and the radar being jammed ($D_{tr} << D_s$), (6.9) is simplified to

$$K_m = \frac{K_j \sigma_{tr}}{G_n A_{r1}} = \frac{K_j \sigma_{tr}}{G_n^2} \tag{6.10}$$

where $G_n = G_m$ is the gain of the repeater receiver antenna. It is assumed that $\gamma_n = 1$. The repeater can be an expendable masking jammer.

6.3.2 Passive Reflectors

The possibility of increasing the RCS of decoys using passive reflectors is based on the specifics of the reflection of an incident plane wave by conducting bodies. The RCS of any body for the given direction is determined using the well-known formula:

$$\sigma = s_2 G \tag{6.11}$$

where $s_2 = \frac{P_2}{p}$ is the ratio of the power (P_2) reflected by the body to the power density (p) of the electromagnetic wave incident on the reflector; and G is the gain of the reflector in the given direction (in the direction towards the observation point).

For flat bodies, as well as for certain other bodies near in their reflection attributes to being flat, the value s_2 is equivalent to the effective area A_r of a certain antenna:

$$A_r = s_2 = \frac{G\lambda^2}{4\pi} \tag{6.12}$$

Substituting G from (6.12) into (6.11), we obtain

$$\sigma = \frac{4\pi}{\lambda^2} s_2^2$$

If an ideally conducting flat plate, the dimensions of which are much greater than the wavelength, is positioned normal to the direction of the incident wave, the RCS is expressed in the form:

$$\sigma_{\max} = \frac{4\pi s^2}{\lambda^2} \tag{6.13}$$

where S is the plate area.

As the orientation of the plate changes, the value of the reflected energy changes quickly. In Figure 6.3 we show the reflection pattern of a metallic plate, the dimensions of which are much greater than the wavelength. Due to the very sharp reflection pattern, a metallic plate is not suitable to increase the RCS of aircraft.

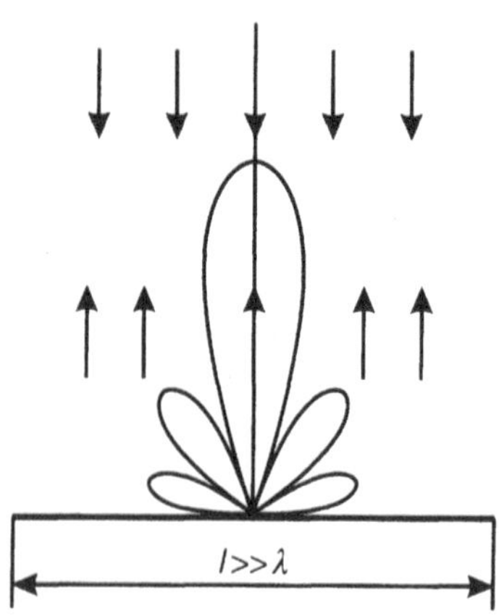

Figure 6.3 The reflection pattern of a metallic plate with dimensions much greater than the wavelength.

Reflectors simulating aircraft must satisfy the following requirements:

- Have a large RCS, but the smallest possible external dimensions and mass;
- Have quite a broad reflection pattern.

So-called corner reflectors of various types, reflectors in the form of Luneburg lenses and Van Atta arrays to a certain degree meet these requirements.

A corner reflector is a rigid structure consisting most often of three mutually perpendicular plates. Depending on the form of the plates, we distinguish (*a*) triangular, (*b*) rectangular and (*c*) round corner reflectors (Figure 6.4). Their maximal RCSs are equal to

$$\sigma_{\triangle} = \frac{4}{3}\pi\frac{a^4}{\lambda^2} \quad \text{(triangular)} \tag{6.14}$$

$$\sigma_{\square} = 12\pi\frac{a^4}{\lambda^2} \quad \text{(square)} \tag{6.15}$$

$$\sigma_{\circ} = 2\pi\frac{a^4}{\lambda^2} \quad \text{(round)} \tag{6.16}$$

Here, *a* is the rib length of the reflector.

Corner reflectors with small dimensions give a high RCS. Thus, when $\lambda = 3$ cm and $a = 50$ cm, $\sigma_{\square} = 2500$ m^2.

The reflection sector of corner reflectors at the half-power level is

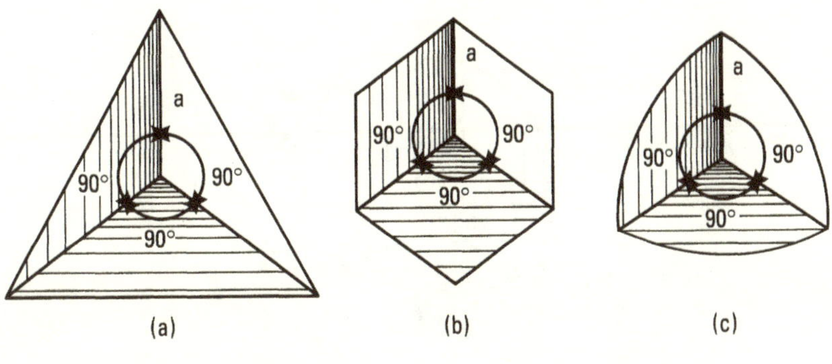

Figure 6.4 (a) Triangular, (b) rectangular and (c) round corner reflectors

approximately 40–50°. In order to increase the reflection sector, several corner reflectors are used, positioned variously in space. For example, the corner reflector shown in Figure 6.5 generates reflection practically in all directions.

The value of the maximum RCS for corner reflectors depends on the precision with which right angles between the plates of the reflector are sustained. A failure to sustain the precision of the angle by only 1° leads to a reduction in the maximum RCS value of a corner reflector by 2–5 times.

Radio countermeasures using corner reflectors against a radar having an antenna with circular polarization may be ineffective. This is explained by the fact that, from the conducting plates of the corner reflector, the ray is reflected an uneven number of times, as a result of which the direction of rotation of the electrical field vector of the reflected signal is reversed (Figure 6.6(a)). The change in polarization of the reflected waves occurring in corner reflectors can be eliminated if one of its plates is covered with a dielectric layer (Figure 6.6(b)) [6]. This issue is discussed in more detail below.

One of the significant drawbacks of corner reflectors is the small width of the reflection pattern at the half-power level. Reflectors implemented with Luneburg lenses have a quite a broad reflection pattern.

A Luneburg lens is a dielectric sphere. The index of refraction of the dielectric (n) in an ideal Luneburg lens depends only on the ratio of the radius within the lens (r) to its external radius (R):

$$n = \sqrt{2 - \left(\frac{r}{R}\right)^2} \tag{6.17}$$

In Figure 6.7(a) we show the ray trajectories in a Luneburg lens. The central ray AO does not undergo refraction in the lens, whereas the

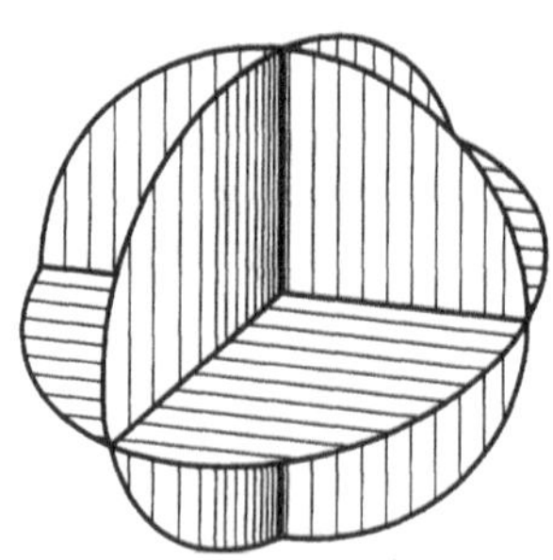

Figure 6.5 A corner reflector reflecting practically in all directions.

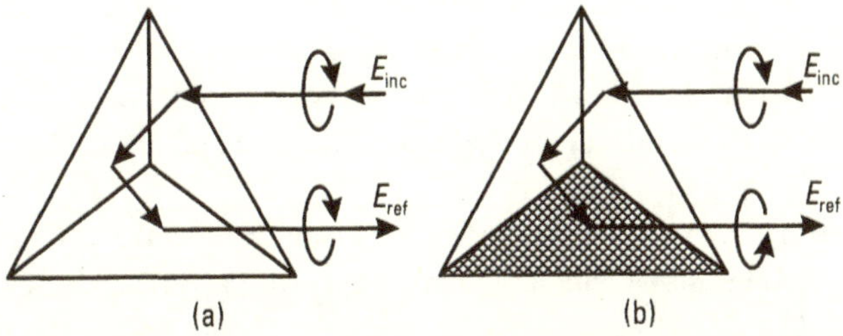

Figure 6.6 (a) Reversed direction of rotation of the electrical field vector of the reflected signal. (b) Elimination of the change in polarization of the reflected waves.

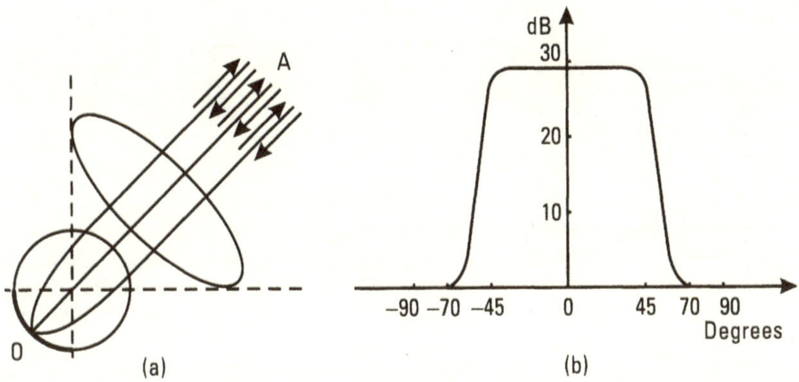

Figure 6.7 (a) Ray trajectories in a Luneburg lens. (b) The width of the reflection pattern at the half-power level.

trajectories of the remaining rays are distorted. As a result, all rays are focused at point *O* on the farther side of the sphere, covered with a metallic film. The point *O*, which is a source of secondary electromagnetic waves, creates an in-phase field distribution at the output of the lens, so that the maximum of the reflection pattern corresponds to the arrival direction of the incident wave.

The maximum RCS of a Luneburg lens can be found by substituting $s = \pi R^2$ in (6.13):

$$\sigma_l = 4\pi^3 \frac{R^4}{\lambda^2} \tag{6.18}$$

The width of the reflection pattern of a Luneburg lens depends on the dimensions of the reflecting (metallic) surface of the sphere. Thus, for a reflecting surface with a dimension of one quarter the surface of the sphere, the width of the reflection pattern at the half-power level is approximately 90° (Figure 6.7(b)). In Figure 6.8, we show the reflection pattern of a Luneburg lens with a 140-degree reflector [6]. The reflection sector of this lens is approximately equal to 140°.

A Luneburg lens does not provide for circular reflection. The latter can be achieved on the basis of this same lens if part of its sphere is encircled by a metallic ring [11].

In Figure 6.9(a) we show a Luneburg lens, omnidirectional in the

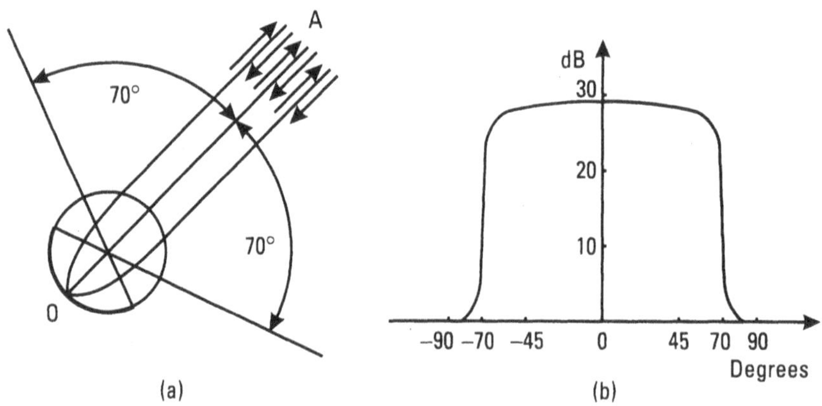

Figure 6.8 (a) A Luneburg lens with a 140 degree reflector and (b) its reflection pattern.

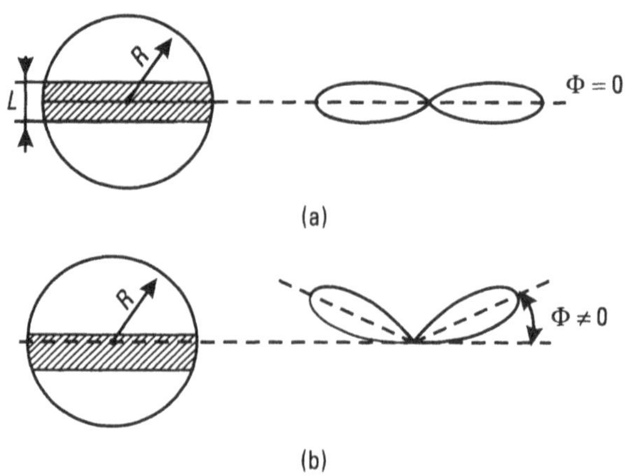

Figure 6.9 A Luneburg lens, omnidirectional in the azimuthal plane, with a reflective ring around the equator: (a) ring around the equator; (b) shifted ring.

azimuthal plane, with a reflective ring around the equator. The position of the metallic ring determines the direction of maximum reflection. Thus, for a ring around the equator, the maximum is located in the equatorial plane (Figure 6.9(a)). If, however, the ring is shifted, the reflection pattern lobe deviates from the equatorial plane (Figure 6.9(b)). The maximum RCS value is determined using the formula:

$$\sigma_k = 4\pi \frac{(\pi R^2 - 2RL)^2}{\lambda^2} \tag{6.19}$$

where R is the radius of the sphere; and L is the width of the metallic ring.

A lens in the form of a dielectric sphere with a metallic ring is somewhat less effective than the Luneburg lens with a reflector considered before. A comparison of (6.18) and (6.19) yields

$$\eta = \frac{\sigma_k}{\sigma_l} = \left(1 - \frac{2L}{\pi R}\right)^2$$

For example, if $L/R = 0.2$, then $\eta \approx 0.9$. Accordingly, the reduction in the RCS σ_k in comparison with an ordinary lens is insignificant.

An increase in the width of the metallic ring leads to the widening of the reflection pattern, but the RCS of the lens is reduced. This contradiction is solved well in a Luneburg lens with a ring in the form of an array of parallel wires wound at an angle of 45° (Figures 6.10 and 6.11). Such a lens is sometimes called a helisphere.

An electromagnetic wave with a linear polarization incident on the helisphere at an angle of 45° passes through the frontal part of the ring and is

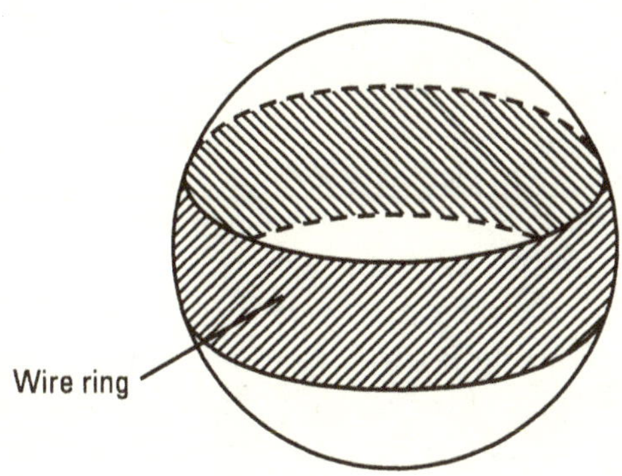

Figure 6.10 An increase in the width of the metallic ring.

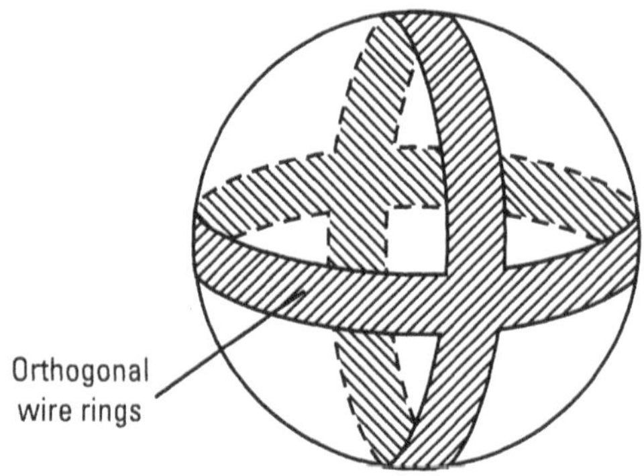

Figure 6.11 A Luneburg lens with a ring in the form of an array of parallel wires wound at an angle.

reflected off the opposite part of the ring. For other polarizations, (vertical, horizontal, circular), there are polarization losses, the maximum of which (when passing through twice) is 6 dB.

To obtain an isotropic pattern, a helisphere with two orthogonal wire arrays is used. By selecting the width (and sometimes the configuration) of the rings, it is possible to obtain a comparatively small heterogeneity of the reflection pattern.

In principle, it is possible to create an ideal lens without metallic rings with isotropic reflection (an Eaton–Lipman lens). For this, the coefficient of the dielectric filler must change according to the law:

$$n = \sqrt{\frac{2R}{r} - 1}$$

where r is the radius within the sphere.

The ray trajectory in a lens with circular reflection is portrayed in Figure 6.12.

Dielectric helispheric lenses are quite heavy. They are difficult to implement due to the necessity of inserting a dielectric with a very large index of refraction into the central part ($r \approx 0$).

Hollow helispheric reflectors have significantly less weight (Figure 6.13(a)). The reflective ring is made from metal and comprises a spherical section. The wire array and the metallic ring are positioned orthogonally. The RCS of such a reflector is 10 dB less than that of an ideal Luneburg lens. The path of the incident and reflected waves are shown in Figure 6.13(b).

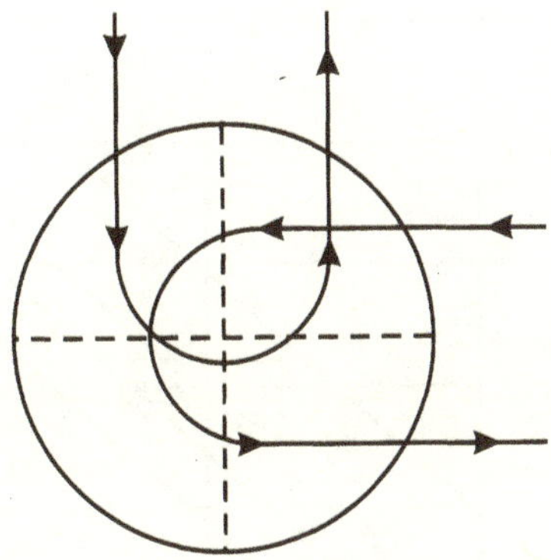

Figure 6.12 The ray trajectory in a lens with circular reflection.

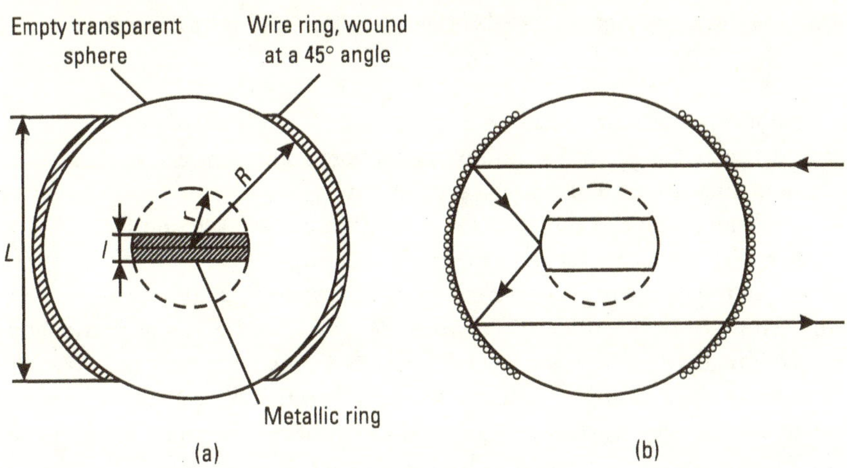

Figure 6.13 (a) A hollow helispheric reflector. (b) The path of the incident and reflected waves.

In Figure 6.14 we show the dependency of the RCS of various reflectors, normalized relative to the RCS of a metallic sphere of the same diameter D, on the parameter D/λ. We lay off the following value along the axis of the ordinates:

$$\alpha = 10 \lg \frac{\sigma}{\sigma_0}$$

where σ_0 is the RCS of a metallic sphere, σ is the RCS of the lens. The graphs

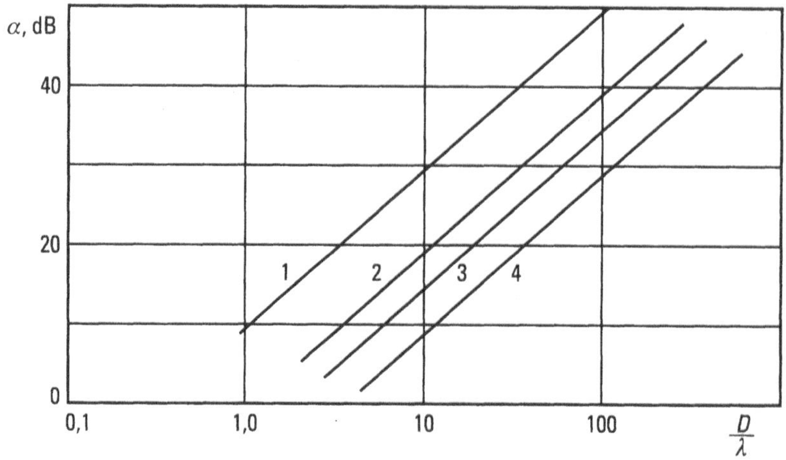

Figure 6.14 The dependency of the RCS of various reflectors on the parameter D/λ.

correspond to: 1— and ideal Luneburg lens; 2 — a helisphere with a ring inside (Figure 6.13); 3 — a hollow helisphere (linear polarization); 4 — a hollow helisphere (circular polarization).

A Van Atta array is, in essence, an antenna array made from a large number of dipoles, spirals and other reflectors (Figure 6.15). The dipoles are located at an equal distance from the axis of symmetry and are connected in pairs using coaxial cables of identical length. An electromagnetic wave received by dipole 1 is reradiated by dipole 6. In turn, dipole 1 reradiates the wave received by dipole 6. The electrical lengths of the antenna feeder system 1–6, as well as of the other dipoles connected in pairs, are identical. Signals, received and reradiated by the dipoles, traverse an identical path. Therefore, the direction of the maximum of the reradiation pattern corresponds to the direction of arrival of the incident wave.

The arrays are designed to reflect waves with any polarization. For this purpose, the dipoles are placed in a metallic screen at various angles (as a rule, each pair forms a 90° angle with the neighboring one).

The RCS of the array formed from n half-wave dipoles located at a distance of $\lambda/2$ from one another and at a distance of $\lambda/4$ from the reflecting screen, can be found using formula [12]:

$$\sigma = 4\pi \frac{S^2}{\lambda^2}\left[\sin\left(\frac{\pi}{2}\cos\Theta\right)\right]^4$$

where Θ is the incidence angle; and S is the array aperture area.

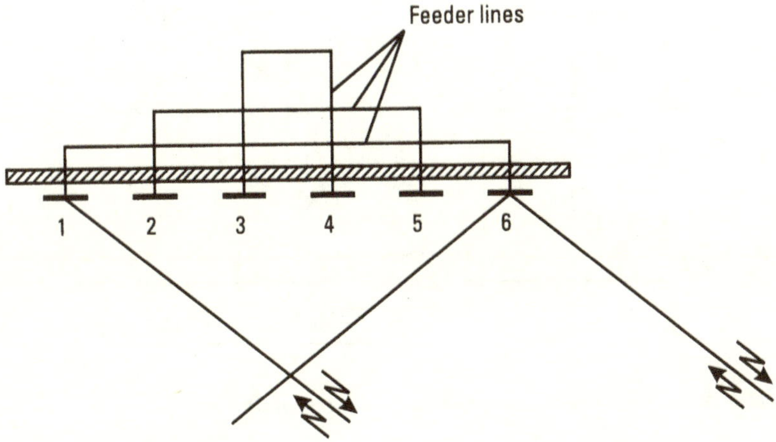

Figure 6.15 A Van Atta array.

Bearing in mind that $S \approx n\lambda^2/4$, we obtain the expression for the maximum RCS of a Van Atta array:

$$\sigma \approx \frac{\pi}{4} n^2 \lambda^2$$

The re-radiated signal may be amplitude modulated. For this purpose, phase shifters are included in the feeder lines connecting the radiators. By making an appropriate change in phase shift, it is possible to obtain the necessary amplitude modulation law in the reradiated signal. Using phase shifters and other components, it is possible to achieve the required radiation parameter randomization in the interests of providing the needed simulation quality.

A Van Atta array can also be made in an active variant (a repeater), where the signal received is amplified in each channel and is reradiated (Figure 6.16). The basic difficulty in making this type of reradiator is isolating the receiving and transmitting sections. The drawback of a Van Atta array is its relatively narrow bandwidth.

In conclusion we would note that the various reflectors considered in this paragraph do not exhaust the manifold devices and techniques for increasing the RCS of false targets. For example, research is being conducted into the possibility of increasing aircraft RCS using space ionization around the false target, in particular, the ionization of the jet engine stream of the decoy missile by adding (injecting) easily ionized elements into the fuel [6].

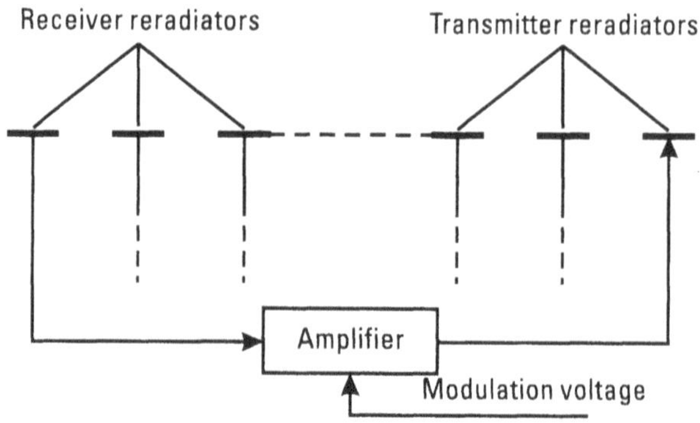

Figure 6.16 A Van Atta array in an active variant (a repeater).

In principle, it is also possible to achieve the needed simulation quality in an analogous manner by using angle reflectors and Luneburg lenses. For this purpose, it is necessary to form the reflected signal using a set of reflectors, separated in space, that form a single system, including components in its high frequency section to provide an ability to modulate and randomize the deception signal. The reflected signal must also ensure that its scattering polarization matrix is identical to the corresponding matrix of the useful signal. In this area, controlled impedance coverings and structures that have appeared in radar technology recently and are associated with the problem of controlling the diffusion of electromagnetic waves, seem very promising [13]. Bearing in mind the assurance of information stability in deception systems in the complex conditions of information conflict, it is convenient to increase their intellectual potential by including knowledge bases containing appropriate adaptation procedures, created on the basis of generalized expert evaluations that have been defined using a large number of typical cases [14], into the automated deception jamming control system.

6.4 Thermal Decoys

Thermal decoys are the most common expendable EW device. They are widely used not only by military, but also by civil aviation flying over a dangerous area. The operations targets of thermal decoys (TD) are thermal (infrared, IR) homing heads (THH) of "air-to-air" and "surface-to-air" class missiles that perform passive homing on aircraft protected by TD. The source of radiant energy, carrying information about the angular coordinates of the aircraft, is thermal (IR) radiation.

The basic irradiators of a turbojet airplane at subsonic speeds are the turbojet engine (TJE) and the jet stream of heated gasses. The most intensive radiation of molecules of a hot gas stream at all temperatures occurs at wavelengths 2.7 and 4.3 μm [15]. The maximum radiation λ_{max} = 2.7 μm, is caused by the overall radiation of water vapor and carbon dioxide gas CO_2. Radiation at a wavelength $\lambda_{max} = 4.3$ μm is caused by CO_2. During intercept guidance of a missile with a THH to supersonic aircraft, the source of radiant energy is their outer covering, heated by the opposing air stream.

Thermal decoys are cartridges (shells), filled with a pyrotechnic compound with a quite high combustion temperature, that are launched (shot) from aircraft in such a fashion that their radiation effects the jamming target. The average combustion time of a TD is up to 6 sec in the infrared range 2–5 μm [1–3].

The energy of the radiant stream of the thermal decoy within the limits of field of sight (the entrance pupil) of the THH must be greater than the radiant stream energy emitted by the aircraft and assure the capture of the missile with a THH at itself. The quantitative evaluation of capture conditions can be determined on the basis of the optoelectronic jamming equation (OEJ), which an analog of the electronic jamming equation. Before writing the OEJ equation, we will mention certain definitions and terms commonly used in optoelectronic technology (when using energy values and units).

Initial in the given case is the radiant stream power $\Phi_{e\ \mathrm{inp}}$, incumbent on the input optics (input pupil) of the radiation receiver of the THH. Specifically the power of the input radiation stream determines the value of the output signal of an optoelectronic receiver [15, 16]. For the THH to be captured by the TD, the following inequality must be fulfilled:

$$\Phi_{e\ \mathrm{inp},\,tt} \geqslant K_j \Phi_{e\ \mathrm{inp},\,ac} \tag{6.20}$$

where $\Phi_{e\ \mathrm{inp},\,tt}$ and $\Phi_{e\ \mathrm{inp},\,ac}$ are the radiant stream powers from the TD and the aircraft at the input to the pupil of the optoelectronic receiver of the THH; K_j is the THH jamming coefficient, determining the degree by which the radiant stream power of $\Phi_{e\ \mathrm{inp},\,tt}$ must exceed $\Phi_{e\ \mathrm{inp},\,ac}$ for there to be a high probability of capture. It is determined experimentally or using modeling.

The order of the value K_j can be determined from expressions (3.83) and (3.84). In the case examined, they are expressed in the following way:

$$\Delta\Theta_s = \Delta\Theta_o \frac{K}{1+K} \tag{6.21}$$

where

$$K = \frac{\Phi_{e\ \mathrm{inp},\,tt}}{\Phi_{e\ \mathrm{inp},\,ac}}$$

Presumably $K_j = 3$–4. This value corresponds to a capture probability no less than 0.6–0.7, if the radiation of the TD affects the input pupil of the optoelectronic receiver.

The relationship (6.20) can be considered to be the optoelectronic jamming equation OEJ. Based on the OEJ equation (6.20), let us determine the required radiation power density $I_{e\ tt}$ for a thermal decoy for a given radiation power density of a turbojet engine or other source of radiation $I_{e\ ac}$.

The radiation power density I_e determines the radiant stream radiated by the source in the given direction. In a spherical system of coordinates [15, 16],

$$I_e = \frac{d\Phi_e}{d\Omega} \, {}^{\mathrm{Wt}}\!/_{\mathrm{ster}} \tag{6.22}$$

The total radiant stream power Φ_e, radiated in the hemisphere by a source with area S is equal to the product of the radiation surface density M_e and the surface area:

$$\Phi_e = M_e S \tag{6.23}$$

In accordance with the Stefan–Boltzmann law for an ideal source (an absolutely black body),

$$M_e = \sigma T^4 \, {}^{\mathrm{Wt}}\!/_{\mathrm{m}^2} \tag{6.24}$$

where σ is the Stefan–Boltzmann constant:

$$\sigma = 5.67 \cdot 10^{-8} \, {}^{\mathrm{Wt}}\!/_{\mathrm{m}^2 \cdot \mathrm{K}^4}$$

T is the absolute temperature in Kelvin.

Assuming the diffuse scattering of radiation and the validity of Lambert's law, and with due consideration to (6.23) and (6.22), we obtain the following expressions, defining the radiation power density of an ideal point source (i.e., quite remote from the radiation receiver):

$$I_e = \frac{1}{\pi} M_e S \cos \Theta$$

or

$$I_e = \frac{1}{\pi} \sigma S T^4 \cos \Theta \, {}^{\mathrm{Wt}}\!/_{\mathrm{ster}} \tag{6.25}$$

Practically speaking, the distribution of the radiation power density is determined experimentally and is represented in the form of radiation contours. In Figure 6.17, we show the radiation contours of an American fighter F-104 with a TJE: (*a*) vertical; (*b*) horizontal. The diffusion contour, allowing for the effect of reflected radiant streams, is indicated using a dotted line [15].

In the Stefan–Boltzmann law, the value M_e defines the integral surface density of radiation, encompassing the entire spectrum range of emissions. The radiation spectral density $M_{e\lambda}$ is determined using Planck's law [15, 16]:

$$M_{e\lambda} = C_1 \lambda^{-5} \left(\exp\left({}^{C_2}\!/_{\lambda T} \right) - 1 \right)^{-1} \tag{6.26}$$

where $C_1 = 2\pi h c^2$; $C_2 = ch/k$; c is the speed of light in a vacuum (m/s); k is Boltzmann's constant ($k = 1.38 \cdot 10^{-23}$ J/K); h is Planck's constant (the quantum constant) ($h = 6.63 \cdot 10^{-24}$ J.S.); and $C_1 = 4.99 \cdot 10^{-24}$ J.m; $C_2 = 1.44 \cdot 10^{-2}$ mK.

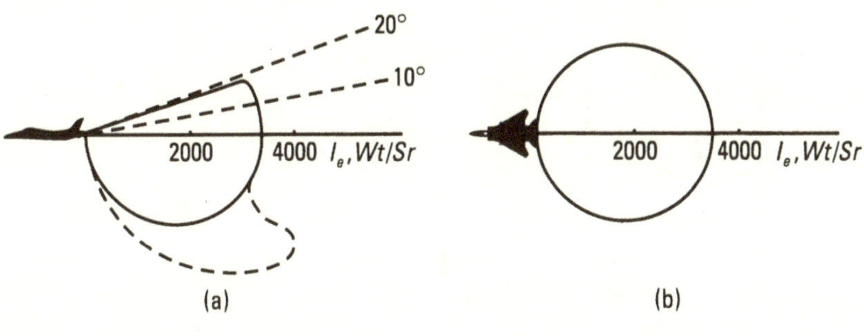

Figure 6.17 The radiation contours of an American fighter F-104 with a TJE: (a) vertical; (b) horizontal.

A graph of the radiation spectral density for an ideal irradiator (an absolutely black body) for three temperature values is given in Figure 6.18 [16]. In this same diagram, a curve is shown using a dotted line to reflect Wien's displacement law ($\lambda_{max} = 2900/T$), according to which, as the absolute temperature rises, the maximum radiation spectral density shifts in the direction of shorter waves.

By definition, the power of the radiant energy stream at the input to the optoelectronic receiver $\Phi_{e\ inp}$, located at a distance D from the point radiation source, is equal to

$$\Phi_{e\ inp} = I_e \Delta\Omega \text{ Wt} \tag{6.27}$$

where $\Delta\Omega = S_{inp}/D_;$ and S_{inp} is the area of the input pupil of the optoelectronic receiver.

The ratio of the power radiation streams for a thermal decoy with an area of S_{tt} and the engine (EN) of a protected aircraft with consideration to (6.27) and (6.25) is expressed in the form:

$$\frac{\Phi_{e\ inp\ tt}}{\Phi_{e\ inp\ ac}} = \frac{I_{e\ tt}}{I_{e\ ac}} = \frac{S_{tt} T_{tt}^4}{S_{en} T_{en}^4} \tag{6.28}$$

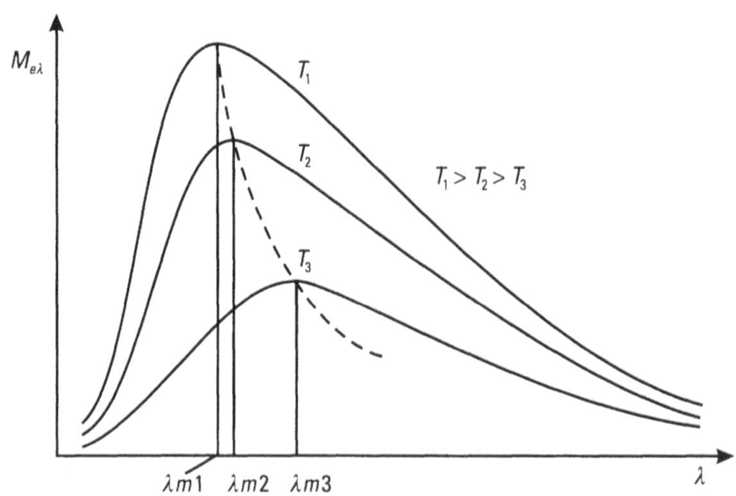

Figure 6.18 A graph of the radiation spectral density for an ideal irradiator for three temperature values.

According to (6.20) and (6.28), the missile is captured by the TD, if the following inequality occurs:

$$S_{tt}T_{tt}^4 \geqslant K_j S_{en} T_{en}^4 \tag{6.29}$$

where S_{en} is the area of the output aperture of the jet engine.

The capture conditions (6.29) must be fulfilled in the infrared radiation range (2.7 and 4.3 μm). As follows from Wien's law, these ranges correspond to engine temperatures of the order 700–1000 K. The combustion temperature of a thermal decoy T_{tt} must be in the range of

$$T_{tt} \geqslant T_{en} \sqrt[4]{K_j \frac{S_{en}}{S_{en}}} \tag{6.30}$$

Assuming the radiating surfaces of the TD and the engine to be round, and the jamming coefficient to be $K_j = 2$, let us write down (6.30) in the following form:

$$T_{tt} \geqslant T_{en} \sqrt[4]{1.4 \frac{d_{en}}{d_{tt}}} \tag{6.31}$$

If we assume diameters of TJE output orifices of the order of ones of meters and launching device diameters in tactical aircraft of the order of hundredths of a meter, then, with the help of Wien's law and the formula $\lambda_{\max} = 2900/T_{tt}$ resulting from it, it is possible to determine that the maximum TD radiation must be in the ultraviolet range.

The circumstance mentioned gives cause to assert that thermal decoys with a small radiation area (small diameter) have a spectrum that differs significantly from that of a TJE, which reduces their information stability, since techniques can potentially be proposed to reduce the degree to which their radiation effects THH [17]. Better from the point of view of providing information stability are TD with a large diameter and a lower radiating surface temperature. An increase in diameter causes problems of a different nature. Certain of them are considered below in conjunction with the particulars of the kinematics and dynamics of decoys launched from an airplane.

To a great extent, the information stability of THH with respect to thermal decoys is attained by increasing the resolution of thermal homing heads. In these conditions, it is necessary to assure the burning TD reaches its nominal temperature in quite a short period of time.

6.5 The Use of Towed and Launched Decoys

6.5.1 Towed Decoys

Towed decoys can be used to thwart a missile or fighter attack during the last phase of guidance (homing). These decoys are towed by a bomber airplane on a thin cable, the length of which can reach several kilometers. When stowed, the decoy is located in a special compartment. At the moment the bomber starts penetrating the most dangerous AAD areas, the decoy is released with the help of a starting device.

The first experience of using towed decoys goes back to the Second World War, when English and American airforces used towed metallic nets as false targets to reduce the AAD effectiveness of the Germans. These nets, creating powerful reflected signals, captured the gun laying stations

A towed decoy is equipped with amplifying repeaters and passive reflectors that increase the value of its RCS to that of the carrier airplane. When necessary, equipment for infrared jamming can be installed in the decoy, as well as sources for its radiation.

Towed decoys can be employed to counteract guidance and homing circuits. Their effectiveness is high if, at the initial moment of missile (or fighter) guidance, the protected airplane and the decoy are represented as a single target (i.e., they are not resolved into angles, ranges and Doppler frequencies). The use of towed decoys has a number of specifics.

The distance of the towed decoy from the airplane basically is determined by the angle and velocity resolution of the system being jammed.

The condition where the airplane and the decoy are not resolved in angle is expressed in the form:

$$L \leqslant \frac{\Delta\Theta_m D}{\sin q} \tag{6.32}$$

where $\Delta\Theta_m$ is the resolution of the victim radar in angle; D is the distance to the victim radar; and q is the decoy aspect ratio.

The condition where the airplane and the decoy are not resolved in Doppler frequency is determined by the passband width of the "velocity gate" ΔF and the difference in Doppler frequencies of the decoy and the airplane Δf:

$$\Delta f \leqslant \Delta F$$

Here,

$$\Delta f = \frac{2}{\lambda}(v_{am} - v_{tr}) = \frac{2\Delta v}{\lambda} \tag{6.33}$$

v_{am}, v_{tr} are accordingly the closing velocities of the attacking missile and the protected airplane and decoy (Figure 6.19).

Conditions (6.32) and (6.33) determine the maximum permissible distance of the decoy from the airplane protected.

In Figure 6.20 we show the resolution areas of the towed decoy and airplane in angle I and the differences in the radial velocities (Doppler

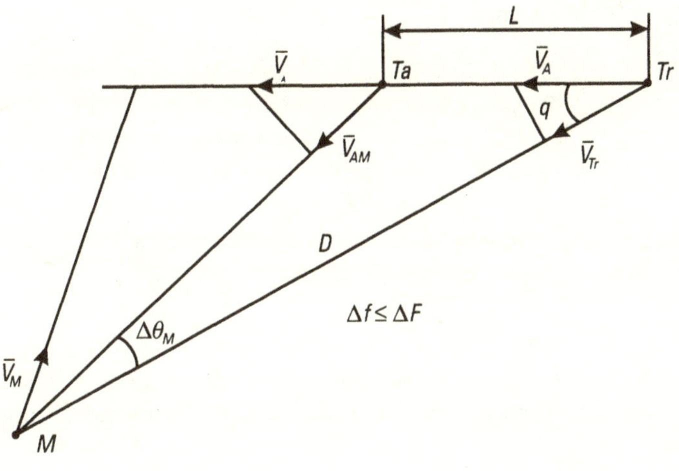

Figure 6.19 A towed decoy.

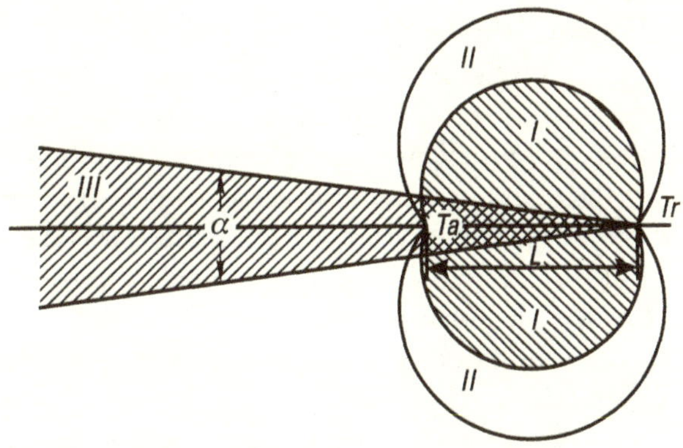

Figure 6.20 The resolution areas of the towed decoy and airplane.

frequencies) II. In the same diagram, we show the area where the decoy is eclipsed by the plane III. The airplane eclipses the decoy if the attacking missile is inside the cone, the angle at the apex of which is equal to

$$a = \text{arc } \sin \frac{k'R}{L}$$

where R is the kill radius of the missile; and k' is the safety factor.

The value of area III is quite significant for attacks from the front hemisphere.

6.5.2 Expendable Decoys

Expendable decoys serve to protect an airplane from an attacking missile (fighter). These decoys have no motors and are active and passive reflectors with a larger RCS than the target airplane they are protecting. A corner reflector, a chaff packet or a burning pyrotechnic compound can serve as a decoy.

The expendable decoy can be locked onto by the tracking system of the missile (fighter), if the following conditions are met:

- The RCS of the decoy is greater than that of the protected airplane;
- The operations time of the false target signal is greater than or equal to the time constant of the tracking system for angle, velocity and distance.

The second condition imposes quite rigid requirements on the design of expendable decoys.

For pulse systems, the length of time that signals from the decoy operate on the homing system is determined by the time the decoy spends in the pulse volume and, for continuous radar, by the time the radial component of the relative velocity of the decoy spends within the limits of the passband of the tracking system for velocity, as well as by the time the decoy spends within the antenna radiation pattern of the radar being jammed. For a thermal decoy, this is the time spent within the limits of the field of sight of the IR radiation receiver.

Let us determine conditions for a pulse automatic target tracking radar to lock onto a expendable decoy.

In order to determine the time the decoy and target remain in the same pulse volume, it is necessary to calculate the trajectory of the false target launched from the airplane when it is in free fall. As is known from a course

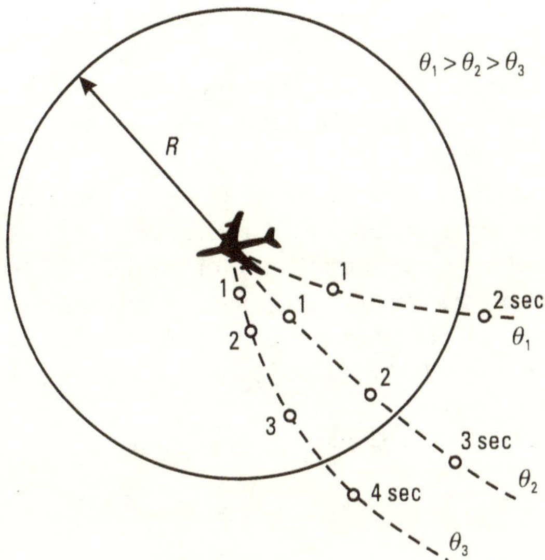

Figure 6.21 The approximate trajectories of falling decoys.

in ballistics, the free-fall trajectory is determined by the characteristic time Θ and the height and velocity the airplane is flying at when the decoy is launched.[1]

In Figure 6.21 we show the approximate trajectories of falling decoys in a mobile system of coordinates, associated with the bomber for various values of the characteristic fall time $\Theta_1 > \Theta_2 > \Theta_3$.

As can be seen from the diagram, the characteristic fall time has a significant influence on the time the airplane and decoy spend in the pulse volume. When $\Theta = \Theta_2$, the time the decoy spends in the volume of a sphere with $R =$ const is greater than when $\Theta = \Theta_1$.

The characteristic fall time of a decoy is determined by the formula, well known in ballistics [18]:

$$\Theta = 20.2 + \frac{id^2}{G} C_x 10^3 \tag{6.34}$$

where $\frac{id^2}{G} C_x 10^3$ is the ballistics coefficient; i is the decoy shape coefficient; d is

1. The characteristic fall time is the time for a body to fall from a height of 2000m in standard atmospheric conditions.

the decoy diameter; G is the mass of the decoy; and C_x is the aerodynamic coefficient.

From (6.34) and Figure 6.21, it follows that, in order to increase the time a decoy spends in the pulse volume, it is necessary to decrease its characteristic fall time Θ.

Since the external dimensions of a decoy are determined by the dimensions of the reflectors installed on it, a decrease in the characteristic time Θ is achieved by increasing the mass of the decoys (see (6.34)). Let us determine, for example, the mass of a decoy comprising a Luneburg lens with a diameter $d = 0.5\text{m}$ ($\sigma \approx 190\ \text{m}^2$, $\lambda = 5$ cm). Let us assume $i = 8$, $C_x = 0.2$. Then the required mass of expendable decoys is approximately

$$G = 500 \text{ kg} \quad \text{if } \Theta = 21 \text{ sec}$$

$$G = 70 \text{ kg} \quad \text{if } \Theta = 26 \text{ sec}$$

The example given shows that expendable decoys must have a significant mass.

In order to increase the time an expendable decoy spends (when its mass is small) in a quite close vicinity to the target, it is necessary to specify its initial velocity in the direction of motion of the aircraft (i.e., to equip the decoy with a start-up motor). Accordingly, its trajectories in a mobile system of coordinates, linked with the bomber, change abruptly. In Figure 6.22 we show the trajectory of a decoy equipped with a start-up motor. Here we also mark, using a dotted line, the trajectory of a decoy without a motor.

To this point in time we have considered that there are no differences in the techniques of jamming guidance and homing circuits using decoys. However, the use of decoys for the guidance circuits of a pulse-type tracking system has important specifics. Basically, these specifics relate to the maximum separation velocity of the decoy from the protected airplane, which in the latter case is determined by the discrete nature in which information about target coordinates is received.

In the simplest pulse-type guidance circuit, data about target coordinates arrives with a certain repetition factor, the value of which depends on the radar antenna rotation velocity (normally a 360° scanning radar). The period for receipt of information in modern systems is evaluated of the order of several seconds.

Releasing the decoy at the moment the protected airplane is irradiated, it is necessary to fulfill conditions (the initial velocity, the average velocity over the radar scan period): the released decoy should not, within one scan

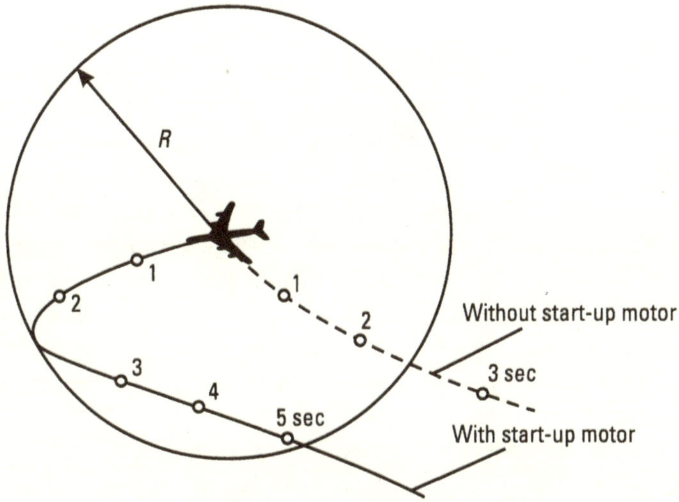

Figure 6.22 The trajectory of a decoy equipped with a start-up motor.

cycle, go beyond the limits of the area in the vicinity of the protected airplane that is attainable by this airplane in accordance with the overload conditions and the possibilities of its power installation for the given time. The absolute value of the initial velocity in the given case is not of any special consequence. It is important that the average velocity during the scan period of the victim radar be less than a certain value.

6.6 Selecting Decoy Launch Time

Depending on the electronic environment taking shape, the launching of decoys is performed either continuously with a certain fixed average rate, or a one-time launch is done in accordance with information arriving from devices indicating the launch or the danger of launch of enemy missiles, or the use of other weapons against the aircraft protected. In the first case, we mentioned, above all, the launch of thermal decoys in the high-threat area where an aircraft may be hit by mobile and other antiaircraft and missile complexes with thermal homing heads. The release rate is selected according to the average fire cycle length of the antiaircraft and missile complex. In the second case, the decoy, including thermal or expendable jammers, is launched according to information received from onboard instruments warning the crew of the threat of launch or the launch of weapons by the enemy attacking.

Information about the threat to the aircraft is normally formed on the basis of observing changes in the signal parameters of the enemy's radio electronic systems. Normally the intensity of the signal and the character of its change, and the arrival angle of radiation (bearing, carrier frequency, duration and pulse repetition frequency) are recorded. Other signal parameters, including their statistical characteristics, can be observed and measured. Measurement (observation) results are compared to corresponding parameter values present a priori in the computer database of the onboard control and information support system. According to the comparison results, a decision is made about the launch of a decoy or an expendable jammer.

Automated decision making circuits of this type have a limited level of information stability. They are based on relatively simple variants of the radio electronic environment and are vulnerable with respect to the simplest enemy countermeasures (i.e., simulated missile launches resulting in premature decoy ejection).

Information stability can be increased by introducing elements of expert evaluation into the automated control system (i.e., by going to intelligent automated control systems). In separate works, in the analogous case, authors refer to automated control systems with artificial intelligence [19]. Practically speaking, intelligent systems are implemented by creating knowledge (procedure) bases in onboard computers. The process of creating the latter is quite onerous [14, 20]. It includes the generalization of results of simulation modeling of typical combat situations, including the results of experimental exercises.

As variants for the definition of procedures, associated with selecting the moment for launching expendable electronic jamming (OEJ) devices, it is possible to use the simplest models of behavioral dynamics of sides in conflict. One of these examples is the episode when there is one decoy on the aircraft being attacked and the attacker has one missile. The enemy attacking determines the optimum moment to launch the missile, and the side being attacked — the optimum moment to release the decoy. The effectiveness indicator is the probability that the aircraft will be hit or the mathematical expectation of losses. The problem can be reduced to a zero-sum matrix game, if the dynamics of the conflict taking place during the time T is considered for discrete moments of time Δt.

Let us designate by $A_i(i = \overline{1, m})$ and $B_j(j = \overline{1, m})$ the pure strategies of the attacking and attacked sides. Strategy A_i means that the side attacked launches a decoy at the moment of time $t_i = i\Delta t$. Strategy B_j

correspondingly means that a missile is launched at the moment of time $t_j = j\Delta t$. The cost function in the given case is the probability P_{ij} that one missile will hit the aircraft being attacked.

According to what has been mentioned, the matrix game being considered can be expressed in the following manner:

$$P\left(A/B\right) = \|P_{ij}\| \quad i = \overline{1, m}; \; j = \overline{1, m} \tag{6.35}$$

where $T = m\Delta t$ is the time of conflict.

Here, P_{ij} is the probability the aircraft will be hit by the missile launched by the enemy at the moment of time $t_j = j\Delta t$ if the side attacked releases its decoy at moment of time $t_i = i\Delta t$:

$$P_{ij} = (1 - P_{rt}(i, j))P_{\text{hit}} \tag{6.36}$$

where $P_{rt}(i, j)$ is the probability the missile will be captured by the decoy (of the expendable jammer) in the case being analyzed; P_{hit} is the probability the aircraft will be hit by one missile in standard conditions.

The variant represented by matrix (6.35) is the simplest and the optimum values t_i and t_j can be determined for one episode. Practically speaking, an optimum solution, even for the simplest situation, can be obtained as a result of analyzing a number of variants of the conflict. In the given case, this means attacking at various course angles and with various available intervals of time T.

In the general case, it should be borne in mind that each of the sides in the conflict plans and conducts measures to reduce the information stability of the opposing side. The number of missiles and decoys can vary. This number depends on the parameters of the battle formation. There can also be other specifics for combat use. All of this stipulates the necessity of developing mathematical models to perform modeling of a broad range of variants and conditions of military operations. The knowledge (procedure) base comprises generalized evaluations and recommendations, obtained by experts on the basis of both logical study of the modeling results and the generalizing of experimental and command headquarters exercises.

References

[1] Paliy, A. I., *Radioe'lektronnaya bor'ba* (*Radio Electronic Warfare*). Moscow: *Voenizdat* (*Military Publishing House*), 1989, pp. 58–98.

[2] Zaporozhets, G. V., *Standartnoe ustroystvo vybrosa sredstv radioe'lektronnoy bor'by samoletov takticheskoy aviatsii VVS SHAA* (*Standard Device for Launching Radio Electronic Warfare Systems from US Airforce Tactical Aircraft*). Moscow: N. E. Zhukovsky Airforce Academy, 1983, 46 pp.

[3] Zaporozhets, G. V., *Sredstva radioe'lektronnoy bor'by samoleta neposredstvennoy aviatsionnoy podderzhki A-10 (SSHA)* (*Electronic Warfare Systems on Board the A-10 Direct Aircraft Support Plane (USA)*). Moscow: N. E. Zhukovsky Airforce Academy, 1982, 28 pp.

[4] Krotov, V., *Sredstva RE'B odnorazovogo ispol'zovaniya VMS zarubezhnykh gosudarstv* (*Single-Use EW Systems in the Navies of Foreign Countries*). Moscow: *Zarubezhnoe voennoe obozrenie* (*Foreign Military Review*), Vol. 5, 1994, pp. 57–60.

[5] Nebabin, V. G., and V. V. Sergeev, *Metody i tekhnika radiolokatsionnogo raspoznavaniya* (*Methods and Technology of Radar Recognition.*). Moscow: *Radio i svyaz'* (*Radio and Communications*), 1984, pp. 7–28, 135–144. English Translation: Artech House, 1995.

[6] Vakin, S. A., and L. N. Shustov, *Osnovy radioprotivodeystviya i radiotekhnicheskoy razvedki* (*The Basics of Radio Countermeasures and Radio Technical Intelligence*). Moscow: *Sov. radio* (*Soviet Radio*), 1968, pp. 9–122, 259–342.

[7] Ostrovityanov, R. V., and F. A. Basalov, *Statisticheskaya teoriya radiolokatsii protyazhennykh tseley* (*The Statistical Theory of Radar for Extended Targets*). Moscow: *Radio i svyaz'* (*Radio and Communications*), 1982, pp. 7–28. English translation: Artech House, 1985.

[8] Ryzhik, I., *Tablitsy integralov, summ, ryadov i proizvedeniy* (*Tables of Integrals, Sums, Series and Products*). Moscow: *Gostekhizdat* (*State Technical Publishing House*), 1948, 400 pp.

[9] Skolnik, M., editor *Spravochnik po radiolokatsii* (*Radar Handbook*). Moscow: *Sov. radio* (*Soviet Radio*), 1976, pp. 193–215.

[10] Varganov, M. E., Yu. S. Zinov'ev, L. Yu. Astanin, et al., *Radiolokatsionnye kharakteristiki letatel'nykh apparatov* (*The Radar Characteristics of Aircraft*)/Under the Editorship of L. T. Tuchkov. Moscow: *Radio i svyaz'* (*Radio and Communications*), 1985, pp. 48–60.

[11] Schrank, H. E., *Vsenapravlennye sfericheskie radiolokatsionnye otrazhateli s vysokoy e'ffektivnost'yu* (*All-Directional Spherical Radar Reflectors with High Effectiveness*). IEEE, Vol. 8, 1965.

[12] *Antennye reshetki* (*Antenna Matrices*). A Review under the Editorship of L. M. Beneson. Moscow: *Sov. radio* (*Soviet Radio*), 1966, pp. 340–355.

[13] Petrov, B. M., and A. I. Semenikhin, *Upravlyaemye impedansnye pokrytiya i struktury* (*Controlled Impedance Coatings and Structures*). Moscow: *Zarubezhnaya radioe'lektronika* (*Foreign Radio Electronics*), Vol. 6, 1994, pp. 9–16.

[14] Klykov, Yu. I., and L. N. Gor'kov, *Banki dannykh dlya prinyatiya resheniya* (*Databanks for Decision Making*). Moscow: *Sov. radio* (*Soviet Radio*), 1980, pp. 15–83.

[15] Lazarev, L. P., *Optiko-e'lektronnye pribory navedeniya letatel'nykh apparatov* (*Optoelectronic Devices for Aircraft Guidance*). Moscow: *Mashinostroenie* (*Machine Building*), 1984, pp. 23–107.

[16] Yakushenkov, Yu. G., *Osnovy optiko-e'lektronnogo priborostroeniya* (*The Basics of Optoelectronic Instrument Building*). Moscow: *Sov. radio* (*Soviet Radio*), 1977, pp. 17–39, 97–147.

[17] Yakushenkov, Yu. G., V. I. Lukantsev, and M. P. Kolosov, *Metody bor'by s pomekhami v optiko-e'lektronnykh priborakh* (*Methods for Countering Interference in Optoelectronic Devices*). Moscow: *Radio i svyaz'* (*Radio and Communications*), 1981, 180 pp.

[18] Kirillov, V. I., *Bombometanie* (*Bombing*). Moscow: *Voenizdat* (*Military Publishing House*), 1960, 210 pp.

[19] Pospelov, D. A., *Situatsionnoe upravlenie* (*Situation Management*). Moscow: *Nauka* (*Science*), 1986, pp. 9–130.

[20] Vakin, S. A., *Radioe'lektronnye sistemy kak ob''ekty radioe'lektronnoy bor'by* (*Radio Electronic Systems as Targets of Electronic Warfare*). Moscow: *Radiotekhnika* (*Radio Technology*), 1994, pp. 40–49.

7

Methods of Reducing Aircraft Detectability and Changing the Electrical Properties of the Environment

7.1 Factors Determining the Complex Nature of the Problem — Possibilities of Reducing the Thermal Detectability of Aircraft

Reducing aircraft detectability is a complex problem, the solution to which is to decrease the radar and thermal detectability of aircraft by using radio and optoelectronic devices to jam defensive sensor systems. Another solution is to change the electrical properties of the medium to increase the attenuation coefficient in the atmosphere. The quantitative evaluations follow from the equations for radar range and radio and optoelectronic jamming.

The radar detection range (D_s) of an aircraft with an RCS of σ_s according to (4.9) is given by the expression:

$$D_s = A\sigma_s^{1/4}10^{-0.05\alpha L_j} \tag{7.1}$$

where

$$A = \left(\frac{P_s G_s^2 \lambda^2}{(4\pi)^3 P_{\text{rec min}}}\right)^{1/4}$$

P_s, G_s, λ, and $P_{\text{rec min}}$ are parameters of the radar: the ERP, the wavelength, and the minimum required power of a reflected signal detected by the receiver; α is the attenuation coefficient in the atmosphere; and L_j is the length of the path with increased absorption.

As follows from (7.1), the detection range D_s is decreased only slowly with a reduction in σ_s. For example, if an aircraft with an RCS of $\sigma_s = 10$ m^2 (10 dBm^2) is detected by a certain radar at a maximum range of $D_{s\max} = 400$ km (56 dB(m)), then one having an RCS of $\sigma_s = 1$ m^2 (0 dBm^2), is detected by the same radar at a distance of 220 km (53.5 dB(m)). The detection of the aircraft at a range of 220 km is sufficient for AAD to solve the problem of its destruction. A reduction in RCS by 30 dB, which gives a value of $\sigma = 0.01$ m^2 (−20 dBm^2), decreases the detection range to $D_s = 70$ km. A final reduction in detection to a value of $D_s = 40$ km (−46 dB(m)) requires a decrease in RCS by 40 dB, which corresponds to a value of $\sigma_s = 0.001$ m^2 (−30 dBm^2).

A reduction of the RCS of an aircraft with an $\sigma_s = 10$ m^2 by 10 dB is achievable in practice while retaining, basically, the classical aerodynamic design of an aircraft. The reduction of RCS to lower values 0.1–0.001 m^2 requires the use of different principles — the newest technologies of design and development of aircraft. At the same time, the required ERP of an onboard jammer on the aircraft decreases proportionally to σ_s. This is confirmed by the jamming equation (4.62), defining the minimum radar detection range of an aircraft ($D_{s\min}$) protecting itself using active jamming:

$$D_{s\min} = \sqrt{\overline{K}_{\text{REJ}}10^{-0.05\alpha L_j}} \tag{7.2}$$

where

$$\overline{K}_{\text{REJ}} = \frac{P_s G_s}{P_j G_j} \frac{\Delta f_j}{\Delta f_s} \frac{K_j \sigma_s}{4\pi\gamma_j} \tag{7.3}$$

$P_j G_j$, Δf_j are the ERP and spectrum width of the jammer; K_j is the jamming coefficient; and γ_j is the polarization coefficient.

Here we understand target detection to mean the determination of its coordinates, including the distance to the target.

From (7.2) and (7.3) it follows that the required jammer ERP for a fixed $D_{s\min}$ changes proportionally to the RCS of the aircraft being protected σ_s. If, for example, the protection of the aircraft with $\sigma_s = 10$ m^2 is provided by a jammer with an ERP of $P_j G_j = 50$W, then for a reduction in RCS to $\sigma_s = 1$ m^2 it is sufficient to have a power of $P_j = 1$W, if $G_j = 5$. In an analogous manner, the expenditure rates for onboard passive and active–passive jamming devices are decreased.

Apparently, with respect to cost/effectiveness, for a given aircraft there exists an optimum relationship between the degree of RCS reduction implemented and the required ERP of the jammer and other indicators of the onboard EW equipment. The problem of counteracting high-precision electronic support measures (ESM) equipment, used with active jammers on the aircraft, can be solved when doing a comprehensive analysis of the entire set of problems and, in particular, by conducting organizational measures when preparing for and waging EW.

The equation (7.2) also indicates the potential for increasing the degree of radar jamming by increasing the absorption coefficient α and the length of the path with increased absorption.

The thermal detectability of aircraft is due to radiation from the engine, the heated gasses of the jet stream at the exhaust output (at a distance not exceeding 2–3m), and the surface that is heated as a result of aerodynamic drag when flying at supersonic speeds. It follows from the optoelectronic jamming equation (6.20) that the reduction of thermal detectability of an aircraft (as in the case with radar detectability) permits a proportional decrease in the radiation intensities of thermal decoys and other OEJ devices.

The potential reduction of the thermal detectability of aircraft depends greatly on the Stefan–Boltzmann law. This law is characterized by expressions (6.24) and (6.25), with which we calculate the radiation power density. According to the law mentioned, the thermal detectability changes proportionally to the fourth power of the temperature of the engine components and gases at the exhaust output. For this reason, even small changes in the engine operating mode have a significant influence on the thermal detectability of aircraft. For example, when changing the engine to the afterburner mode of operation, the exhaust temperature is increased by a factor of 1.4–1.5, which, according to (6.22), leads to an increase in radiation density of 4–5. A reduction in radiation is achieved by converting a turbojet engine to using serpentine input and exhaust ducts. Particularly, in the tactical U.S. Airforce F-117A fighter, two nonafterburning engines with serpentine ducts are installed [1]. The maximum of the radiation pattern for the aircraft can be oriented in a less dangerous direction. Also, ability of water vapor and carbon dioxide to absorb infrared radiation at wavelengths 2.7 and 4.3 çm should be kept in mind.

Along with the problem of decreasing radar and thermal detectability of aircraft in the dynamics of flight, the problem of concealing aircraft at airports and other bases is becoming a high priority, because of the appearance of intelligence-and-strike complexes and other systems of high-precision weapons.

7.2 Modern Technologies for the Development of Aircraft with Low Radar Detectability and Problems of EW Dynamics

As the experience of implementation of "STEALTH" technology shows, the aerodynamic and electrodynamic design of aircraft with an RCS not exceeding 0.1–0.01 m^2 over a broad band of carrier frequencies, is possible at the present time only through synthesis, combining methods of computational aerodynamics and applied electrodynamics with the application of modern computers. The ultimate result of synthesis is the definition of rational aerodynamic and electrodynamic designs that ensure a level of backscattering of a radar signal that does not exceed a specified value within a certain solid angle. The power density of backscattering (the Umov–Poynting modulus vector), observed in a given direction for a fixed electromagnetic field density illuminating an aircraft unambiguously determines its RCS in the given direction. When we speak of backscattering, we normally presume that the radar receiver and transmitter are collocated.

Aircraft, including those implemented in the "STEALTH" program, have a quite complex form. Backscattering is a superposition of fields reflected from a comparatively large number of components. The geometric size of each of the components exceeds by many times (by two or three orders of magnitude) the radar wavelength in the centimeter bands, and, by smaller factors in the decimeter bands. According to the basic concept of "STEALTH", the shape of each component, its position and its orientation are determined with due consideration to aerodynamic limitations, in such a way as to minimize the backscattered power density. The latter is calculated by solving problems of the electromagnetic field diffraction on elements of corresponding shapes and sizes.

Backscattering from diffraction on the given component is characterized by its envelope and initial phase. The average value of the backscattered power density over the RF cycle determines the RCS value corresponding to the given component. The sum of RCS components for individual elements comprises the RCS of the aircraft as a whole. When summing up the RCS elements, it is necessary to consider that, under flight conditions, the envelopes and initial phases of the backscattering of individual elements is random. Moreover, the randomization of backscattering is also implemented deliberately with the purpose of reducing, when all other conditions are equal, the radar detection probability of the signal reflected from the aircraft. This follows from the expressions (1.4), (1.5) and (1.8), which determine the detection probabilities by receivers optimized according to the Neyman–Pearson criteria, for a totally known signal and for a signal with a random initial phase and envelope.

In order to obtain the required degree of reduction in backscattering, calculations of rational shapes and aerodynamic design of aircraft are used in addition to radio absorbing materials (RAM) and coverings and controllable impedance structures that permit adaptive control of scattering. The necessity of this is stipulated by limitations in the aerodynamic design of an aircraft and its propulsion system. The use of RAM and adaptive scattering coverings provides for a compromise between the demands of aerodynamics and electrodynamics. We should separately note the problem of reducing detectability of antenna devices in onboard radio equipment of aircraft. This issue is examined in the material that follows.

To the extent we can judge from material published on the B-2 and F-117A aircraft, developed and deployed in the arsenal of the U.S. Air Force at the present time and implemented using "STEALTH" technology, RCS values for the B-2 are $\sigma_s = 1$ m^2, and for the F-117A — from the front hemisphere $\sigma_s = 0.01$ m^2, and from other aspects $\sigma_s = 0.025$ m^2 [1, 2]. American specialists consider it possible to attain an RCS of the order of 0.001 m^2 [3]. This latter circumstance resulted in unfounded assertions about the undesirability of placing onboard devices for generating active jamming on aircraft implemented using "STEALTH" technology. Concerning this issue, it should be noted that such assertions are metaphysical in nature. They don't take into account the specifics of the developmental dialectics of EW systems and techniques as an element of armed conflict. We confirm what has been mentioned using individual examples.

Aerodynamic and electrodynamic aircraft designs developed using "STEATH" technology have been accomplished using approximate methods of solving the corresponding aerodynamics diffraction problems. Their strict solutions for the majority of the scattering elements entering in the aerodynamic designs are unknown, and calculations for known solutions require unacceptably large amounts of machine time to perform. As is testified to by materials published [4], the solution to scattering problems in the process if implementing "STEALTH" technology was performed using an approximation of the physical diffraction theory developed by P. Ya. Ufimtsev [5] and providing sufficient precision in asymptotic approximations for practical purposes. Specifically, it is assumed that the parameter $\lambda/a(\lambda/a << 1)$ will be small, where λ is the wavelength and a is a certain characteristic size for the scattering element, for example, in the case of diffraction from a disk — its radius. It should be borne in mind that, when using other known methods, the solution to diffraction problems is found in asymptotic approximation [6].

We may assume that the degree of RCS reduction in aircraft achieved using "STEALTH" technology applies to the centimeter and decimeter wavebands. In the meter waveband, especially in those cases where there are components of the aircraft of sizes that are half-wave multiples of the radar, the RCS of the aircraft can increase significantly. The angular coordinate resolution problem that arises in the meter waveband is solved using synthesized antenna radar.

The detection of aircraft with low detectability can be performed successfully using multistatic radar. Aerodynamic and electrodynamic designs synthesized using "STEALTH" technology provide for the reduction of backscattering in a limited range of angles. The scattered radiation in other directions can significantly exceed the backscattering. The well-known Babinet principle from optics should not be forgotten, according to which the substitution of a transparent screen in a certain aperture with a supplementary (nontransparent) one does not change the Fraunhofer diffraction picture in the focal plane of the lens with the exception of the focus itself. There are also other possibilities of increasing the radar effectiveness for aircraft with low detectability [7].

It would seem to be convenient to use cost/effectiveness criteria when solving the problem of reducing radar detectability, in specific media, by determining the relationship between the RCS and the composition and parameters of the onboard jamming complex, bearing in mind the necessity of counteracting not only radar, but also high-precision AAD ESM systems.

7.2.1 Radio Absorptive Coatings

Let us consider the reflection of a plane electromagnetic wave from an infinite ideally conducting surface coated with a substance characterized by a complex dielectric (ε') and magnetic (μ') permittivity (Figure 7.1):

$$\varepsilon' = \varepsilon_r' + j\varepsilon_\kappa' \tag{7.4}$$

$$\mu' = \mu_r' + j\mu_\kappa' \tag{7.5}$$

Here, $\varepsilon = \varepsilon_0\varepsilon$ is the dielectric permittivity of the coating (in free space $\varepsilon' = \varepsilon_0$); $\varepsilon_r'/\varepsilon_0 = \varepsilon_r$ is the relative dielectric permittivity of the coating; $\varepsilon_\kappa'/\varepsilon_0 = \varepsilon_\kappa$ is the imaginary part of the dielectric permittivity resulting from dielectric losses and the electrical conductance of the surface;

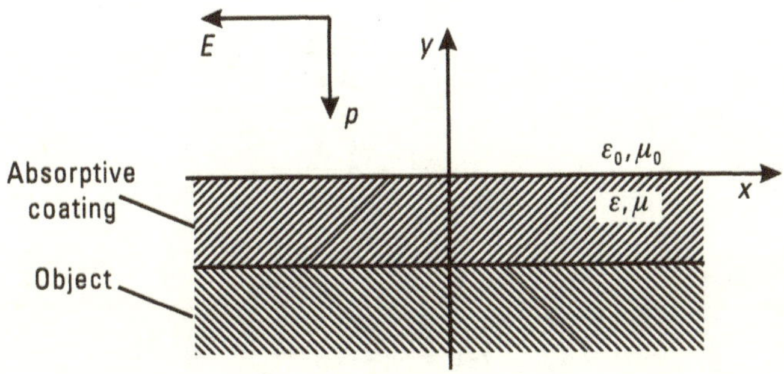

Figure 7.1 The reflection of a plane electromagnetic wave from an infinite ideally conducting coated surface.

$\mu' = \mu_0\mu$ is the magnetic permittivity of the coating (in free space $\mu' = \mu_0$); $\mu_r'/\mu_0 = \mu_r$ is the relative magnetic permittivity of the coating; $\mu_\kappa'/\mu_0 = \mu_\kappa$ is the imaginary part of the magnetic permittivity resulting from losses.

Let us determine the value of parameters μ' and ε' of the absorptive coating for which the reflection coefficient from the boundary ($y = 0$) is equal to zero. First we write the expression for the complex reflection coefficient for a plane wave from the boundary of two media:

$$R = \frac{\bar{z} - z_0}{\bar{z} + z_0} \tag{7.6}$$

where z_0 is the wave impedance of free space:

$$z_0 = \sqrt{\frac{\mu_0}{\varepsilon_0}} = 120\pi \tag{7.7}$$

$\bar{z}$ is the wave impedance of the absorptive coating:

$$\bar{z} = \sqrt{\frac{\mu'}{\varepsilon'}} \tag{7.8}$$

Substituting (7.7) and (7.8) into (7.6), we obtain

$$R = \frac{1 - \sqrt{\dfrac{\varepsilon' \mu_0}{\mu' \varepsilon_0}}}{1 + \sqrt{\dfrac{\varepsilon' \mu_0}{\mu' \varepsilon_0}}} \tag{7.9}$$

Considering that

$$\sqrt{\varepsilon\mu} = n + ik \tag{7.10}$$

where n is the index of refraction, and k is the attenuation coefficient of the medium, then (7.9) can be transformed to two expressions equivalent in form, more precisely:

$$R = \frac{\mu - n - ik}{\mu + n + ik} \tag{7.11}$$

$$R = \frac{n + ik - \varepsilon}{n + ik + \varepsilon} \tag{7.12}$$

Representing μ and ε in a complex form:

$$\mu = \mu_r + j\mu_k \tag{7.13}$$

$$\varepsilon = \varepsilon_r + j\varepsilon_k \tag{7.14}$$

it is possible to transform (7.11) and (7.12) to the following format:

$$R = \frac{(\mu_r^2 + \mu_k^2) - (n^2 + k^2) + j2(n\mu_k = k\mu_r)}{(\mu_r + n)^2 + (\mu_k + k)^2} \tag{7.15}$$

$$R = \frac{(n^2 + k^2) - (\varepsilon_r^2 + \varepsilon_k^2) + j2(\varepsilon_r k - n\varepsilon_k)}{(n_\varepsilon r)^2 + (k + \varepsilon_r)^2} \tag{7.16}$$

In both cases, the reflection coefficient from the boundary of two media is a complex number. It is equal to zero when both the real and imaginary parts are also equal to zero. The denominators in (7.15) and (7.16) are positive numbers, which permits us to write the following equalities, specifying for (7.15) that $R = 0$ if

$$\mu_r^2 + \mu_k^2 = n^2 + k^2 \tag{7.17}$$

$$n\mu_k = k\mu_r \tag{7.18}$$

When R is determined by expression (7.16), R = 0 if

$$n^2 + k^2 = \varepsilon_r^2 + \varepsilon_k^2 \tag{7.19}$$

$$\varepsilon_k n = k\varepsilon_r \tag{7.20}$$

It follows from (7.17) and (7.19) that the absorptive material can be either a magnetodielectric, the complex permittivity μ of which satisfies the conditions:

$$|\mu|^2 = |\sqrt{\varepsilon\mu}|^2 \tag{7.21}$$

or a dielectric with complex permittivity satisfying analogous conditions:

$$|\varepsilon|^2 = |\sqrt{\varepsilon\mu}|^2 \tag{7.22}$$

Besides this, the conditions (7.18) and (7.20) must also be fulfilled. In the first case, (expression (7.15) equal to zero) the surface absorbs magnetic energy, and in the second (expression (7.16) equal to zero) electrical energy is absorbed (Ohmic losses).

The thickness of the absorptive coating didn't enter into formulas (7.15) and (7.16). In the given case it can be assumed to be comparatively large. In practice, however, it is necessary to provide for a sufficient degree of attenuation of radiation by coatings of limited thickness. When doing this, segments with a wave impedance of the order of $Z = 377$ ohm (free space) and of the order of $Z = 0$ ohm (metallic surface) should be matched. Practically speaking, the problem is solved by using multilayer absorptive coatings in which each of the layers can be considered to be a wave impedance transformer providing matching of adjacent layers. The input impedance Z_{inp} of a certain ith layer of thickness d_i is determined from the expression:

$$Z_{\text{inp}} = \sqrt{\mu_i / \varepsilon_i}\; th\left(\sqrt{\varepsilon_i \mu_i}\, d_i k_0\right) \quad (i = \overline{0, m}) \tag{7.23}$$

where μ_i and ε_i are the complex magnetic and dielectric permittivities of the absorptive material in the ith layer determined using (7.13) and (7.14); $k_0 = 2\pi/\lambda$ is the wave number in free space. The number of absorptive layers in the coating, in principle, can be quite large.

Speaking of the thickness of the RAM, it is necessary to bear the following in mind. When a plane wave is reflected from a metallic surface, a standing wave of the electrical field is formed with an antinode at a distance of $\lambda_\varepsilon/4$ from the conductor (λ_ε is the wavelength in the dielectric). For this reason, the thickness of a dielectric RAM coating must be selected so that the antinode region of the electrical field is located in the coating. The most intensive absorption takes place precisely in this region of the dielectric. For the purpose of maximizing this absorption, short metallic fibers are introduced in the region of the dielectric corresponding to the antinode.

In the case of RAM with magnetic losses (ferromagnetics), the thickness of the covering can be comparatively small, since the magnetic field antinode occurs directly on the metallic surface.

At the present time RAMs are being developed on the basis of both analytical and experimental methods [8–11]. The simplest is the analysis method [10]. In this case, additional elements that conduct electricity and contain magnets are introduced in the composite materials that modify the complex dielectric and magnetic permeabilities, as well as the conductance of the mixtures. The short metallic fibers located in the mixture are reradiators (scatterers) of electromagnetic energy and can comprise up to 3% of the mass. Particles of magnetite are sprayed as magnet-containing additives [10].

On the basis of the variant adopted for the distribution of electromagnetic parameters in the coating structure, its electrodynamic model is justified and calculations are performed to determine reflection or transmission coefficients in a specified frequency band.

The problems of synthesis are more complex, involving the construction of the coating so as to provide for the minimum reflection coefficient in the given frequency band for minimum thickness and mass. Problems of synthesis also involve the determination of the electromagnetic parameters of the component parts of the ingredient materials and their distribution within the thickness of the layer, providing for optimum characteristics for the RAM for a given frequency band and a maximum permissible reflection coefficient [9, 10].

The synthesis of the layered nonhomogeneous structure of a RAM with optimum parameters can be made using the solution of the electrodynamic problem of the propagation of radio waves in a layered absorbing medium, the properties of which constantly change in the direction z (Figure 7.2). The complex index of refraction of the medium $n(z)$ changes from the

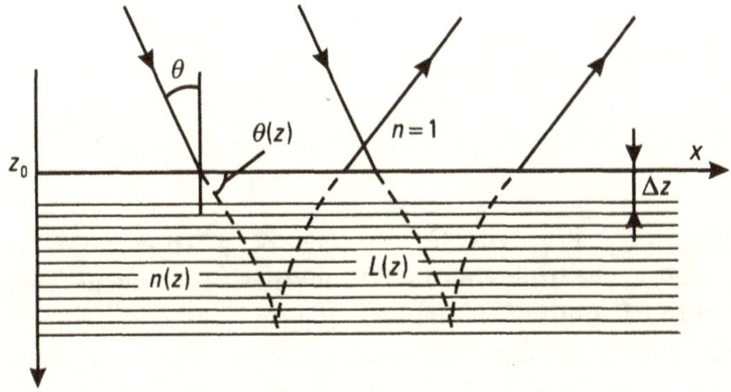

Figure 7.2 The synthesis of the layered nonhomogeneous structure of a RAM with optimum parameters.

value $n = 1$ when $z = z_0$ to the value $n(z) = \infty$ when $z = \infty$. It is shown that when a plane wave passes through the layered inhomogeneous medium, the current value of the reflection coefficient $R(z)$ satisfies a Riccati differential equation of the following form [11]:

$$\frac{dR}{dt} + 2i\beta R - \chi(1 - R^2) = 0 \tag{7.24}$$

Here, depending on the field polarization relative to the plane of incidence,

$$\beta(z) = \beta_{\perp} = \kappa_0\sqrt{n^2(z) - \sin^2\Theta(z)}$$

$$\beta(z) = \beta_{\parallel} = \frac{\kappa_0}{n^2(z)}\sqrt{n^2(z) - \sin^2\Theta(z)}$$

$$\chi_{\perp} = \frac{1}{2\beta_{\perp}}\frac{d\beta_{\perp}}{dz} \qquad \chi_{\parallel} = \frac{n^2(z)}{2\beta_{\parallel}}\frac{d\beta_{\parallel}}{dz}$$

where β is the phase factor (the phase scale) occurring in layer z; χ is a coefficient taking into consideration the change in the phase factor (the phase scale) in the layer with coordinate z (gradient β); and $\kappa_0 = 2/\pi\lambda$ is the wave number in free space.

The equation (7.24), assuming slow changes in parameters of the medium ($\chi = \text{grad}\,\beta << 1$) and equality of the field to zero for $z \to \infty$, is solved by iteration. For the mth approximation, the following iteration formula is obtained:

$$R_m = e^{-jL(z)} \int_z^{\infty} \chi(z)\left(1 - R_{m-1}^2(z)\right) e^{jL(z)} dz \qquad m = 1, 2, 3 \ldots$$

where $L(z) = 2\int_{z_0}^{z} \beta(z)dz$ is the phase length; when $m = 1$, $R_0 = 0$.

In essence, the problem of RAM synthesis for a small reflection level consists of the definition and implementation in practice of an optimal law for changing the phase factor gradient grad$\beta(z)$ considering the decrease in the complex amplitude of the incident wave as z increases. At the same time, the thickness of the absorptive coating is also minimized. RAMs of the type considered are called gradient RAMs. Normally real absorbing materials and coatings of the gradient type are implemented as a set of layers with constant parameters approximating the required law for the change of the complex index of refraction. Other ways of solving the problems of analyzing and synthesizing an ideal RAM structure are also known [9, 12].

The required attenuation of the reflected electromagnetic wave and the corresponding maximum permissible value of the reflection coefficient R are determined by the specific conditions of solving the problem of reducing radar detectability. As concerns "STEALTH" technology, this is a result of the requirement to reach a compromise between the requirements of aerodynamics and electrodynamics. In the case of concealing aircraft at airports and other surface targets, it is sufficient to reduce the RCS to a level ensuring the "matching" of the target detectability to that of the surrounding area. Below, as examples, we provide the characteristics of specific types of RAM [8, 10, 11].

Composite RAM can be developed on the basis for ferrite–rubber mixtures with the addition of metallic fibers. One of the RAM samples [10] provides for an absorption of more than 15 dB in the 8–26 GHz range and more than 20 dB at a frequency of 4.5 GHz. The possibility of using composite materials of this type at frequencies of 30, 60 MHz and 1 GHz are shown.

Composite RAM using fibers from graphite of carbon, boron, polyethylene and polyvinylchloride with a ferrite filler also permits significant decrease in aircraft detectability. Thus, at a frequency of 11.6 GHz, a three-layer RAM with an overall thickness of 10.7 mm provides for a reflection coefficient of the order of −16 dB at angles of incidence from 0 to 30 degrees. High-quality RAM with increased resistance to fire provides for a reflection coefficient of −30 to −50 dB over a broad band of frequencies from 100 MHz to 40 GHz. A new trend in the development of RAM is a coating developed by a group of specialists from the Center for Molecular Electronics at the Pittsburgh Carnegie-Mellon University (USA) on the basis of Shiffer retinil salts, which, according to the opinion of its creators, can be applied with a brush like paint to the object to be protected. Shiffer retinil salts are much lighter than ferrites, but are comparable in their ability to absorb HF and UHF radiation. We shall note, for example, that a mass of 1 m^2 of ferrite absorbent of the brand NZ-1 with a thickness of 2.4 cm comprises about 250 kg [11].

7.2.2 Interference and Diffusion Radio Absorbing Coatings

The reduction of the reflection coefficient in interference coatings is achieved as a result of the absorption of electromagnetic energy in the material of the coating, as well as the partial cancellation of the reflected signal due to interference. The simplest interference coating is a fine homogeneous layer of ferrite with thickness *d*, placed on a metallic substrate (Figure 7.3). The coating is matched to free space by selecting a definite relationship of the value of magnetic losses in the coating, its thickness *d* and the wavelength λ. According to (7.6) and with consideration to (7.24), the reflection coefficient R, from such a coating, is determined using the following expression:

$$R = \frac{z_{inp} - 1}{z_{inp} + 1} = \frac{\sqrt{\mu/\varepsilon}\, th\left(\sqrt{\varepsilon\mu}k_0 d\right) - 1}{\sqrt{\mu/\varepsilon}\, th\left(\sqrt{\varepsilon\mu}k_0 d\right) + 1} \tag{7.25}$$

where μ and ε are determined using (7.13) and (7.14); $k_0 = 2\pi/\lambda$ is the wave number, and λ is the wavelength in free space.

If magnetic losses are quite large (i.e., $\mu_2 >> \mu_1$) and the thickness of

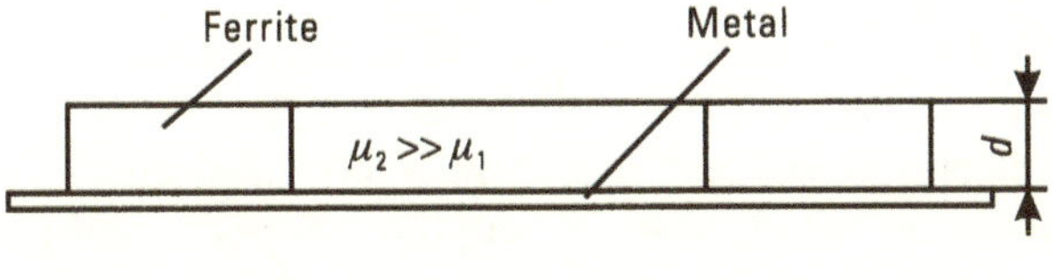

Figure 7.3 The simplest interference coating is a fine homogeneous layer of ferrite with thickness *d*, placed on a metallic substrate.

the coating is much less than the wavelength ($d << \lambda$), then, in the first approximation, (7.25) becomes

$$R = {(\mu_r k_0 d - 1)}\Big/{(\mu_r k_0 d + 1)}$$

The reflection coefficient is equal to zero when

$$\mu_r = \frac{\lambda}{2\pi d} \qquad (7.26)$$

From (7.26), it follows that the RAM being considered is quite wideband in nature.

Ferromagnetic materials of the type NZ-1 and NZ-2 from the Emerson company are intended for the attenuation of reflections in the waveband from units of centimeters up to 6–8m. The thickness of coating NZ-1 is $d = 2.4$ cm, and of NZ-2 is $d = 0.6$ cm.

Interference coatings are implemented on a dielectric basis with the inclusion in the antinode region of an electrical field of conducting fibers. The antinodes are located at a distance from the metallic substrate which is a multiple of an uneven number of quarters of the wavelength in the dielectric.

In the practice of developing anechoic chambers, as well as for the solution of specific problems of antenna technology and antiradar concealment, various radio absorbing materials of the diffusive type are used. The surface of such materials is distinguished by high geometric heterogeneity. A typical example is the pyramidal diffusive RAM, the surface of which is a set of pyramids with a square base and a constant distance between the summits of adjacent pyramids, which is equal to l. In the long-wave part of the band, when $l >> \lambda$, the coating acts as a gradient RAM. In the short-wave part of the band ($l << \lambda$), multiple wave reflection occurs from the edges of adjacent pyramids, which significantly increases the effectiveness of the coating such that the reflection coefficient R may possibly not exceed -50 dB.

For concealment in ground-based conditions, absorbing coatings are made in the form of hair, rubber or wooden mats, saturated with a mixture of neoprene (a type of rubber) and soot. Such mats with a thickness of several centimeters are capable of reducing the reflected signal power by 13–17 dB. Regarding materials at hand, coatings from wet hay and grass can be used with success.

7.2.3 Controlling the Scattering of Radio Waves

A significant reduction in the RCS of the target can be achieved in principle by controlling the parameters of the scattered field. The objective of such control is to change the properties of the target as a reradiating source to such a degree that the energy scattered in the required direction is minimum [9, 13–16].

One of the first suggested techniques of controlling scattering was the attachment of a complex load to the reflecting object. This technique has a certain similarity with the techniques described above for decreasing RCS using antiradar coatings. However, the difference in principle is that a complex load is attached to a local region whose dimensions are significantly less than those of the target as a whole. The loaded region in the particular case can be a gap with concentrated or distributed loads.

Along with reduction of backscattering, scattering control is performed to develop false targets and controlled reflectors and in concealment using "bright spots."

In Figure 7.4 we show target *Ta* with coupling aperture *s*, loaded for a complex load. The target is illuminated by transmitter *A*, and the receiving of scattered radio waves is performed at point *B*. The secondary field at receiving point *B* can be represented as the result of superposition of two fields. One of them is the field of the nonloaded body *Ta*, and the second is the field of the loaded aperture *s*. It is worthwhile mentioning that, in view of the small area of the aperture *s*, the overall configuration of the target and its area can be considered to be unchanged.

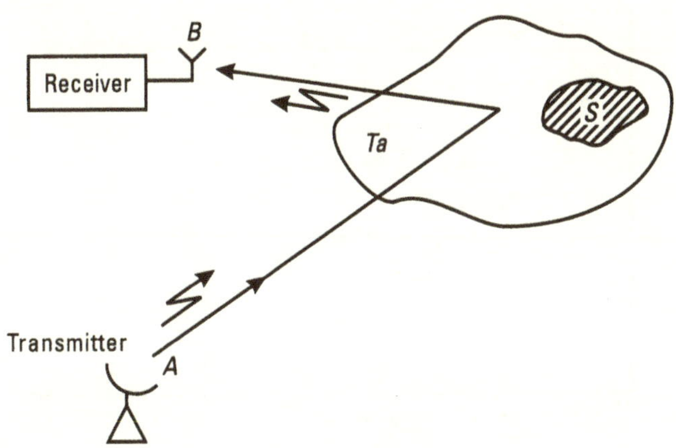

Figure 7.4 Target *Ta* with coupling aperture *s*, loaded for a complex load.

The scattered field of aperture s is determined by the form of the aperture and the parameters of the load impedance. By adjusting these parameters, it is possible to change the distribution of the amplitude and phase of the field scattered by the gap and, as a result, to achieve the required decrease in the resulting field at receiving point B. The relative RCS change of a loaded target can be evaluated using [16]

$$\frac{\sigma_{ut}}{\sigma_{lt}} = (1 - \alpha)^2 \left| \frac{Z + \dfrac{Z_A}{1 - \alpha^*}}{Z + Z_A} \right| \tag{7.27}$$

where σ_{ut} is the RCS of the nonloaded target; σ_{lt} is the RCS of the loaded target; Z_A is the equivalent complex load from the side of its attachment points in the absence of illumination by transmitter A; Z is the complex load; and α^* is the function of the separation of the transmitter from the receiver and shape of the target, as well as the place of the gap on the target and the nature of the load.

In Figure 7.5 we show the qualitative dependency of σ_{lt}/σ_{ul} on the reactive load $Z = jx$. The minimum RCS of the loaded target approximately corresponds to the reactive load:

$$Z = -\text{Im}[Z_A|1 - \alpha^*|]$$

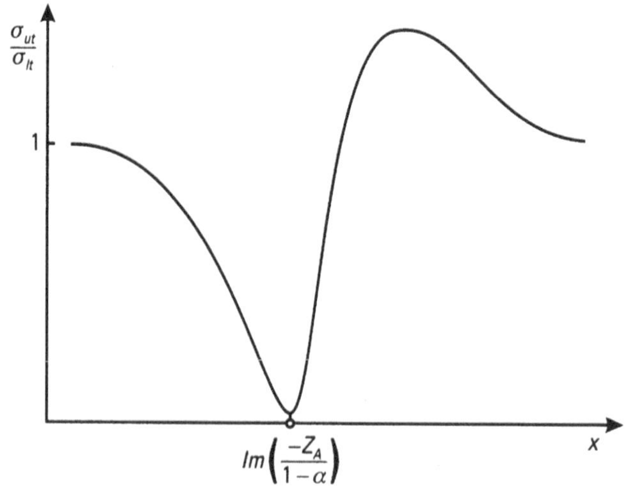

Figure 7.5 The qualitative dependency of σ_{lt}/σ_{ul} on the reactive load $Z = jx$.

The finding of the shape functions (7.27) for complex targets poses insurmountable mathematical difficulties. However, for bodies of simple shape (dipoles, spheres), solutions have been obtained in the form of graphs.

Calculations show that the RCS of a thin dipole, by connecting a complex load, can be reduced by 20–35 dB. Physically this is explained by the detuning of the dipole introduced by the reactive load.

In Figure 7.6 we show the dependency of the relative RCS value of the loaded dipole on the scattering angle Θ.

The dipole was loaded with an inductive load $Z_L = j600$. Here also we show the analogous dependency in the case of a nonloaded dipole $Z = 0$ (the length of the dipole is $l = 0.43\lambda$ and the width $b = 0.043\lambda$).

Changes in the parameters of the complex load can be achieved by attaching concentrated or distributed reactances, implemented in the form of various cavities (for example, ring gaps). In Figure 7.7 we show the relative RCS values of a loaded sphere, recorded experimentally [17]. The complex load was a gap, and the size and nature of the load were regulated by

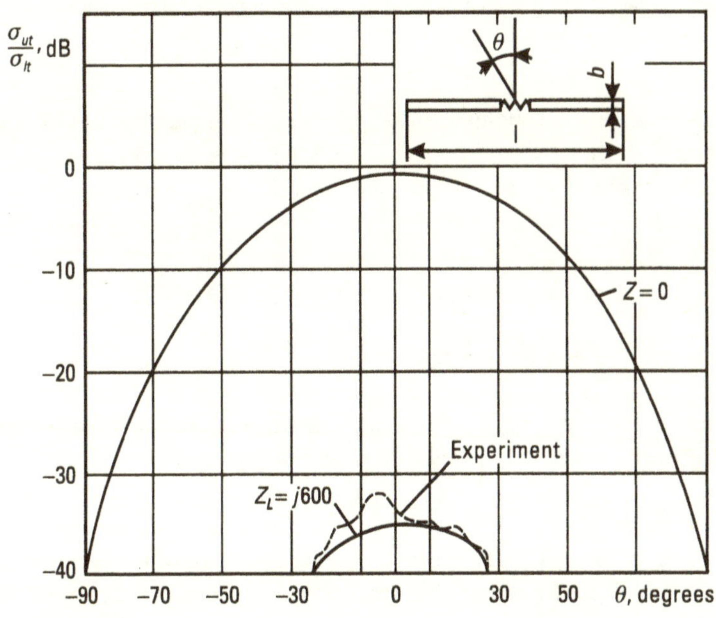

Figure 7.6 The dependency of the relative RCS value of the loaded dipole on the scattering angle Θ.

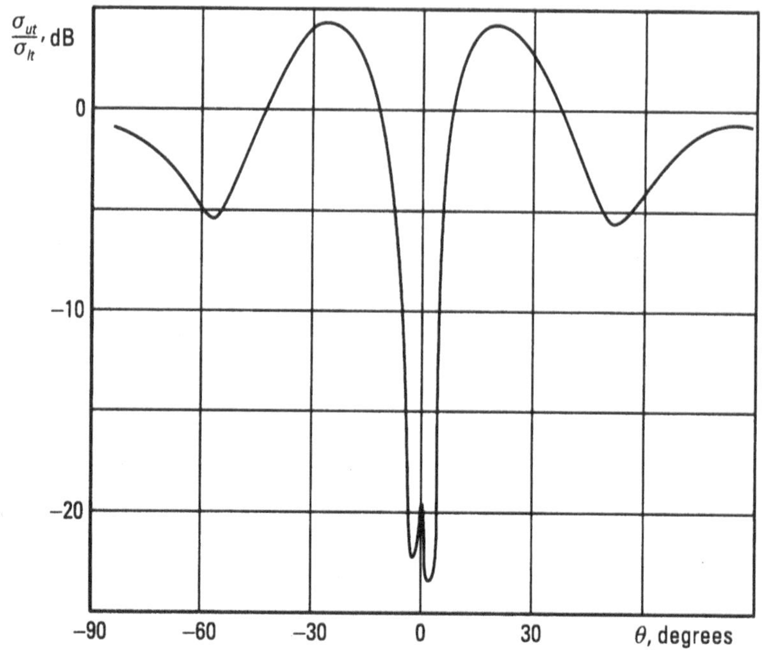

Figure 7.7 The relative RCS values of a loaded sphere, recorded experimentally.

changing its depth. In the given case, the depth of the gap was changed by substituting shorting discs. In Figure 7.8 we present the geometry of the surrounding sphere. A disc with the following parameters was experimentally studied:

$$R = 42.25 \text{ mm} \quad d = 1.6 \text{ mm} \quad \Theta = 90^\circ,$$

$$2\pi r/\lambda = 4.28 \quad f = 51.36 \text{ GHz}$$

The dependency of the RCS of loaded scatterers on the reradiation angle (Figures 7.6 and 7.7) shows the possibility of reducing the RCS by 20–35 dB.

At present, the known principles for the practical implementation of scattering control systems presume the use of impedance coatings, one variant of which is the intelligent reflecting structure [14, 15]. The basis of controlled impedance coatings (or structures) are controlled components. As a rule, these are ferrites, the parameters of which depend on the magnetizing current. The latter is determined by instructions generated

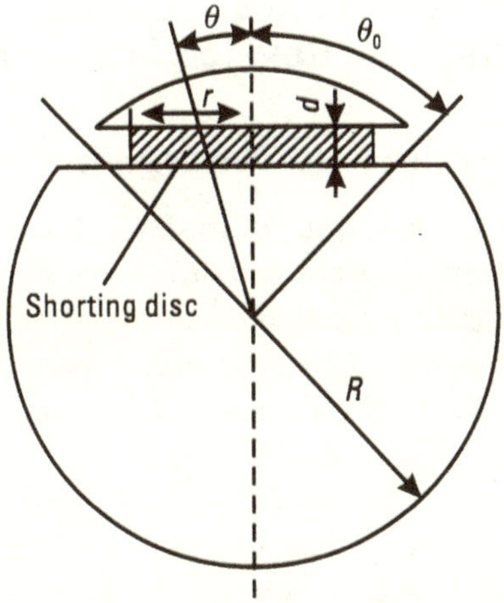

Figure 7.8 The geometry of the surrounding sphere.

by a microprocessor which is part of the impedance coating system. In order to adapt to the environment and to preclude unauthorized access to control of the ferrite components, in other words, to ensure the information stability of the scattering control system in the dynamics of EW, it is necessary to provide for appropriate data- and knowledge bases in the microprocessor, determining the intelligence level of the structure. An information support system must also be provided for, including above means for ESM, which in principle can automate the decision making process for scattering parameters depending on the evolving radio environment. Anisotropy and nonlinear properties of the magnetized ferrites permit the deliberate automatic change of the reflective properties of the impedance surface over a quite broad range of values.

Analyzing the question of intelligent coatings used in military technology, we should keep in mind that the characteristics of the coating can change depending on the meteorological and combat conditions. In the mass media, cases have been noted where rain and snow have influenced the detectability of "STEALTH" airplanes. It can be assumed that such effects increase as the level of "STEALTH" technology increases and the RCS of aircraft correspondingly decreases. This circumstance once again stresses the requirement to solve the problem of aircraft detectability by an optimum combination of its RCS value and the ERP of onboard jamming systems.

7.3 Potential for Reducing Radar Detectability of Aircraft Antennas — Optimum Gain for Jammer Antennas

7.3.1 Potential for Reducing Radar Detectability of Aircraft Antennas

The number of antennas for various purposes onboard a modern aircraft is quite large. Many of them, if special measures are not taken, can prove to be unmasking elements for an aircraft built using "STEALTH" technology. Indeed, let us assume that the RCS of the aircraft has been reduced to a level of -10 dBm^2 and that there is a linear antenna installed on it with a length of the order of 0.60–0.65m. The RCS of this antenna, considered as a half-wave reradiator, for a meter-range radar ($\lambda = 1.2$–1.3m) when polarizations match, is $\sigma = 1.1$–1.6 dBm^2 (i.e., it exceeds the RCS of the aircraft itself by more than an order of magnitude).

If the RCS of the aircraft is reduced to a lower value, for example, to a value of -20 dBm^2, then such a target can be unmasked by a linear antenna that is only 10 cm in length, the RCS of which for a radar on the 20 cm range is $\sigma_A = 4.7$ dBm^2, which exceeds the original aircraft RCS by 15.3 dB.

These examples show that without a solution to the problem of reducing radar detectability of onboard antennas, work on reducing the detectability of aircraft practically speaking makes no sense. The problems become more acute as the level of detectability of the aircraft as a whole decreases. In turn, the problem of reducing the radar detectability of aircraft antennas requires the development of new antenna technologies.

At the present time, a comparatively large number of works are known devoted to solving the problem of reducing the detectability of aircraft antennas [18]. In the first approximation, they can be summarized as follows.

The reduction of the number of antennas on board the aircraft by replacing most of them with a multifunction antenna system using a conformal phased antenna array (PAA). It is assumed that such an antenna provides a solution to the problems of radar, radio communications, radio navigation and, possibly, jamming. The external surface of a PAA is selected so as to minimize the backscattering of external radar signals. One of the problems in a multifunctioning PAA is providing isolation between its concurrently functioning subsystems for various purposes.

Certain proposals solving particular problems are of interest. For example, it has been proposed to cover radar antennas with a large metallic screen with gaps cut in its surface. By selecting parameters, it is possible to limit the polarization and frequency passband of the screen to only a given radar. For external radiation of a different frequency and polarization, the antenna with this screen appears as an object with low backscattering. It is asserted that the antenna RCS in this case can be reduced approximately by 30 dB.

Reduction of the radar detectability of linear antennas in the meter and decameter ranges is achieved by using ferrite coatings with a high magnetic permittivity μ, applied to the metallic base from which the antenna is made. At the same time, electrical characteristics are retained and the RCS in the centimeter waveband is reduced.

An effective means of decreasing the radar detectability of linear antennas is the use of reactive loads, as discussed earlier. It would seem to be possible to synthesize linear antennas with reactive loads included in them that minimize the backscattered power while retaining their basic electrical characteristics.

One decreasing radar detectability of large aircraft antennas is the use of plasma generated in a space covered with a cowl, within which the atmosphere pressure has been reduced. In order to reduce significantly backscattering at frequency f, the density of electrons in the plasma (N el/cm^3) must be greater than or equal to the electron concentration N_0 determined by the plasma frequency f_0 [19]:

$$f_0 = 8960\sqrt{N_0} \tag{7.28}$$

In the case of counteracting a radar in the 10-cm band ($f_0 = 3 \times 10^9$ Hz), it is necessary to ensure $N_0 = 10^{15}$ el/cm^3. This is a quite high concentration. All the same, an onboard radar in the 3-cm band with a somewhat reduced ERP (P_sG_s) can still operate because the critical electron concentration is 1.1×10^{16} el/cm^3, an order of magnitude higher than for a radar in the 10-cm band. In the general case, there exists a problem controlling the concentration of electrons under the cowl in the dynamics of onboard radar operations.

An original suggestion is to decrease the radar detectability of antennas by using a controlled multilayer sheet applied directly to the antenna surface. Such a sheet is a set of flat controlled and uncontrolled layers. It is proposed to use powerful p-i-n diodes and ferroelectric ceramic capacitors as control elements. The uncontrolled layers can be lossless homogeneous dielectrics or two-dimensional periodic arrays. Such sheets are analogous to the controlled impedance structures mentioned earlier.

As one of the more radical suggestions, it is possible to consider the miniaturization of reradiators in the centimeter waveband. It is suggested to use ferrite elements as PAA reradiators. In particular, in receiving operations mode, it was suggested to use parametric amplifiers made from ferrites.

It should be noted once again that, without solving the problems of the detectability of onboard antennas, just as without a comprehensive analysis, it would not seem to be possible to come up with a satisfactory solution to the problem of aircraft detectability as a whole. The key factor here is the cost/effectiveness criterion.

It would seem that the solution of the problem only by reducing RCS is not very promising. As was noted earlier, it can be solved only by a combination of decreasing the RCS and determining the optimum value of the jammer ERP, which is determined as the product of the jammer power P_j and the gain of its antenna G_j.

7.3.2 Optimum Transmitting Antenna Gain in an Aircraft (Helicopter) Jammer

The effectiveness of radar jamming depends on the ERP of the active jammer. An erroneous opinion has been formed that almost limitless increase in jammer ERP is possible using highly-directional antennas (for example, a PAA with a large number of elements and a large aperture). However, limitations exist on the gain of a jammer antenna, associated with the effect of antenna size on the radar detectability of the aircraft being protected. When the gain of the antenna G_j increases, as is known, the jammer ERP increases. However, the increase in G_j is a result of an increase in antenna size, which, in its turn, causes an increase in the radar detectability of the aircraft due to the increase in the contribution of the RCS of the antenna σ_A to the total RCS of the aircraft σ_{Ta}.

When the equivalent RCS of the antenna σ_A becomes commensurate with the RCS of the aircraft σ_{Ta}, further increase in the jammer ERP due to the increase in the antenna gain does not lead to an increase in jamming effectiveness. Moreover, jamming effectiveness, when the value of G_j is too large, can be significantly reduced. Let us determine the optimum value of the transmitting antenna gain G_{opt}, for which we use the self-screening jamming equation:

$$K = \frac{P_j G_j 4\pi D_j^2}{P_s G_s \sigma_{Ta\Sigma}} \tag{7.29}$$

Here, P_j is the jammer power; G_j is the jammer antenna gain; P_s is the radar transmitter power; G_s is the radar transmitting antenna gain; D_j is the distance between the radar and the jammer; and $\sigma_{Ta\Sigma}$ is the total RCS of the aircraft.

The total RCS of the aircraft on which the jammer is installed can be determined using the expression:

$$\sigma_{Ta\Sigma} = \sigma_{Ta} + \sigma_A \tag{7.30}$$

where σ_{Ta} is the RCS of the aircraft in the absence of a jammer antenna; and σ_A is the RCS of the jammer antenna.

The RCS of the aircraft σ_{Ta} depends on the type of aircraft and is usually determined experimentally. The RCS of the jammer antenna σ_A depends on the effective antenna area A and the degree of its matching to free space and the load (the output stage of the transmitter). In accordance with the reciprocity theorem, the transmitting antenna, having aperture S_A, can be replaced by an equivalent receiving antenna, loaded by the complex load Z_l (Figure 7.9).

When a plane wave is incident on an antenna with field intensity $\overline{E}$, the power of the field received by the antenna is equal to

$$P_A = pA \tag{7.31}$$

where $p = \frac{\langle |\overline{E}^2| \rangle}{240\pi}$ is the power density of the incident wave; A is the effective area of the antenna aperture; and $\langle |\overline{E}^2| \rangle$ is the mean square electrical field intensity.

The effective area of the antenna aperture is linked with the antenna gain by the expression:

$$A = \frac{\lambda^2}{4\pi} G_j \tag{7.32}$$

The power delivered to the load is equal to

$$P_l = \zeta p A \tag{7.33}$$

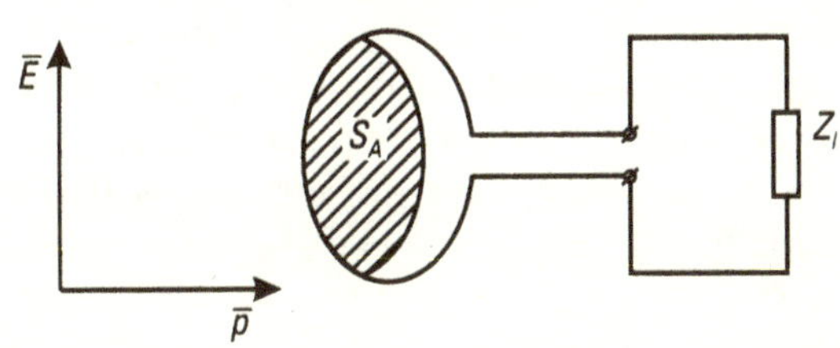

Figure 7.9 An equivalent receiving antenna, loaded by the complex load Z_l.

where

$$\zeta = \frac{4R_l R_a}{|Z_a + Z_l|} \tag{7.34}$$

$$Z_a = R_a + jX_a \tag{7.35}$$

$$Z_l = R_l + jX_l \tag{7.36}$$

R_l, R_a are the active impedances of the load and the antenna; and X_l, X_a are the reactive impedances of the load and the antenna.

The power scattered by the antenna is determined using the formula:

$$P_{\text{dif}} = P_a - P_l \tag{7.37}$$

On the other hand, this power is linked to the full power intercepted by the antenna by

$$P_{\text{dif}} = \rho P_a \tag{7.38}$$

where ρ is the reflection coefficient, characterizing the share of power being reradiated by the antenna into free space. From (7.31), (7.33), (7.37), and (7.38) we find

$$\rho = 1 - \zeta = 1 - \frac{4R_l R_a}{(X_a + X_l)^2 + (R_a + R_l)^2} \tag{7.39}$$

Reflections from the antenna are absent when it is ideally matched to the load, when the following conditions are fulfilled:

$$X = -X_l \quad R_l = R_a \tag{7.40}$$

However, in practice conditions (7.40) are not met and for real antennas the reflection coefficient has a value of $\rho = 0.1$–0.7 depending on the polarization, frequency and arrival direction of the illuminating wave. In the book [20] are reviewed the dependencies of the reflection coefficient ρ on certain parameters of the incident wave for an antenna system comprising an antenna array consisting of matched metal–dielectric waveguides (Figure 7.10).

In Figure 7.11 we show the typical dependency of the reflection coefficient on the incidence angle of the wave Θ for an array with

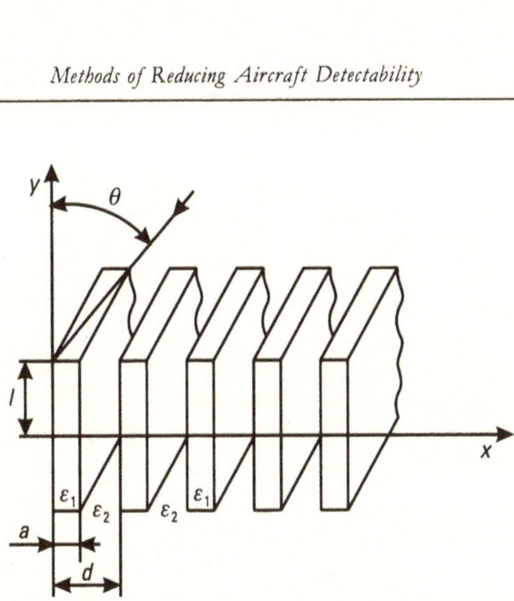

Figure 7.10 An antenna system comprising an antenna array consisting of matched metal–dielectric waveguides.

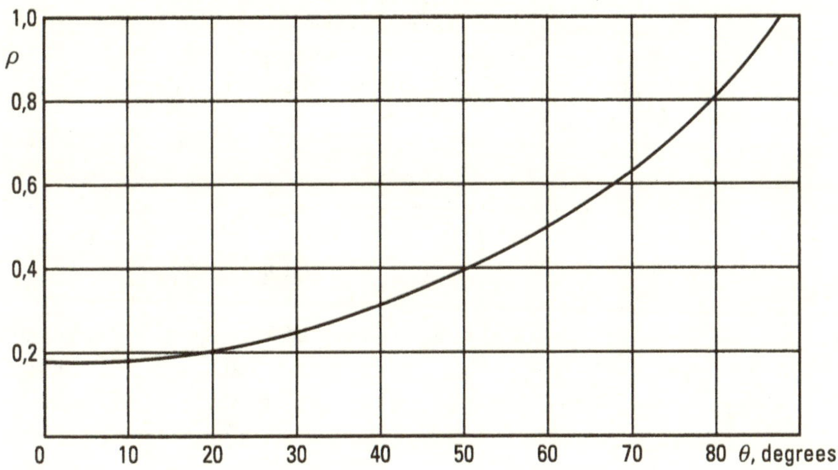

Figure 7.11 The typical dependency of the reflection coefficient on the incidence angle of the wave Θ for an array.

parameters ($a/\lambda = 0.22$; $d/\lambda = 0.55$; $l/\lambda = 0.2$; $\varepsilon_1/\varepsilon_2 = 6$). The dependencies have been calculated for the frequency f_0 at which the antenna is best matched with free space. When mismatched by $\delta f = f - f_0$, a quadratic dependency of the increase of ρ on δf is observed:

$$\rho = \rho_0\left[1 + (\delta f/f_0)^2\right] \tag{7.41}$$

where ρ_0 is the reflection coefficient when $f = f_0$.

Let us replace the jammer antenna with the equivalent plate having a reflecting area of S, proportional to the effective absorption area of the antenna A:

$$S = \rho A \tag{7.42}$$

The RCS of such a plate for monostatic scattering is equal to

$$\sigma = 4\pi S^2/\lambda^2 \tag{7.43}$$

Substituting (7.42) into (7.43), we obtain

$$\sigma_A = (4\pi/\lambda^2)\,(\rho^2 A^2) \tag{7.44}$$

From (7.32) and (7.44), we find

$$\sigma_A = (\rho^2\lambda^2 G_n^2)/4\pi \tag{7.45}$$

Taking into consideration (7.29), (7.30) and (7.45), we write the jamming equation in the following form:

$$K = \frac{4\pi P_j D^2 G_j}{P_s G_s\left(\sigma_{Ta} + \rho^2\lambda^2 G_j^2 \Big/ 4\pi\right)} \tag{7.46}$$

In Figure 7.12 we show the dependency of the jamming/signal ratio given at the input to the receiver k_{rec} on the antenna gain $G_j = 3$ cm, $\lambda = 10$ cm and $\lambda = 30$ cm. The jamming/signal ratio k_{rec} given is equal to

$$K_{\text{rec}} = \frac{G_j}{1 + \left(\rho^2\lambda^2 G_j^2 \Big/ 4\pi\sigma_{Ta}\right)}$$

From Figure 7.12 it follows that, in the centimeter range, when G_j increases to a value of $G_{j1} = 100$, the jamming/signal ratio at the input to the receiver being jammed grows linearly. When $G_j > G_{j1}$, significant differences begin to manifest themselves in the dependencies of K_{rec} on G_j for various wavelengths λ.

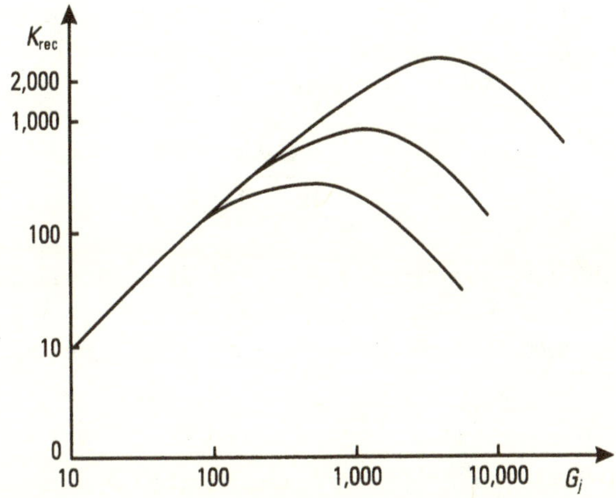

Figure 7.12 The dependency of the jamming/signal ratio given at the input to the receiver k_{rec} on the antenna gain G_j.

From (7.46) we find the optimum value $G_{j\ \mathrm{opt}}$ at which $K = K_{\max}$:

$$G_{j\ \mathrm{opt}} = 2\sqrt{\pi\sigma_{Ta}} \Big/ \rho\lambda \tag{7.47}$$

The expression (7.47) permits us to justify the requirements on the value for the gain of the jammer transmitter antenna. As follows from (7.47), the maximum permissible antenna gain $G_{j\ \mathrm{opt}}$ depends most significantly on the degree to which it is matched to free space ρ and the wavelength λ. To a lesser degree, $G_{j\ \mathrm{opt}}$ depends on the RCS of the aircraft (proportional to $\sqrt{\sigma_{ta}}$).

The RCS of an optimum jammer antenna system from (7.45) and (7.47) is equal to

$$\sigma_{A\ \mathrm{opt}} = \sigma_{Ta} \tag{7.48}$$

Substituting (7.47) into (7.46), we find the maximum value of the jamming/signal ratio at the input to the radar receiver being jammed:

$$K_{\max} = \frac{4\pi^{3/2} P_j D_j^2}{\lambda\rho P_s G_s \sqrt{\sigma_{Ta}}} \tag{7.49}$$

Equating the right part of (7.49) to the jamming coefficient, we determine the minimum value of jammer transmitter power required to jam the radar down to the specified minimum jamming range $D_{j\,\min}$:

$$P_{J\,\min} = \frac{K_j P_s G_s \lambda \rho \sqrt{\sigma_{Ta}}}{4\pi^{3/2} D_{j\,\min}^2} \tag{7.50}$$

The expression (7.50) introduces the idea of adaptive control of jammer power. If there are a priori dependencies of the coefficient ρ and the target RCS on the direction towards the victim radar and on the wavelength λ:

$$\rho = \rho(\Theta) \quad \sigma_{Ta} = \sigma_{Ta}(\Theta) \tag{7.51}$$

then the law for controlling jammer power has the form:

$$P_{j\,\min} = k_1 \rho(\theta, \lambda) \sqrt{\sigma_{Ta}(\theta, \lambda)} \tag{7.52}$$

where

$$k_1 = \frac{K_j P_s G_s \lambda}{4\pi^{3/2} D_{j\,\min}^2} \tag{7.53}$$

The analysis performed on the effect of parameters of the jammer antenna system on jamming effectiveness has shown that, when developing jammer antenna systems, it is necessary to ensure that they are matched to free space. When the match is poor ($\rho = 0.5$–0.9), it is impossible to achieve a high jammer ERP by simply increasing antenna gain.

7.4 Methods of Changing the Electrical Properties of the Environment

7.4.1 Local Changes in the Electrical Properties of the Environment to Decrease Radar Detectability

At the present time, two EW trends are known associated with changing the electrical properties of the environment:

- Local changes in the propagation conditions of radio waves to decrease radar detectability;
- The use of high-altitude nuclear explosions for the purpose of globally disrupting systems of radio communications and radio navigation.

The physical principles of radio countermeasures using artificial ionization of space are based on the phenomena of absorption, reflection and refraction of electromagnetic waves in plasma.

Consider the phenomena of refraction and reflection. As is known from electrodynamics, the reflection of electromagnetic waves occurs in all cases when the macroscopic parameters of heterogeneity $n = \sqrt{\varepsilon\mu}$ and the conductance σ differ from the corresponding parameters of the environment in which radio waves propagate. The refraction of radio waves occurs for this same reason. Thus, to ensure a significant reflection of radio waves using ionized formations, corresponding local change in the macroscopic parameters of the environment μ and ε are necessary. The use of the macroscopic parameters mentioned as electrical characteristics of the environment is permissible, if the average distance between the particles forming the environment is much less than the wavelength ($d << \lambda$) (i.e., when the environment seems to be continuous for the incident and propagating waves).

Errors in the determination of the direction of arrival of the radio wave occur when the local heterogeneity, is itself heterogeneous (i.e., its refraction coefficient is a function of coordinates (Figure 7.13)).

Errors in determining the direction can, in principle, also occur with constant electrical parameters of local heterogeneity, if this heterogeneity has a suitable form (for example, nonrectangular).

The index of refraction of an ionized environment (local heterogeneity), ignoring the Earth's magnetic field, is calculated approximately using the following formula:

$$n = \sqrt{1 - 81 {}^{N}/_{f^2}}$$

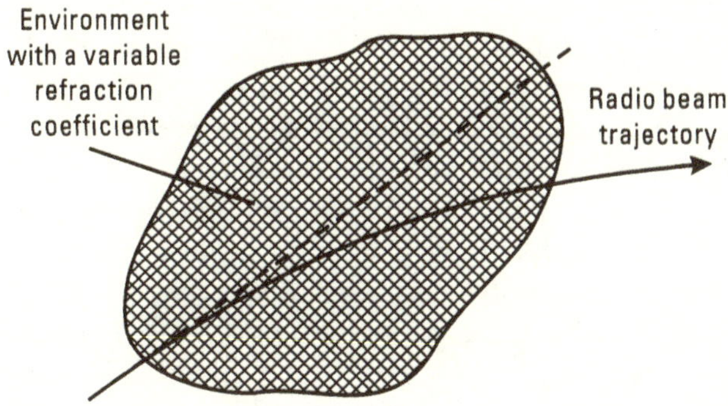

Figure 7.13 Errors in the determination of the direction of arrival of the radio wave.

Here, f is the carrier frequency in Hertz; and N is the electron density per cubic meter.

When the concentration of electrons is quite high, radio waves can be fully reflected from the ionized region (full internal reflection). The critical frequency corresponding to the full reflection of radio waves is determined from the condition $n = 0$, whence it follows that

$$F_{cr} = 9N^{1/2}$$

corresponds to the plasma frequency f_0 determined by (7.28).

Thus, to obtain full reflection of waves with a carrier frequency f, it is necessary to have an electron density:

$$N = f_{cr}^2/81 \text{ el/m}^3$$

For example, for a wave $\lambda = 3$ cm we obtain $N = 10^{18}$ el/m^3, or $N = 10^{12}$ el/cm^3.

To ensure the specified concentration of electrons (N), it is necessary to have powerful sources of ionization. The required electron production rate of the source of ionization is equal to $I = \alpha N^2$, where α is the electron recombination rate, equal at the Earth's surface to approximately $\alpha = 10^9$ cm^3/sec·el.

At $N = 10^{12}$ el/cm^3 and $\alpha = 10^{-9}$ cm^3/sec·el, the ionization rate must be $I = 10^{15}$ el/cm^3·sec. In other words, to create an ionized region with a concentration of 10^{12} electrons per cubic centimeter, the ionization source must create in 1 sec 10^{15} electrons per cubic centimeter. Such a high concentration of electrons can be created by rapid burning of large quantities of easily ionized elements, for example, cesium, or with high energy electron beams or several laser beams of equal wavelength [21]. Let us note that the threshold concentration of electrons necessary for visual detection of ionization, is equal to 10^{17}–10^{18} electrons per cubic meter.

When nuclear ammunition explodes, in the epicenter of the explosion large concentrations of electrons are formed. However, as a result of their recombination, the concentration of electrons quickly decreases with time so that the interference of a nuclear explosion on radar in the centimeter range is quite short lived. High-altitude nuclear explosions have a more significant effect on the propagation of waves in the meter range.

Consider the absorbing properties of locally ionized media. The mechanism of absorbing radio waves in an ionized region can be explained in the following manner. Free electrons under the influence of an electric field of an incident wave experience forced oscillations with a frequency, equal to the carrier frequency of electromagnetic waves. In the process of oscillating motions, the electrons collide with neutral molecules, atoms and ions and increase their kinetic energy. Thus, electromagnetic field energy is converted into thermal energy in the medium.

The absorbing property of an ionized region is characterized by the absorption coefficient (dB/km):

$$\beta = 1.8 \times 10^{-2} N\nu/(\omega^2 + \nu^2) \tag{7.54}$$

where N is the electron density per m^3; ν is the number of collisions of electrons with other particles (ions, atoms, and gas molecules) per second; and ω is the angular frequency.

From (7.54) it follows that the absorption coefficient has maximum at a certain collision frequency. Using the rule for finding the maximum of a function, we obtain that $\beta = \beta_{max}$ at $\omega = \nu$.

The collision frequency ν is proportional to the air density. For this reason there is a certain interval of atmospheric altitudes beyond which the attenuation of radio waves is small. Calculations and experimental studies show that the attenuation of radio waves is greatest in a 16-km band with a center located at an altitude of approximately 72 km. The number of collisions at an altitude of 72 km is approximately equal to $\nu = 6 \times 10^6$ collisions per second. For signals with carrier frequencies $f > 5$ MHz, the value ω^2 in (7.54) is significantly greater than ν due to which

$$\beta = 0.45 \times 10^{-2} N\nu/f^2 \tag{7.55}$$

The attenuation of shorter waves does not increase at low altitudes because the lifetime of electrons is reduced and the initial assumptions on which formula (7.54) is based become invalid.

Using (7.55), it is possible to determine the required electron density providing for the specified attenuation β in dB/km:

$$N = 2.2\beta f^2 \times 10^3 \ \nu^{-1} \ \text{el/m}^3$$

To obtain, for example, an attenuation of $\beta = 10$ dB/km at an altitude of 72 km for a wavelength $\lambda = 3$ cm, the ionized region must have an electron density $N = 0.37 \times 10^{18}$ el/m^3. At the present time, such a high concentration for large extents can be created only for a very short time using high energy electron beams or lasers.

Let us evaluate the potential for implementing a local region of atmospheric ionization to reduce aircraft radar detectability. Let us determine, in the first approximation, the required ERP of appropriate ionization sources.

The ERP of an ionization source (a relativistic electron accelerator, laser etc.) can be considered to be sufficient if it ensures the required degree of power reduction ΔP_s of radar signals reflected from aircraft in a specified direction. The value ΔP_s is determined by the product of the absorption coefficient in plasma (β dB/m) and the doubled equivalent thickness of the plasma layer l_{pe},m:

$$\Delta P_s = \beta 2 l_{pe} \quad \text{dBWt} \tag{7.56}$$

According to (7.55),

$$\Delta P_s = -.45 N_e \nu 2 l_{pe} / f^2 \quad \text{dBWt} \tag{7.57}$$

where N_e is the electron density per cm^3.

The equivalent thickness of the plasma layer l_{pe} is determinable under the assumption that the electron density in the coordinate z follows a parabolic curve [22]:

$$N_e(z) = N_{em}(1 - z^2/z_0^2) \tag{7.58}$$

Here, N_{em} is the maximum electron density ($1/\text{cm}^3$); and z_0 is the full thickness of the plasma layer ($N_e(z) = 0$ at $z = z_0$).

Hence,

$$l_{pe} = \int_0^{Z_0} \left(1 - {z^2}\Big/{z_0^2}\right) dz = {}^2\!/_3 z_0 \tag{7.59}$$

The value we are seeking, the degree of power reduction for the reflected signal ΔP_s, is determined by

$$\Delta P_s = 0.6 N_{em} \nu z_0 / f^2 \quad \text{dB} \tag{7.60}$$

As an example, let us consider the instance when $f_s = 3 \times 10^9$ Hz, ($\lambda = 0.1$m); $\nu = 10^7$ Hz. The full thickness of the plasma layer $z_0 = 1$m. We shall determine the electron density for a critical frequency $f_s = 3 \times 10^9$ Hz:

$N_{em} = 1.10 \times 10^{11}$ 1/cm^3. The attenuation of the reflected radar signal is then $\Delta P_s = 0.073$ dB. A significant reduction in detectability ($\Delta P_s =$ 73 dB) occurs, if the electron density in the plasma increases by two orders of magnitude ($N_{em} = 1.1 \times 10^{13}$ 1/cm^3) and the full thickness of the plasma increases by one order of magnitude ($z_0 = 10$m).

The examples considered show that, to implement the required reduction in aircraft radar detectability, a quite high electron density is required when flying at a sufficiently high altitude ($\nu = 10^7$ 1/s). If we bear in mind that the performance coefficient of modern ionizers does not surpass 0.1, then the required powers for primary current sources must comprise hundreds of kW.

7.4.2 The Effect of Nuclear Explosions on the Operations of Radio Systems

Explosions at altitudes less than 16 km do not result in lengthy ionization, so they do not have a significant effect on the work of radio systems. In the case of surface (underground or underwater) explosions regions may be formed in which intense absorption and reflection of radio waves occur. However, the effect of absorption and reflection of radio waves in this case is not associated with the ionization of space, but with the presence of local heterogeneity of the environment, having a high concentration of particles of solid substances and water thrown into the atmosphere. The effective influence on the operations of radio systems in the centimeter range due to the products of the explosion can be exerted only during the initial stage of the explosion.

The duration of ionized regions depends on the altitude and the force of the explosion, the time of day etc. At high altitudes (> 40–50 km) quite stable regions are formed with relatively high concentrations of electrons. In the case of the explosion of nuclear ammunition at high altitudes, ionized regions with a small concentration of electrons (10^{10}–10^{11}) can exist for several minutes and even for hours.

In the first approximation, it is possible to distinguish ionized regions of two types.

First of all, there are regions of slow electrons, formed as a result of ionization of the environment, primarily, thermal x-rays. These regions have comparatively limited sizes, tens and hundreds of kilometers, and the concentration of electrons in them decreases approximately according to the law:

$$N = 10^{13} t^{-1} \quad e/m^3$$

Here, t is the time in seconds.

The development of the region in time, after its formation, proceeds basically according to the laws for diffusion.

Secondly, there are regions of fast (relativistic) electrons (β-particles), irradiated by radioactive products of fission. Fast electrons are caught up by the Earth's magnetic field, in which case the ionization of space on the scale of the Earth assumes a global character (and not local, as was stated earlier). Consider this issue in more detail, recalling several initial physical prerequisites.

A charge e, moving with velocity $\bar{v}$ in a magnetic field with intensity $\overline{H}$, is acted upon by a Lorentz force $\overline{F}$ determined by the formula:

$$\overline{F} = e[\bar{v}\overline{H}]$$

This force, all other conditions being equal, is proportional to the speed of the particles. If the magnetic field is homogeneous (i.e., its force lines are parallel to one another), then the trajectory of the charged particle, having entered the magnetic field with a velocity of $\bar{v}$ is a cylindrical spiral with a constant pitch, wound around a force line. The radius of the cylinder (radius of gyration) is determined using the formula:

$$r = mv(eH)$$

In a heterogeneous field the trajectory of motion of the charged particle becomes much more complex. Particularly, as the charged particle moves in the heterogeneous magnetic field in the direction of increasing field intensity, the charged particle will be acted upon by a force pushing it into a region of lower field intensity values (in the direction of motion of the particle, the force lines of the magnetic field converge) [23]. The appearance of this force is a result of the magnetic field component H_2 (Figure 7.14), the magnitude of which increases together with the heterogeneity of the field (the greater the angle γ). If the field H were heterogeneous, then all its force lines would be parallel to the component $H_1(\gamma = 0)$ and the component H_2 would be equal to zero. The heterogeneity of the field generates the component H_2.

Let us review a simpler case: the motion of an electron along a circular orbit in a heterogeneous field (Figure 7.14). The component H_1 of magnetic field H generates a centripetal force providing the electron with a rotational movement. The component H_2, acting at point O on the electron, moving with velocity v, generates a force $F(H_2)$, pushing the particle into the region of the minimum field. The presence of this force in a

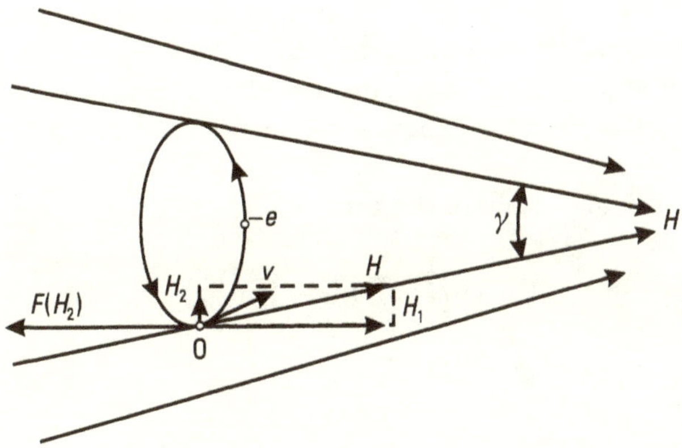

Figure 7.14 The trajectory of motion of a charged particle in a heterogeneous field.

heterogeneous magnetic field and the heterogeneity itself lead to a significant transformation of the spiral trajectory.

In Figure 7.15 we depict the approximate trajectory of an electron in a heterogeneous magnetic field. As the electron progresses into the region of greater concentration of magnetic force lines, the pitch of the spiral and the radius of gyration decrease.

Since the Lorentz forces in the first approximation don't change the absolute value of the particle velocity, then, as the longitudinal movement of the electron slows down due to the braking force, the velocity component v_1, perpendicular to H_1, will of necessity increase. More in-depth studies indicate that the ratio v_1^2/H can be considered to be

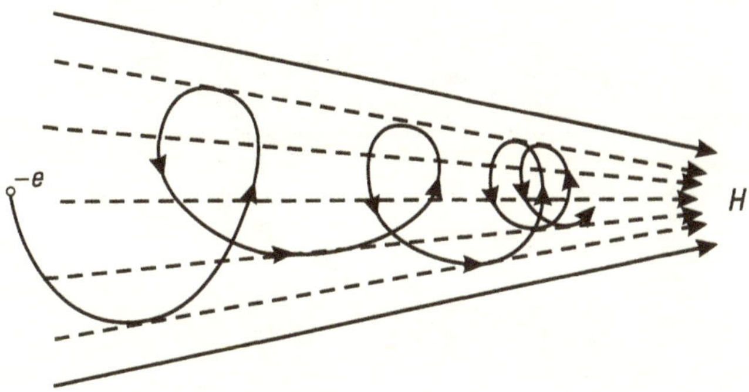

Figure 7.15 The approximate trajectory of an electron in a heterogeneous magnetic field.

approximately constant during the entire time the particle is in motion. This circumstance permits us to find the angle α between the velocity vector of the particle and the direction of the force line at the given moment of time, if the initial angle α_0 is known at the moment the particle enters the magnetic field.

Accordingly, we have (Figure 7.16)

$$(\sin^2 \alpha_0)/H_0 = (\sin^2 \alpha)/H \tag{7.61}$$

Hence,

$$\sin \alpha = \sin \alpha_0 (H/H_0)^{1/2}$$

For the specified angle α_0, let us determine the intensity of the magnetic field at which forward movement ceases into the region of greater concentration of magnetic force lines ($\sin \alpha = 1$):

$$H = H_0/(\sin^2 \alpha_0)$$

The electron cannot penetrate into the region of greater magnetic field intensity. Having reached the indicated region of the magnetic field, the electron begins moving in the opposite direction. Thus, the region of high concentration of magnetic force lines can play the role of a magnetic mirror of sorts.

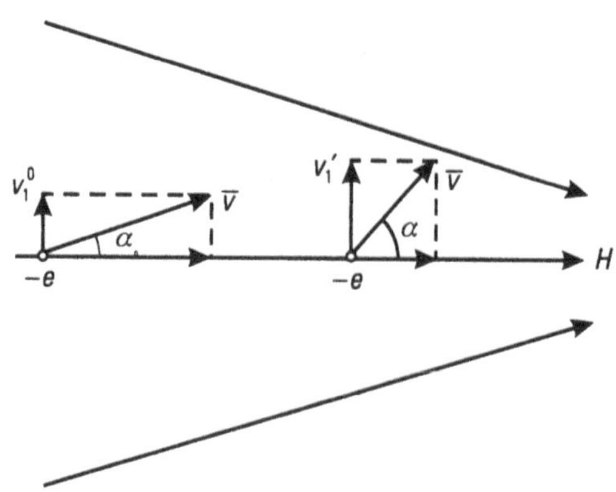

Figure 7.16 Finding the angle α between the velocity vector of the particle and the direction of the force line.

The magnetic field of the Earth has two regions of high concentration of magnetic force lines: the north and south magnetic poles (Figures 7.17 and 7.18).

If the maximum intensities of the magnetic field at the poles are identical (H_{max}), then the possibility of reflecting an electron from the magnetic mirror is determined by angle α_0.

All particles for which $\sin \alpha_0 > (H_0/H_{max}^{1/2})$ will be reflected from the magnetic mirrors formed at the poles.

Thus, the magnetic field of the Earth, for the particles mentioned, in fact serves as a "magnetic trap." The regions in space, in which the reflection of electrons from "magnetic mirror" occurs, are called conjugate points (A_1 and A_2 in Figure 7.18). The conjugate points can in principle also not coincide with the magnetic poles. The formula (7.61) permits us to find the conjugate points in other sectors of space near the Earth.

Electrons, caught up by the magnetic field of the Earth, moving along the magnetic force lines, concurrently experience a "magnetic drift" from west to east. The "magnetic drift" of electrons is caused by a decrease in magnetic field intensity for altitude.

The physical reason for the "magnetic drift" of electrons due to the decrease in the field along the radius is illustrated in Figure 7.19. In Figure 7.19 we depict the direct current magnetic field I, decreasing along the radius R, flowing along a direct wire. If an electron (or a positive ion) rotates near the force lines A and B, then the radius of gyration of its trajectory in sectors

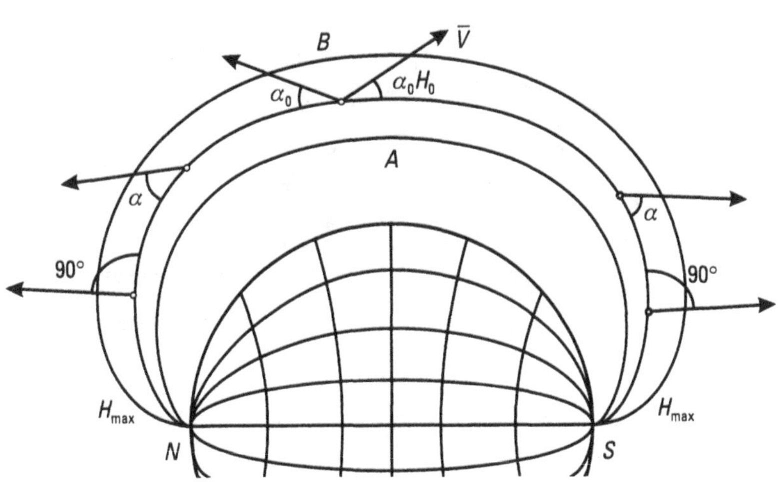

Figure 7.17 The magnetic field of the Earth.

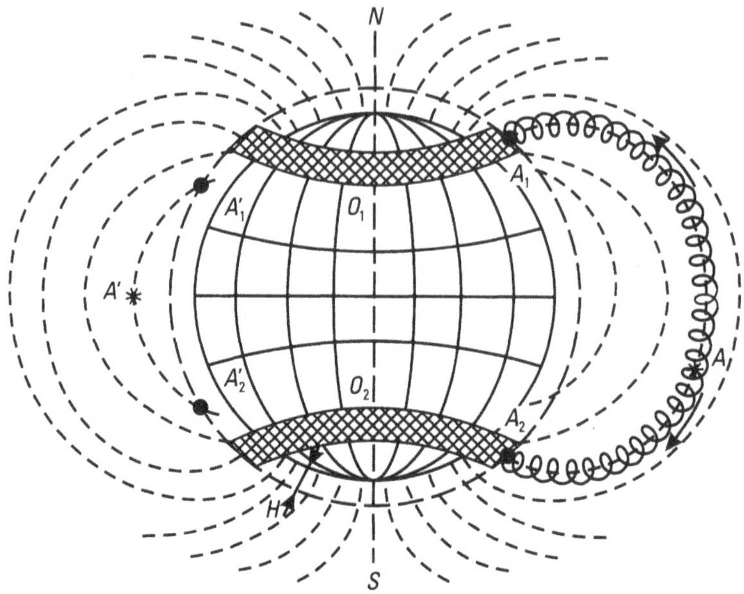

Figure 7.18 The magnetic field of the Earth: the north and south magnetic poles.

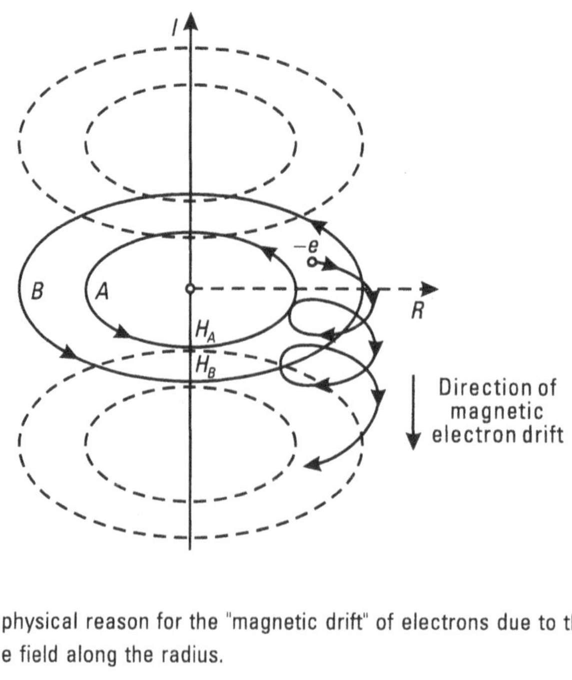

Figure 7.19 The physical reason for the "magnetic drift" of electrons due to the decrease in the field along the radius.

A and B will vary. In sector A, the radius of gyration will be less than in sector B with a lesser magnetic field intensity. After completing the first loop, the electron, moving as shown in Figure 7.19, will end up somewhat lower than its initial point of motion. During subsequent revolutions it will constantly drift in the direction, opposite to the direction of the current I generating the given magnetic field. From here it directly follows that, in conditions on Earth, electrons, moving along force lines, "shift" from west to east. Accordingly, presumably charged particles "drift" from east to west. An analogous phenomenon occurs in the Earth's so-called radiation belts.

Thus, due to the quick electrons formed during a high-altitude nuclear explosion, the ionization of space near the Earth assumes a global character. However, the density of electrons on the average turns out to be low, and the effects described have no noticeable influence on the operation of radar systems, with the exception, perhaps, of radar of the meter range, working in the region where conjugate points are formed. As concerns communications and radio navigation equipment, especially in the short wave and middle wave bands, then here there can be serious disruptions in operations for a comparatively long time.

The lifetime of global ionized regions consisting of quick electrons depends on the height of conjugate points above the Earth. If the conjugate points are located in a region of increased atmospheric density, then the loss of electrons occurs more quickly due to absorption by neutral molecules and positive ions located in the region of magnetic mirrors. The absorption of quick electrons leads to the ionization of space in the vicinity of the conjugate points. The loss of electrons outside the conjugate points, as a rule, is insignificant. When nuclear ammunition was exploded at an altitude of 480 km (operation "Argus"), the formation of clearly expressed layers with a thickness of approximately 100 km was observed. The ionized regions remained for several days [24, 25].

References

[1] Alekseev, S., *Amerikanskiy malozametnyy istrebitel' F-117A* (*The American F-117A Fighter with Low Detectability*). Moscow: *Zarubezhnoe voennoe obozrenie* (*Foreign Military Review*), Vol. 1, 1993, pp. 40–45.

[2] Afinov, V., *Sredstva RE'B strategicheskoy aviatsii VVS SSHA* (*EW Systems on US Airforce Strategic Aircraft*). Moscow: *Zarubezhnoe voennoe obozrenie* (*Foreign Military Review*), Vol. 3, 1994, pp. 35–45.

[3] Anan'in, E'. V., R. G. Vaksman, and Yu. M. Petrakov, *Metody snizheniya radiolokatsionnoy zametnosti* (*Methods for Reducing Radar Detectability*). Moscow: *Zarubezhnaya radioe'lektronika* (*Foreign Radio Electronics*), Vol. 4/5, 1994, pp. 5–21.

[4] Erofeev, Yu. N., Moscow: Weekly "*Nedelya*" ("*Week*"), Vol. 37, November 1997.

[5] Ufimtsev, P. Ya., *Metod kraevykh voln v fizicheskoy teorii difraktsii* (*The Method of Edge Waves and The Physical Theory of Diffraction*). Moscow: *Sov. radio* (*Soviet Radio*), 1962, pp. 24–48, 194–255.

[6] Shtager, E. A., *Metody rascheta radiolokatsionnykh kharakteristik ob''ektov, nakhodyashchixsya vblizi zemnoy ili morskoy poverkhnosti* (*Methods for Calculating the Radar Characteristics of Targets Located near the Surface of the Earth or Sea*). Moscow: *Zarubezhnaya radioe'lektronika* (*Foreign Radio Electronics*), Vol. 4/5, 1994, pp. 22–39.

[7] Bochkarev, A. M., and M. N. Dolgov, *Radiolokatsiya malozametnykh letatel'nykh apparatov* (*Radar Detection of Aircraft with Low Detectability*). Moscow: *Zarubezhnaya radioe'lektronika* (*Foreign Radio Electronics*), Vol. 2, 1989, pp. 3–17.

[8] Pirumov, V. S., A. G. Alekseev, and B. V. Ayzikovich, *Novye radiopogloshchayushchie materialy i pokrytiya* (*New Radio Absorptive Materials and Coatings*). Moscow: *Zarubezhnaya radioe'lektronika* (*Foreign Radio Electronics*), Vol. 6, 1994, pp. 2–7.

[9] Ayzikovich, B. V., V. Yu. Lapshin, et al., *Metody analiza i rascheta radiolokatsionnykh kharakteristik poglotiteley E'MV* (*Methods for the Analysis and Calculation of Radar Characteristics of Absorbents of Electromagnetic Waves*). Moscow: *Zarubezhnaya radioe'lektronika* (*Foreign Radio Electronics*), Vol. 4/5, 1994, pp. 41–53.

[10] Alimin, B. F., *Sovremennye razrabotki poglotiteley E'MV i radiopogloshchayushchie materialy* (*Modern Developments in Electromagnetic Wave Absorbents and Radio Absorptive Materials*). Moscow: *Zarubezhnaya radioe'lektronika* (*Foreign Radio Electronics*), Vol. 2, 1989, pp. 75–82.

[11] Mitsmakher, M. Yu., and V. A. Torgovanov, *Beze'xovye kamery SVCH* (*UHF Anechoic Chambers*). Moscow: *Radio i svyaz'* (*Radio and Communications*), 1982, pp. 83–96.

[12] Alimin, B.F., and V. A. Torgovanov, *Metody raschetov poglotiteley e'lektromagnitnykh voln* (*Calculation Methods for Electromagnetic Wave Absorbants*). Moscow: *Zarubezhnaya radioe'lektronika* (*Foreign Radio Electronics*), Vol. 3, 1976, pp. 29–57; Vol. 8, 1976, pp. 60–80.

[13] Vakin, S. A., and L. N. Shustov, *Osnovy radioprotivodeystviya i radiotekhnicheskoy razvedki* (*The Basics of Radio Countermeasures and Radio Technical Intelligence*). Moscow: *Sov. radio* (*Soviet Radio*), 1968, pp. 9–122, 259–342.

[14] Petrov, B. M., and A. I. Semenikhin, *Upravlyaemye impedansnye pokrytiya i struktury* (*Controlled Impedance Coatings and Structures*). Moscow: *Zarubezhnaya radioe'lektronika* (*Foreign Radio Electronics*), Vol. 6, 1994, pp. 9–16.

[15] Golik, A. M., Yu. A. Kleymanov, and D. A. Gromov, *Intellektual'nye antennye reshetki* (*Intelligent Antenna Matrices*). Moscow: *Zarubezhnaya radioe'lektronika* (*Foreign Radio Electronics*), Vol. 4–5, 1992, pp. 3–8.

[16] Skandler, J., et al., *Upravlenie rasseyaniem e'lektromagnitnoy e'nergii putem podklyucheniya kompleksnoy nagruzki* (*Controlling Diffusion of Electromagnetic Energy by Connecting a Complex Load*). IEEE, Vol. 8, 1965.

[17] Liepa, V., and T. Senior, *Izmenenie rasseyaniya ot sfery putem podklyucheniya reaktivnoy nagruzki* (*Changing Diffusion from a Sphere by Connecting a Reactive Load*). IEEE, Vol. 8, 1965.

[18] Mikhaylov, G. D., V. I. Sergeev, et al., *Metody i sredstva umen'sheniya radiolokatsionnoy zametnosti antennykh sistem* (*Methods and Means for Reducing Radar Detectability in Antenna Systems*). Moscow: *Zarubezhnaya radioe'lektronika* (*Foreign Radio Electronics*), Vol. 4/5, 1994, pp. 54–59.

[19] Smirnov, B. M., *Vvedenie v fiziku plazmy* (*Introduction to the Physics of Plasma*). Moscow: *Nauka* (*Science*), 1990, pp. 15–102.

[20] Zakhar'ev, L. N., and A. A. Lemanskiy, *Rasseyanie voln "chernymi" telami* (*The Diffusion of Waves by "Black" Bodies*). Moscow: *Sov. radio* (*Soviet Radio*), 1972, 228 pp.

[21] Plokhikh, A. P., and N. A. Vazhenin, *Metody i sredstva modifikatsii sredy pri nablyudeniyakh vozdushnykh ob"ektov* (*Methods and Means of Modifying the Environment when Observing Airborne Objects*). Moscow: *Zarubezhnaya radioe'lektronika* (*Foreign Radio Electronics*), Vol. 9, 1992, pp. 29–56.

[22] Chernyy, F. B., *Rasprostranenie radiovoln* (*Propagation of Radio Waves*). Moscow: *Sov. radio* (*Soviet Radio*), 1962, pp. 89–147, 309–326.

[23] Artsimovich, L. A., *E'lementarnaya fizika plazmy* (*Elementary Physics of Plasma*). Moscow: *Atomizdat* (*Atomic Publishing House*), 1963, pp. 132–160.

[24] *Vozdeystvie yadernykh vzryvov na radiosvyaz' i radiolokatsiyu* (*The Influence of Nuclear Explosions on Radio Communications and Radar*). Moscow: *Novosti zarubezhnoy voennoy radioe'lektroniki* (*News from Foreign Military Radio Electronics*), Vol. 14(206), 1962, 48 pp.

[25] *Operaciya "Argus"* (*Operation "Argus"*). Moscow: *Atomizdat* (*Atomic Publishing House*), 1960, 116 pp.

About the Authors

Sergei Aleksandrovich Vakin

S. A. Vakin was born on August 20, 1920 in the village of Afanasovo in the Ulyanovsk District of the Kaluga Region. In 1945, he graduated from the N. E. Zhukovsky Airforce Engineering Academy. In 1955 he was appointed head of the scientific and research laboratory of the Radar Department. In 1951 he defended his candidate's dissertation on onboard radar. In 1958 he founded the Electronic Warfare Department and became its first head. He defended his doctoral thesis in 1970 and has been a professor since 1973. In 1974 he completed senior officer management courses at the General Staff Military Academy.

The basic subject of his scientific activity has been the development of the theoretical foundations for the jamming of electronic systems; ways of using onboard electronic warfare (EW) systems during combat; and the evaluation of their effectiveness in penetrating antiaircraft defenses. He was the first to develop methods to determine the optimum and rational parameters of jamming signals in airborne EW systems. He is the founder of Russian EW theory. For his work in these areas, as well as his active participation in the creation and testing of principally new models of onboard jamming systems, he was awarded the title of "Honored Scientist and Engineer of the Russian Federation" in 1983. He was the first to propose and develop the theoretical bases for generating informationally stable masking jamming signals.

He is the founder and permanent head of the scientific school of information stability of electronic systems during the conducting of EW. He has personally trained and educated 50 candidates of engineering. Five of them have subsequently become doctors. He has published two monographs, he has more than 90 scientific works and 18 inventions, and he has personally carried out more than 30 scientific and research projects. He is involved in active social and scientific work; he is a member of four dissertation councils. He has been rewarded by the government on several occasions. He has three decorations and 15 medals. In 1998 he was granted the title of "Honorary Professor of the N. E. Zhukovsky Airforce Engineering Academy."

His main scientific works are in the area of UHF radio physics and theoretical issues related to systems engineering applicable to problems of Electronic Warfare. He has authored articles in *Reports of the USSR Academy of Science* (1952, Physics) and the journal *Radiotekhnika* (*Radio Engineering* — 1958–1998). Together with L. N. Shustov, he has published *Fundamentals of Electronic Countermeasures and Radio Technical Intelligence* (M. Soviet Radio Publishing House, 1968).

Lev Nikolaevich Shustov

L. N. Shustov was born on March 13, 1933 in the city of Tula. In 1956 he graduated from the USSR Airforce Higher Aviation Engineering Military School in Riga, Latvia, where he majored in the Application of Electronic Airforce Systems. He served in airforce engineering units in various management positions. Since 1960 he has worked in the Electronic Warfare (EW) Department at the N. E. Zhukovsky Airforce Engineering Academy. In 1982 he defended his doctoral thesis on the theory of space and time jamming and its applications. Since 1984 he has been a professor in the EW Department at the Academy. He has capably managed the department's staff, first as deputy head, and then as head of the department.

A leading scientist in the area of EW, he is the author of fundamental textbooks and works in this area. He has published two monographs that have been translated in many countries of the world, which are advanced with respect to military engineering. He has more than 115 printed works and is the author of 15 inventions.

He is the Russian founder of the theory of space and time jamming. He developed and elaborated the theory of constructing EW complexes. He created the scientific school in this area and continues on as its irreplaceable leader to this day. He has trained and educated 11 candidates of engineering.

Over the entire duration of his career, he has conducted extensive scientific work. He has organized a number of scientific and research project laboratories in prospective areas of EW systems and techniques of their application, including the functional destruction of electronic systems for various purposes.

He has been rewarded by the government, the Ministry of Defense, the Commander-in-Chief of the Airforce and the commanding officers of the academy and the department on several occasions. For his outstanding contribution to the development of science and technology, he was awarded the Order of the Red Star and 11 medals.

At present, he is working as a professor in the EW Department. He is conducting extensive social work teaching young people. He is a member of two specialized committees.

Robert H. Dunwell

Robert H. Dunwell graduated from the University of California at Los Angeles (1972). From 1974 to 1993 he worked for IBM Corporation in Europe and the Russian Federation, working in systems engineering and image processing systems. Since 1994 he has been the General Director, DBT Ltd, Moscow, Russian Federation.

Index

Page numbers in *italics* refer to figures.

Recent Titles in the Artech House Radar Library

David K. Barton, Series Editor

Advanced Techniques for Digital Receivers, Phillip E. Pace

Airborne Pulsed Doppler Radar, Second Edition, Guy V. Morris and Linda Harkness, editors

Bayesian Multiple Target Tracking, Lawrence D. Stone, Carl A. Barlow, and Thomas L. Corwin

CRISP: Complex Radar Image and Signal Processing, Software and User's Manual, August W. Rihaczek, et al.

Design and Analysis of Modern Tracking Systems, Samuel Blackman and Robert Popoli

Digital Techniques for Wideband Receivers, Second Edition, James Tsui

Electronic Intelligence: The Analysis of Radar Signals, Second Edition, Richard G. Wiley

Electronic Warfare in the Information Age, D. Curtis Schleher

EW 101: A First Course in Electronic Warfare, David Adamy

Fundamentals of Electronic Warfare, Sergei A. Vakin, Lev N. Shustov, and Robert H. Dunwell

High-Resolution Radar, Second Edition, Donald R. Wehner

Introduction to Electronic Warfare, D. Curtis Schleher

Microwave Radar: Imaging and Advanced Concepts, Roger J. Sullivan

Millimeter-Wave and Infrared Multisensor Design and Signal Processing, Lawrence A. Klein

Modern Radar System Analysis, David K. Barton

Multitarget-Multisensor Tracking: Applications and Advances Volume III, Yaakov Bar-Shalom and William Dale Blair, editors

Principles of High-Resolution Radar, August W. Rihaczek

Radar Cross Section, Second Edition, Eugene F. Knott, et al.

Radar Evaluation Handbook, David K. Barton, et al.

Radar Meteorology, Henri Sauvageot

Radar Reflectivity of Land and Sea, Third Edition, Maurice W. Long

Radar Resolution and Complex-Image Analysis, August W. Rihaczek and Stephen J. Hershkowitz

Radar Signal Processing and Adaptive Systems, Ramon Nitzberg

Radar System Performance Modeling, G. Richard Curry

Radar Technology Encyclopedia, David K. Barton and Sergey A. Leonov, editors

Range-Doppler Radar Imaging and Motion Compensation, Jae Sok Son, et al.

Theory and Practice of Radar Target Identification, August W. Rihaczek and Stephen J. Hershkowitz

For further information on these and other Artech House titles, including previously considered out-of-print books now available through our In-Print-Forever® (IPF®) program, contact:

Artech House
685 Canton Street
Norwood, MA 02062
Phone: 781-769-9750
Fax: 781-769-6334
e-mail: artech@artechhouse.com

Artech House
46 Gillingham Street
London SW1V 1AH UK
Phone: +44 (0)20 7596-8750
Fax: +44 (0)20 7630-0166
e-mail: artech-uk@artechhouse.com

Find us on the World Wide Web at:
www.artechhouse.com

www.ingramcontent.com/pod-product-compliance
Lightning Source LLC
LaVergne TN
LVHW052248150926
848178LV00001B/8
* 9 7 8 1 5 8 0 5 3 0 5 2 1 *